PRIOR INFORMATION IN LINEAR MODELS

PRIOR INFORMATION IN LINEAR MODELS

Helge Toutenburg
Institute for Mathematics,
the Academy of Sciences
of the German Democratic Republic

JOHN WILEY & SONS
Chichester · New York · Brisbane · Toronto · Singapore

Library of Congress Cataloging in Publication Data:

Toutenburg, Helge.
Prior information in linear models.

(Wiley series in probability and mathematical
statistics. Tracts in probability and statistics)
Includes index.
1. Mathematical statistics. 2. Linear models
(Statistics) I. Title. II. Series.
QA276.T67 519.5 81-14653
ISBN 0 471 09974 0 AACR2

British Library Cataloguing in Publication Data:

Toutenburg, Helge
Prior information in linear models.
(Wiley series in probability and mathematical statistics).
1. Regression analysis
I. Title
519.5′36 QA278.2

ISBN 0 471 09974 0

Photo Typeset by
Macmillan India Ltd., Bangalore

Printed at Pitman Press, Bath.

Contents

Preface

This book describes how prior information can be used in estimating coefficients of linear models. Prior information here does not imply a Bayesian approach, but includes, for example, equations and inequalities as well as prior estimates on the unknown parameters.

The literature in this area is so extensive that any attempt at a complete coverage would lead either to a multi-volume work or to a mere catalogue. Thus we do not attempt such completeness, but concentrate instead on several aspects which so far have not been handled in depth in any other book. These include the robustness of minimax estimators, restricted least squares and two-stage least squares estimators, and mixed estimators which allow for part of the prior information to be incorrect.

The book is devoted to estimation and model choice, and references are also made to prediction as considered by Bibby and Toutenburg (1978). In order to make the book more readable, proofs have been omitted where they are obvious. The mathematical background which we assume of the reader is described at the beginning of Chapter II.

This book is based upon a lecture course given by the author at the Humboldt University in Berlin (GDR). The author would like to thank James Cameron of John Wiley and Sons; Professor Matthes; the Royal Society; the Open University; and the London School of Economics who supported some of this work. Thanks are also due to John Bibby for his comments and criticisms, which are much appreciated.

Berlin, January 1981 HELGE TOUTENBURG

I Introduction

1.1 THE BACKGROUND OF THE BOOK

Although in statistical practice many models and related methods for estimation and prediction have been developed and applied, at the present time methods of linear regression are of central relevance. The reason is that these can enable a sufficient approximation of various processes in economics and other sciences, although of course the usefulness of the linear model depends on the used information. If only the sample information is available, then the estimators due to Gauss or Aitken are most favourable; however, often one can give *auxiliary* information which improves the linear model. Thus estimation and model choice under linear restrictions, or under prior knowledge of inequalities about the parameters, may be used as an alternative to other statistical methods as, for example, nonlinear regression. *Prior information contains a component of uncertainty*; thus one is led to questions regarding the robustness of the proposed restrictive (or restricted) estimators. Some of these questions will be answered within this book, in which we confine ourselves to a classical (i.e. non-Bayesian) approach.

1.2 THE STRUCTURE OF THE BOOK

Chapter II presents a brief survey of certain basic ideas of the linear model: descriptive regression, the maximum likelihood principle, and simple test procedures. Further, we investigate the relation between the model with uncorrelated disturbances and various standard models with correlated disturbances.

Section 2.3 is devoted to *linear restrictions* and related problems such as multicollinearity, testing linear hypotheses, and estimation under linear restrictions. The results on optimal estimation give the lower bounds of the quadratic risk for the possible linear set-ups (heterogeneous and homogeneous, biased and unbiased) of an estimator. As a basis for the following chapters, the various well-known measures of efficiency of an estimator such as mean-square error, scalar mean square error, and the general quadratic risk are introduced.

In Chapter III we look at the problem of *mixed estimation*. We introduce linear stochastic restrictions of type $\mathbf{r} = \boldsymbol{R}\boldsymbol{\beta} + \boldsymbol{\phi}$, which arise in many practical situations. The mixed estimation procedure due to Theil and Goldberger (1961) is more efficient than the unrestricted least-squares estimator (LSE) where disturbances variance σ^2 is known, though to ensure its *practicability* one must have auxiliary information on σ^2. It is proposed to use natural bounds on σ^2 which guarantee that the resulting modified mixed estimator dominates the LSE. In order to make the linear restrictions more realistic, in Section 3.3 we investigate situations where the prior information is incorrect in the mean. Thus we are led to a biased estimator which dominates the LSE under certain conditions depending on the degree of incorrectness. That idea enables us to interpret the *ridge* estimator as a special mixed estimator. A method is given for finding a risk-minimizing value of k, and a stepwise test procedure supports the process of model building by finding the optimal k. This approach gives an idea of the interval of k in which the mixed estimator has the best performance.

Also in Chapter III, a method for testing the null hypothesis H_0: 'prior and sample information are in agreement' is proposed. Based on the biased restricted estimator, and using three mean-square error criteria for comparing estimators, there is given an F-test to check the validity of a specified linear restriction. Finally, some alternative estimators are developed which mix exact and stochastic restrictions.

Chapter IV is devoted to a more sensitive type of prior information—the *inequality restrictions*—which may be written as a *concentration ellipsoid.* Using the minimax principle we derive biased estimators—the family of *minimax-linear estimators* (MILE)—which optimally use this prior information. The MILE in every case dominates the unrestricted LSE. On the other hand, using the familiar MSE-risk leads to conditions for superiority which are comparable with those connected with the ridge estimator. A wide discussion is devoted to the robustness of the MILE, where we look at various aspects of *incorrect specified prior information.* A simulation experiment investigates the stability of the MILE in terms of its dependence on the size and the location of a prior region.

Chapter V introduces some basic results on *prediction* in linear models to an extent which is relevant for problems in model choice. The most important case, that of reducing the number of included exogeneous variables, is included in the problem of deciding between an unrestricted model and a model under linear restrictions. Thus we are led again to a comparison of biased and unbiased estimators or predictors. As the main point of interest we define two-stage estimators which decide between the alternative models on the basis of an optimal region for the restricted estimator. This may be understood as a *pre-testing procedure* using initial estimates of the parameters.

In Chapter VI we investigate the use of prior information of standard type in order to identify stochastic regressors for single-equation models as well as *simultaneous-equation models of econometrics.* We propose an econometric estimator, based on a mathematical interpretation of the familiar 2 SLSE, which takes into account linear restrictions on the parameters of a structural equation and can be shown to dominate the 2 SLSE.

The two appendices concern matrix algebra and matrix differentiation.

Full references for the book appear in the Bibliography, which is followed by a subject index and a glossary of abbreviations and notation.

II General Linear Model

2.1 GENERAL LINEAR MODEL

2.1.1 Descriptive Regression

In economics, as well as in other sciences such as physics, chemistry, biology, or agriculture, we are often confronted with the problem of finding a relationship between a family of variables. This interdependency may be derived from theory and/or may be based on empirical investigations. As a fundamental premise we have to assume that the essential phenomena of the underlying relation can be reproduced in principle. In this way we can hope that our empirical and theoretical investigations allow us to formulate a mathematical model which will be the starting point for further considerations in statistical inference, as for instance prediction.

First we want to describe empirically, without reference to a statistical or probabilistic model, the 'naive' analysis of statistical dependence within a given family of observed (i.e. quantitative) variables. Examples:

(i) The quantity of a commodity demanded by a consumer depends on the consumer's income and on the price of the commodity.
(ii) The cost of producing a commodity depends on the quantity produced and on the prices of the raw material and other production factors.
(iii) Individual consumption is related to disposable income.

In general we have a family of $K+1$ quantitative real-valued variables $Y, X_1, \ldots, X_K$, such as volume (output) of production, income, consumption, etc.

We assume that we have $T \geq K$ jointly observed values of these variables which can be arranged in the empirical data matrix

$$\begin{pmatrix} y_1 & X_{11} & \ldots & X_{K1} \\ y_2 & X_{12} & \ldots & X_{K2} \\ \vdots & \vdots & & \vdots \\ y_t & X_{1t} & \ldots & X_{Kt} \\ \vdots & \vdots & & \vdots \\ y_T & X_{1T} & \ldots & X_{KT} \end{pmatrix} = (\mathbf{y}, X). \tag{2.1.1}$$

It is usual in practice to interpret the index t as time, or as the period t. This interpretation is taken for convenience and does not restrict the application of empirical regression to time series data. The style of writing the data matrix (1) indicates that the variable Y depends on the variables collected in $\boldsymbol{X}$. Later we shall define Y as *endogenous* and $X_1, \ldots, X_K$ as *exogenous*.

If there are only two variables (Y and X_1, i.e. $K = 1$) the empirical data can be exhibited graphically in a *bivariate scattergram* (see Figure 2.1.1). As pointed out by Till (1973, p. 303) and by Bibby and Toutenburg (1978, p. 41), fitting a straight line to a set of measurements is a common procedure in many branches of science; the scattergram is in fact the fundamental analytic procedure for the detection of statistical dependence.

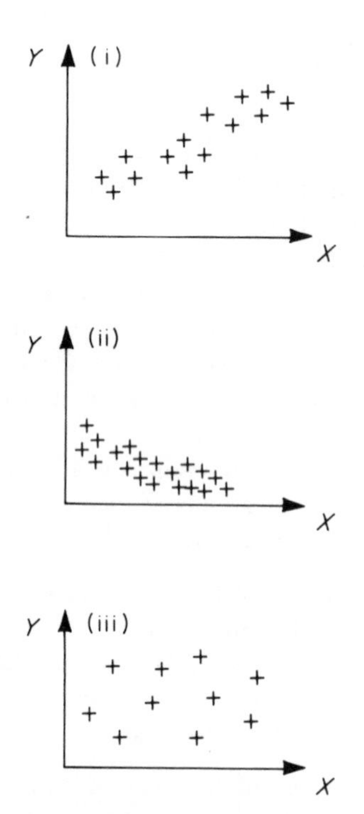

Figure 2.1.1 Bivariate scattergram. (i) Weak positive correlations; (ii) negative correlation; (iii) without correlation

Of course, in many situations a function sufficiently complicated could be found such that it passes through every point of the scattergram. It is clear from analysis that, for instance, a polynomial of sufficiently high degree can be chosen. In general we have a function $Y = f(X_1, \ldots, X_K)$ which will ensure a curve which fits the data. The function $f(.)$ is called the *empirical regression*. Since in practice we are interested in a simple type of function which approximates the conjectured statistical dependence sufficiently, we often concentrate on the linear function

$$Y = f(X_1, \ldots, X_K) = X_1\beta_1 + \ldots + X_K\beta_K \tag{2.1.2}$$

where the β_i are unknown regression coefficients. If linearity seems to be inappropriate there often exist adequate transformations which will realize a linear set-up. Thus linearity is much more useful than is apparent at first sight (for a detailed discussion see Schönfeld, 1973), and seems to be a *reasonable compromise between practicability and mathematical optimality.*

Having chosen the type of function the second problem is to determine numerical values for the coefficients $\beta_i (i = 1, \ldots, k)$ such that the function (2) gives a sufficient or, if possible, the best fit of empirical data. It is thus necessary to find a criterion for determining the *goodness of fit*. Any such criterion has to take into account the deviations between the observed values y_t and the fitted values $\hat{y}_t = \sum_{i=1}^{K} x_{ti} b_i$. Therefore we have to define the regression discrepancies $e_t (t = 1, \ldots, T)$ by

$$y_t - \hat{y}_t = e_t \tag{2.1.3}$$

or, equivalently,

$$y_t = \sum x_{it} b_i + e_t \qquad (t = 1, \ldots, T). \tag{2.1.4}$$

This gives in matrix form

$$\mathbf{y} = \boldsymbol{X}\mathbf{b} + \mathbf{e}. \tag{2.1.5}$$

Given the data matrix (1), the vector of regression discrepancies $\mathbf{e}$ is a function of the chosen numerical value of $\mathbf{b}$. The choice of $\mathbf{b}$ therefore directly influences the goodness of data fitting. In measuring the goodness there exist possible alternatives. We know for instance

$$\max_t |e_t|, \ \sum_{t=1}^{T} |e_t|, \ \text{ or } \ \sum_{t=1}^{T} e_t^2 = \mathbf{e}'\mathbf{e}. \tag{2.1.6}$$

Choosing the numerical values of $b_i (i = 1, \ldots, K)$ such that the third function in (6) is minimized gives the *principle of least squares*. If there are no restrictions on the coefficients β_i of (2), the set B of admissible vectors $\boldsymbol{\beta}$ is $B = E^K$ (the K-dimensional Euclidean space).

Principle of least squares The vector $\mathbf{b} \in E^K$ gives the best fit of data in the sense of the least squares measure if

$$\min_{\boldsymbol{\beta} \in E^K} \mathbf{e}'\mathbf{e} = \min_{\boldsymbol{\beta} \in E^K} (\mathbf{y} - \boldsymbol{X\beta})'(\mathbf{y} - \boldsymbol{X\beta}) = (\mathbf{y} - \boldsymbol{X}\mathbf{b})'(\mathbf{y} - \boldsymbol{X}\mathbf{b}) \tag{2.1.7}$$

where the data matrix $(\mathbf{y}, \boldsymbol{X})$ is assumed to be fixed.

As $S(\boldsymbol{\beta}) = (\mathbf{y} - \boldsymbol{X\beta})'(\mathbf{y} - \boldsymbol{X\beta})$ is a real-valued nonnegative quadratic function, the existence of a finite minimum is assured. To get the solution we write

$$S(\boldsymbol{\beta}) = \mathbf{y}'\mathbf{y} + \boldsymbol{\beta}'\boldsymbol{X}'\boldsymbol{X\beta} - 2\boldsymbol{\beta}'\boldsymbol{X}'\mathbf{y}$$

and differentiate with respect to $\boldsymbol{\beta}$ (see Theorems A.50 and A.52) which gives

$$\frac{\partial S(\boldsymbol{\beta})}{\partial \boldsymbol{\beta}} = \boldsymbol{X}'\boldsymbol{X\beta} - \boldsymbol{X}'\mathbf{y}.$$

Equating to zero we get the so-called normal equation for the solution $\mathbf{b}$, namely

$$X'X\mathbf{b} = X'\mathbf{y}. \tag{2.1.8}$$

Assumption Let rank $X'X = p(0 \leq p \leq K)$. Without loss of generality let the first p columns $X_1, \ldots, X_p$ of the $T \times K$-matrix X be linearly independent.

We can then rearrange X and $\mathbf{b}$ as

$$X = (X_1, X_2) \quad \text{and} \quad \mathbf{b} = \begin{pmatrix} \mathbf{b}_1 \\ \mathbf{b}_2 \end{pmatrix}$$

where

$$\underset{T,p}{X_1} = (X_1, \ldots, X_p), \qquad \underset{T,(K-p)}{X_2} = (X_{p+1}, \ldots, X_K)$$

$$\mathbf{b}_1' = (b_1, \ldots, b_p), \qquad \mathbf{b}_2' = (b_{p+1}, \ldots, b_K).$$

As rank X = rank $X_1 = p$, the columns of X_2 are linearly dependent on X_1, i.e. there exists a matrix L such that $X_2 = X_1 L$. The normal equation (8) then becomes

$$\begin{pmatrix} X_1'X_1 & X_1'X_1L \\ L'X_1'X_1 & L'X_1'X_1L \end{pmatrix} \begin{pmatrix} \mathbf{b}_1 \\ \mathbf{b}_2 \end{pmatrix} = \begin{pmatrix} X_1'\mathbf{y} \\ L'X_1'\mathbf{y} \end{pmatrix}. \tag{2.1.9}$$

The first p rows of this equation give

$$X_1'X_1\mathbf{b}_1 + X_1'X_1L\mathbf{b}_2 = X_1'\mathbf{y}. \tag{2.1.10}$$

The rows $p+1, \ldots, K$ of (9) are the linear combination of (10). Therefore any solution of (10) solves the last rows equation in (9), also.

As rank $X_1 = p$, the inverse $(X_1'X_1)^{-1}$ exists. Premultiplying (10) by $(X_1'X_1)^{-1}$ gives

$$\mathbf{b}_1 = (X_1'X_1)^{-1}X_1'\mathbf{y} - L\mathbf{b}_2.$$

The $(K-p)$-vector $\mathbf{b}_2$ may be chosen arbitrarily so that we have a so-called $(K-p)$-dimensional manifold of solutions.

If $p = K$ (i.e. the matrix X is of full column rank) we have the *unique* solution

$$\mathbf{b} = (X'X)^{-1}X'\mathbf{y} \tag{2.1.11}$$

which is called the *empirical least squares estimator* (ELSE).

If in general rank $X = p \leq K$, two solutions $\mathbf{b}$ and $\mathbf{b}^*$ of the normal equation span the same response surface (also called the LS hyperplane), i.e. we have

$$X\mathbf{b} = X\mathbf{b}^*.$$

The proof is simple. As $\mathbf{b}$ and $\mathbf{b}^*$ are solutions of the normal equation we have

$$X'X\mathbf{b} = X'\mathbf{y} \quad \text{and} \quad X'X\mathbf{b}^* = X'\mathbf{y}$$

and therefore

$$X'X(\mathbf{b} - \mathbf{b}^*) = \mathbf{0}.$$

This gives

$$X(\mathbf{b}-\mathbf{b}^*)=\mathbf{0}$$

which completes the proof.

Thus we have

$$S(\mathbf{b})=(\mathbf{y}-X\mathbf{b})'(\mathbf{y}-X\mathbf{b})=(\mathbf{y}-X\mathbf{b}^*)'(\mathbf{y}-X\mathbf{b}^*)=S(\mathbf{b}^*).$$

Notation The estimated vector of dependent variables $\hat{\mathbf{y}}=X\mathbf{b}$ is called the empirical response surface. The difference $\mathbf{y}-X\mathbf{b}$ is denoted by $\hat{\mathbf{e}}$ and is called the vector of *estimated residuals.*

We have the relation (often called analysis of variance)

$$\mathbf{y}'\mathbf{y}=\hat{\mathbf{y}}'\hat{\mathbf{y}}+\hat{\mathbf{e}}'\hat{\mathbf{e}} \tag{2.1.12}$$

which follows from

$$\mathbf{b}'X'X\mathbf{b}=\mathbf{b}'X'\mathbf{y} \quad \text{((8) premultiplied by } \mathbf{b}')$$

and

$$S(\mathbf{b})=\hat{\mathbf{e}}'\hat{\mathbf{e}}=(\mathbf{y}-X\mathbf{b})'(\mathbf{y}-X\mathbf{b})=\mathbf{y}'\mathbf{y}-\mathbf{b}'X'\mathbf{y}=\mathbf{y}'\mathbf{y}-\hat{\mathbf{y}}'\hat{\mathbf{y}}.$$

Thus the sum of the squares of the observed values $\mathrm{SS}(\mathbf{y})$ is given by the components $\mathrm{SS}(\hat{\mathbf{y}})$ and $\mathrm{SS}(\hat{\mathbf{e}})$; $\mathrm{SS}(\hat{\mathbf{y}})$ is due to regression and $\mathrm{SS}(\hat{\mathbf{e}})$ is the residual term not explained by regression.

Based on (12) it is possible to define a measure of goodness-of-fit for the empirical regression. Let

$$R^2=\frac{\mathrm{SS}(\hat{\mathbf{y}})}{\mathrm{SS}(\mathbf{y})}=1-\frac{\mathrm{SS}(\hat{\mathbf{e}})}{\mathrm{SS}(\mathbf{y})}. \tag{2.1.13}$$

R^2 is called the *coefficient of multiple determination,* and clearly $0 \le R^2 \le 1$. The nearer R^2 is to 1 the better is the fit of the empirical regression to the points $(\mathbf{y}, X)$. If $R^2=0$ we cannot believe in the linear model used. So R^2 measures the percentage of the variation in the dependent variables explained jointly by the independent variables.

Frequently, in economic research, one is faced with the situation of having to determine, on the basis of the sample information, a subset of all the suggested independent variables, assuming that the functional form of dependence is given. R^2 can be used as a criterion for the comparison of the goodness of fit between two alternative sets of independent variables.

As pointed out in Huang (1970, p. 81), R^2 is a nondecreasing function of the number of regressors included in the model. So R^2 must be corrected for this number of degrees of freedom, giving a measure

$$\bar{R}^2=1-(1-R^2)\frac{T-1}{T-K}$$

where K is the number of variables included. (For a more detailed theoretical justification of $\bar{R}^2$ see Barten, 1962.)

2.1.2 The Classical Regression Model

We now leave the idea of descriptive regression and consider $\boldsymbol{\beta}$ as the unknown parameter of a linear regression process which is under investigation. So $\boldsymbol{\beta}$ is no longer chosen according to a fixed idea of data fitting but has to be estimated.

The standard model with which most of this book is concerned is given by

$$\mathbf{y} = \boldsymbol{X\beta} + \boldsymbol{\varepsilon}, \quad \boldsymbol{\varepsilon} \sim (\mathbf{0}, \boldsymbol{\Sigma}) \tag{2.1.14}$$

where the symbols have the following meanings:

$\mathbf{y}$	the $T \times 1$ vector of the dependent variable, is random and observable
$\boldsymbol{X}$	the $T \times K$ matrix of explanatory variables, is fixed and observable
$\boldsymbol{\beta}$	the $K \times 1$ vector of regression coefficients, is fixed and unobservable
$\boldsymbol{\varepsilon}$	the $T \times 1$ vector of disturbances, is random and unobservable
$\boldsymbol{\Sigma}$	the $T \times T$ dispersion matrix, is fixed and unobservable.

The notation $\boldsymbol{\varepsilon} \sim (\mathbf{0}, \boldsymbol{\Sigma})$ is an abbreviation born in recent years and it means that $\boldsymbol{\varepsilon}$ is distributed with expectation $E\boldsymbol{\varepsilon} = \mathbf{0}$ and dispersion (covariance) $E\boldsymbol{\varepsilon\varepsilon}' = \boldsymbol{\Sigma}$. Only in those cases where we are interested in testing statistical hypotheses does any special type of distribution have to be assumed, additionally. Rao (1971) gave a further abbreviation of the model (1) by introducing the triplet $(\mathbf{y}, \boldsymbol{X\beta}, \boldsymbol{\Sigma})$.

The type of the dispersion matrix $\boldsymbol{\Sigma}$ often determines the name of the specified regression as well as the methods to be applied for estimation of the parameters. The most simple case is that of *uncorrelated* disturbances, which is expressed by $\boldsymbol{\Sigma} = \sigma^2 \boldsymbol{I}$. The corresponding model (14) is called classical regression, or regression with *homoscedastic* distributed disturbances. If the matrix $\boldsymbol{X}$ is of full rank K, then the well-known standard estimator of $\boldsymbol{\beta}$ is the *ordinary least squares estimator* (OLSE).

$$\mathbf{b}_0 = (\boldsymbol{X}'\boldsymbol{X})^{-1}\boldsymbol{X}'\mathbf{y} \tag{2.1.15}$$

(which coincides with the empirical least squares estimator (11)).

If rank $\boldsymbol{X} < K$ no regular inverse of $(\boldsymbol{X}'\boldsymbol{X})$ exists. In this case a method using generalized inverses may be employed.

To calculate the OLSE $\mathbf{b}_0$, both algebraical and geometrical methods, for instance projection matrices, are possible (see Seber 1966). Another method, used in this book, is the derivation of linear estimators by minimizing quadratic risk functions.

2.1.3 Linear Estimators

$\mathbf{y}$ is a random vector. Let $\boldsymbol{C}$ and $\mathbf{c}$ be respectively a matrix and a vector of the relevant size, then the vector

$$\hat{\boldsymbol{\beta}} = \boldsymbol{C}\mathbf{y} + \mathbf{c} \tag{2.1.16}$$

may be called a linear estimator of $\boldsymbol{\beta}$. $\hat{\boldsymbol{\beta}}$ is called a *homogeneous* estimator if $\mathbf{c} = \mathbf{0}$; otherwise $\hat{\boldsymbol{\beta}}$ is called a *heterogeneous* estimator. Clearly the OLSE $\mathbf{b}_0$ given in (15) is a special case of a homogeneous estimator, obtained by putting $\boldsymbol{C} = (\boldsymbol{X}'\boldsymbol{X})^{-1}\boldsymbol{X}'$ and $\mathbf{c} = \mathbf{0}$ in (16).

2.1.4 Mean Square Error

If a linear estimator of type (16) is given, its *mean square error* (MSE) is defined by

$$\text{MSE}\,(\hat{\boldsymbol{\beta}}) = E(\hat{\boldsymbol{\beta}}-\boldsymbol{\beta})(\hat{\boldsymbol{\beta}}-\boldsymbol{\beta})'. \tag{2.1.17}$$

As

$$\hat{\boldsymbol{\beta}}-\boldsymbol{\beta} = (\boldsymbol{C}\boldsymbol{X}-\boldsymbol{I})\boldsymbol{\beta}+\mathbf{c}+\boldsymbol{C}\boldsymbol{\varepsilon}, \tag{2.1.18}$$

$$E(\hat{\boldsymbol{\beta}})-\boldsymbol{\beta} = \text{bias}\,\hat{\boldsymbol{\beta}} = (\boldsymbol{C}\boldsymbol{X}-\boldsymbol{I})\boldsymbol{\beta}+\mathbf{c} \tag{2.1.19}$$

and

$$V(\hat{\boldsymbol{\beta}}) = \sigma^2\boldsymbol{C}\boldsymbol{C}' \tag{2.1.20}$$

we have

$$\text{MSE}\,(\hat{\boldsymbol{\beta}}) = \sigma^2\boldsymbol{C}\boldsymbol{C}' + [(\boldsymbol{C}\boldsymbol{X}-\boldsymbol{I})\boldsymbol{\beta}+\mathbf{c}][(\boldsymbol{C}\boldsymbol{X}-\boldsymbol{I})\boldsymbol{\beta}+\mathbf{c}]'. \tag{2.1.21}$$

This expression is the well-known formula MSE = variance + (bias)^2 in its matrix form.

We need to distinguish between the MSE (21), which is of matrix type, and another definition of a *scalar mean square error* (SMSE):

$$\text{SMSE}\,(\hat{\boldsymbol{\beta}}) = E(\hat{\boldsymbol{\beta}}-\boldsymbol{\beta})'(\hat{\boldsymbol{\beta}}-\boldsymbol{\beta}). \tag{2.1.22}$$

Applying the trace operator we have immediately

$$\text{SMSE}\,(\hat{\boldsymbol{\beta}}) = \text{tr}\,E(\hat{\boldsymbol{\beta}}-\boldsymbol{\beta})(\hat{\boldsymbol{\beta}}-\boldsymbol{\beta})' = \text{tr MSE}\,(\hat{\boldsymbol{\beta}}). \tag{2.1.23}$$

Both definitions will be used in this book.

If a given estimator $\hat{\boldsymbol{\beta}}$ can be proved to be homogeneous and unbiased, i.e. if $\mathbf{c} = \mathbf{0}$ and the well-known *condition for lack of bias*

$$\boldsymbol{C}\boldsymbol{X}-\boldsymbol{I} = \boldsymbol{0} \tag{2.1.24}$$

holds, then we have MSE $(\hat{\boldsymbol{\beta}}) = V(\hat{\boldsymbol{\beta}}) = \sigma^2\boldsymbol{C}\boldsymbol{C}'$.

This result is true, for example, for the special case of OLSE (15):

$$\text{MSE}\,(\mathbf{b}_0) = V(\mathbf{b}_0) = \sigma^2(\boldsymbol{X}'\boldsymbol{X})^{-1}. \tag{2.1.25}$$

In this book we shall seek estimators that minimize the value of (21), or of certain functions of it.

In general our purpose is to improve estimation under squared error loss which measures the loss of an estimator $\hat{\boldsymbol{\beta}}$ of $\boldsymbol{\beta}$ by

$$L(\hat{\boldsymbol{\beta}}, \boldsymbol{A}) = (\hat{\boldsymbol{\beta}}-\boldsymbol{\beta})'\boldsymbol{A}(\hat{\boldsymbol{\beta}}-\boldsymbol{\beta}) \tag{2.1.26}$$

where $\boldsymbol{A}$ is a positive definite symmetric matrix of corresponding order $K \times K$.

Note Throughout the book we shall use the abbreviations

$$\boldsymbol{A} > \boldsymbol{0} \text{ if } \boldsymbol{A} \text{ is positive definite}$$

and

$$\boldsymbol{A} \geq \boldsymbol{0} \text{ if } \boldsymbol{A} \text{ is nonnegative definite.}$$

Note, moreover, that all matrices considered in this book are real.

Given (26), the risk of an estimator $\hat{\boldsymbol{\beta}}$ is defined as

$$\begin{aligned} R(\hat{\boldsymbol{\beta}}, \boldsymbol{A}) &= E(L(\hat{\boldsymbol{\beta}}, \boldsymbol{A})) \\ &= E(\hat{\boldsymbol{\beta}} - \boldsymbol{\beta})' \boldsymbol{A} (\hat{\boldsymbol{\beta}} - \boldsymbol{\beta}). \end{aligned} \tag{2.1.27}$$

Applying the trace operator we have a connection between the risk and the (matrix-valued) mean square error of an estimator, namely

$$R(\hat{\boldsymbol{\beta}}, \boldsymbol{A}) = \operatorname{tr} \boldsymbol{A} E(\hat{\boldsymbol{\beta}} - \boldsymbol{\beta})(\hat{\boldsymbol{\beta}} - \boldsymbol{\beta})' = \operatorname{tr} \boldsymbol{A} \operatorname{MSE}(\hat{\boldsymbol{\beta}}). \tag{2.1.28}$$

This formula and Theorem A.15 together show the equivalence of minimizing the risk $R(\hat{\boldsymbol{\beta}}, \boldsymbol{A})$ or the MSE $(\hat{\boldsymbol{\beta}})$ for estimators $\hat{\boldsymbol{\beta}}$ which are in a fixed class of possible estimators.

Theorem 2.1 (Theobald, 1974)

Let two estimators $\boldsymbol{\beta}_1^*$ and $\boldsymbol{\beta}_2^*$ be given. The following two statements are then equivalent:

(a) $\quad \operatorname{MSE}(\boldsymbol{\beta}_1^*) - \operatorname{MSE}(\boldsymbol{\beta}_2^*) \geq \boldsymbol{0}$

(b) $\quad R(\boldsymbol{\beta}_1^*, \boldsymbol{A}) - R(\boldsymbol{\beta}_2^*, \boldsymbol{A}) \geq 0$ for all $\boldsymbol{A} \geq \boldsymbol{0}$.

Proof By (28) we have

$$\begin{aligned} R(\boldsymbol{\beta}_1^*, \boldsymbol{A}) - R(\boldsymbol{\beta}_2^*, \boldsymbol{A}) &= \operatorname{tr} \boldsymbol{A}\{\operatorname{MSE}(\boldsymbol{\beta}_1^*) - \operatorname{MSE}(\boldsymbol{\beta}_2^*)\} \\ &= \operatorname{tr} \boldsymbol{A}\Delta, \text{ say.} \end{aligned}$$

By Theorem A.15 we have $\operatorname{tr} \boldsymbol{A}\Delta \geq 0$ for all $\boldsymbol{A} \geq \boldsymbol{0}$ if and only if $\Delta \geq 0$.

The Canonical Form

To simplify considerations in the linear model, the so-called canonical form of regression (see, for example, Liski, 1979, p. 10) is often used. Since the $K \times K$-matrix $\boldsymbol{X}'\boldsymbol{X}$ is symmetric, there exists a $K \times K$ orthogonal matrix $\boldsymbol{P}$ such that

$$\boldsymbol{P}'\boldsymbol{P} = \boldsymbol{P}\boldsymbol{P}' = \boldsymbol{I}$$

and

$$\boldsymbol{P}'\boldsymbol{X}'\boldsymbol{X}\boldsymbol{P} = \boldsymbol{\Lambda} = \operatorname{diag}(\lambda_1, \ldots, \lambda_K)$$

where $\lambda_i (\mathrm{i} = 1, \ldots, K)$ are the eigenvalues of $\boldsymbol{X}'\boldsymbol{X}$ (see Theorem A.2). If rank

$X = K$, all λ_i are positive. The model (14) can be rewritten as

$$\begin{aligned} y &= XPP'\beta + \varepsilon \\ &= \tilde{X}\tilde{\beta} + \varepsilon \end{aligned} \tag{2.1.29}$$

where $\tilde{X} = XP$, $\tilde{\beta} = P'\beta$, and the columns of $\tilde{X}$ are orthogonal. The elements of $\tilde{\beta}$ are called the *regression coefficients of the principal components.* The scalar mean square error (22) is invariant with respect to the following operation.

Let $\hat{\beta}$ be a linear estimator and $\beta^* = P'\hat{\beta}$. Then we have

$$\begin{aligned} \text{SMSE}(\beta^*) &= E(P'\hat{\beta} - P'\beta)'(P'\hat{\beta} - P'\beta) \\ &= E(\hat{\beta} - \beta)'PP'(\hat{\beta} - \beta) \\ &= \text{SMSE}(\hat{\beta}). \end{aligned} \tag{2.1.30}$$

On the other hand we have for the matrix-valued mean square error:

$$\begin{aligned} \text{MSE}(\beta^*) &= E(P'\hat{\beta} - P'\beta)(P'\hat{\beta} - P'\beta)' \\ &= P'\,\text{MSE}(\hat{\beta})P. \end{aligned} \tag{2.1.31}$$

By (28) we get as the corresponding relation for the risk:

$$R(\beta^*, A) = \text{tr}\, P'AP\,\text{MSE}(\hat{\beta}). \tag{2.1.32}$$

For the OLSE $\mathbf{b}_0$ (15) we find

$$\begin{aligned} \text{SMSE}(\mathbf{b}_0) &= \sigma^2\, \text{tr}\,(X'X)^{-1} \\ &= \sigma^2\, \text{tr}\, \Lambda^{-1} = \sigma^2 \sum_{i=1}^{K} \lambda_i^{-1} \end{aligned} \tag{2.1.33}$$

and by using (31)

$$R(\mathbf{b}_0, A) = \sigma^2\, \text{tr}\, PAP'\Lambda^{-1}. \tag{2.1.34}$$

2.1.5 Admissibility of Estimators

If a well-defined class of estimator functions, as, for example, the class of linear homogeneous unbiased estimators, $\hat{\beta} = C\mathrm{y}$, $CX = I$, and a risk function are given, the best estimator can be calculated by minimizing the risk over that class of estimators. The resulting estimator would be called the R-optimal (e.g. homogeneous unbiased) estimator.

On the other hand, if two estimators are given explicitly, the problem of comparing the goodness (in some sense) of these two estimators arises. So we are led to the following definitions (see also Ferguson, 1967, p. 54).

Definition 2.1 An estimator $\hat{\beta}_1$ is said to dominate an estimator $\hat{\beta}_2$ under the quadratic loss $L(\hat{\beta}, A)$ (see (26)) if

$$R(\hat{\beta}_2, A) - R(\hat{\beta}_1, A) \geq 0 \quad \text{for all } \beta \tag{2.1.35}$$

and

$$R(\hat{\boldsymbol{\beta}}_2, A) - R(\hat{\boldsymbol{\beta}}_1, A) > 0 \qquad \text{for some } \boldsymbol{\beta}. \tag{2.1.36}$$

Following (28) and Theorem A.15 we may conclude that an estimator $\hat{\boldsymbol{\beta}}_1$ dominates an estimator $\hat{\boldsymbol{\beta}}_2$ if the difference of matrices

$$\text{MSE}(\hat{\boldsymbol{\beta}}_2) - \text{MSE}(\hat{\boldsymbol{\beta}}_1) \geq \boldsymbol{0} \quad \text{for all } \boldsymbol{\beta} \tag{2.1.37}$$

and

$$\text{MSE}(\hat{\boldsymbol{\beta}}_2) - \text{MSE}(\hat{\boldsymbol{\beta}}_1) \neq \boldsymbol{0} \qquad \text{for some } \boldsymbol{\beta}. \tag{2.1.38}$$

The relations (35) and (36) between two estimators $\hat{\boldsymbol{\beta}}_1$ and $\hat{\boldsymbol{\beta}}_2$ clearly depend on the matrix A of the loss function (26). In other words, the use of different weight matrices A leads to different loss functions, and therefore to different risks. It may happen that $\hat{\boldsymbol{\beta}}_1$ dominates $\hat{\boldsymbol{\beta}}_2$ with respect to a quadratic loss function $L(\hat{\boldsymbol{\beta}}, A_1)$ but not with respect to $L(\hat{\boldsymbol{\beta}}, A_2)$ with $A_1 \neq A_2$ (see Liski, 1979, p. 13).

The concept of dominating defines a relation between two given estimators. To find out which estimator is best in some sense leads to the concept of *admissibility*.

Definition 2.2 An estimator is said to be admissible with respect to the risk $R(\hat{\boldsymbol{\beta}}, A)$ if no estimator exists which dominates it.

Clearly, admissibility is related to a given class of estimators: in general to the class of all linear estimators $\hat{\boldsymbol{\beta}} = C\mathbf{y} + \mathbf{c}$.

Admissibility can be defined also with respect to the mean square error (17) of $\hat{\boldsymbol{\beta}}$.

Definition 2.3 An estimator $\hat{\boldsymbol{\beta}}$ is said to be admissible with respect to the mean square error if for all estimators $\tilde{\boldsymbol{\beta}}$

$$\text{MSE}(\tilde{\boldsymbol{\beta}}) - \text{MSE}(\hat{\boldsymbol{\beta}}) \geq \boldsymbol{0} \quad \text{for all } \boldsymbol{\beta}$$

and

$$\text{MSE}(\tilde{\boldsymbol{\beta}}) - \text{MSE}(\hat{\boldsymbol{\beta}}) \neq \boldsymbol{0} \quad \text{for some } \boldsymbol{\beta}.$$

By Theorem 2.1, both concepts of admissibility are equivalent if and only if the estimator $\hat{\boldsymbol{\beta}}$ is admissible with respect to $R(\hat{\boldsymbol{\beta}}, A)$ for all $A \geq \boldsymbol{0}$. To ensure this, the admissibility of $\hat{\boldsymbol{\beta}}$ with respect to $R(\hat{\boldsymbol{\beta}}, I)$ is sufficient.

Theorem 2.2 (Shinozaki, 1975)

Let $\hat{\boldsymbol{\beta}}$ be admissible with respect to $R(\hat{\boldsymbol{\beta}}, I)$. Then $\hat{\boldsymbol{\beta}}$ is admissible under

$$R(\hat{\boldsymbol{\beta}}, A) \text{ for any } A \geq \boldsymbol{0}.$$

For the proof the reader is referred to Shinozaki, 1975.

2.1.6 Gauss–Markov Theorem

In Section 2.1.3 we found the OLSE $\mathbf{b}_0 = (X'X)^{-1}X'\mathbf{y} = C_0\mathbf{y}$ to be a special homogeneous linear estimator which is, moreover, unbiased by $C_0X = I$ and $\mathbf{c} = \mathbf{0}$.

We shall now consider in a more general way the problem of best linear unbiased estimation of $\boldsymbol{\beta}$ in the model (14) in which X is assumed to be of full rank. Based on (16), our starting point is the linear homogeneous estimator $\hat{\boldsymbol{\beta}} = C\mathbf{y}$ in which C is to be optimized with respect to the quadratic risk

$$R(\hat{\boldsymbol{\beta}}, I) = E(\hat{\boldsymbol{\beta}} - \boldsymbol{\beta})'(\hat{\boldsymbol{\beta}} - \boldsymbol{\beta}). \tag{2.1.39}$$

Using the homogeneous linear set-up gives

$$\hat{\boldsymbol{\beta}} - \boldsymbol{\beta} = (CX - I)\boldsymbol{\beta} + C\boldsymbol{\varepsilon}. \tag{2.1.40}$$

The demand of unbiasedness of $\hat{\boldsymbol{\beta}}$, i.e. $E\hat{\boldsymbol{\beta}} = \boldsymbol{\beta}$, is equivalent to

$$CX = I \tag{2.1.41}$$

i.e. written row by row

$$\mathbf{c}_i X = \mathbf{e}_i \quad (i = 1, \ldots, K) \tag{2.1.42}$$

where $\mathbf{e}_i$ and $\mathbf{c}_i$ are the ith rows of I and C, respectively.

Therefore the risk (27) becomes

$$R(\hat{\boldsymbol{\beta}}, I) = \sigma^2 \operatorname{tr} CC'. \tag{2.1.43}$$

In this way (43) and (41) result in the optimization problem

$$\min_C \{\sigma^2 \operatorname{tr} CC' \mid CX = I\} \tag{2.1.44}$$

or equivalently to

$$\min_C \{\sigma^2 \operatorname{tr} CC' - 2 \sum_{i=1}^{K} \boldsymbol{\lambda}_i'(\mathbf{c}_i X - \mathbf{e}_i)'\}. \tag{2.1.45}$$

Here $\boldsymbol{\lambda}_i$ is a $K \times 1$-vector of Lagrange multipliers and

$$\underset{K \times K}{\Lambda} = \begin{pmatrix} \boldsymbol{\lambda}_1' \\ \vdots \\ \boldsymbol{\lambda}_K' \end{pmatrix}.$$

Differentiating (45) with respect to C and Λ (see Theorems A.53 and A.51) yields the normal equations

(i) $$\sigma^2 C - \Lambda X' = \mathbf{0}$$

(ii) $$CX - I = \mathbf{0}$$

which give the solution

$$\hat{\Lambda} = \sigma^2 (X'X)^{-1}$$

and therefore the optimal $\hat{C} = C_0$

$$C_0 = (X'X)^{-1}X'. \tag{2.1.46}$$

Thus we get as the solution of the optimization problem (45) the OLSE $\mathbf{b}_0 = C_0\mathbf{y}$ (15), which clearly is unbiased by $C_0X = (X'X)^{-1}.\ X'X = I$ and has dispersion

$$V(\mathbf{b}_0) = E(\mathbf{b}_0 - \boldsymbol{\beta})(\mathbf{b}_0 - \boldsymbol{\beta})' = \sigma^2(X'X)^{-1}. \tag{2.1.47}$$

This result is summarized in the following well-known theorem.

Theorem 2.3 (Gauss–Markov Theorem)

In the classical linear regression model

$$\mathbf{y} = X\boldsymbol{\beta} + \boldsymbol{\varepsilon}, \quad \boldsymbol{\varepsilon} \sim (\mathbf{0}, \sigma^2 I), \quad \text{rank } X = K \tag{2.1.48}$$

the OLSE $\mathbf{b}_0$ (15) with dispersion matrix $V(\mathbf{b}) = \sigma^2(X'X)^{-1}$ is the *best linear unbiased estimator* (BLUE) of $\boldsymbol{\beta}$.

The property of $\mathbf{b}_0$ to be 'best' is related to the risk function (45) or—due to unbiasedness—to optimality in the following sense.

Theorem 2.4

Let $\tilde{\boldsymbol{\beta}} = \tilde{C}\mathbf{y}$ be any linear homogeneous estimator of $\boldsymbol{\beta}$ with dispersion $V(\tilde{\boldsymbol{\beta}})$, $\mathbf{d}$ a fixed nonzero $K \times 1$-vector and $\text{var}(\tilde{\beta}_i)$ and $\text{var}(\mathbf{b}_{0i})$ the ith diagonal elements of $V(\tilde{\boldsymbol{\beta}})$ and $V(\mathbf{b}_0)$, respectively. The following relations are then equivalent:

(i) $$V(\tilde{\boldsymbol{\beta}}) - V(\mathbf{b}_0) \geq \mathbf{0}$$

(ii) $$\text{var}(\tilde{\beta}_i) - \text{var}(\mathbf{b}_{0i}) \geq 0 \quad (i = 1, \ldots, K)$$

(iii) $$\mathbf{d}'V(\tilde{\boldsymbol{\beta}})\mathbf{d} - \mathbf{d}'V(\mathbf{b}_0)\mathbf{d} \geq 0.$$

Proof The proofs of (ii) and (iii), as well as the proof of the equivalence of (i), (ii), and (iii), are left to the reader.

We now prove (i).

Without loss of generality we may write

$$\tilde{C} = C_0 + D = (X'X)^{-1}X' + D.$$

As $\tilde{\boldsymbol{\beta}}$ is assumed to be unbiased we have

$$\tilde{C}X = C_0X + DX = I, \quad \text{i.e. } DX = \mathbf{0}.$$

Using this, we get the dispersion of $\tilde{\boldsymbol{\beta}}$ as

$$\begin{aligned} V(\tilde{\boldsymbol{\beta}}) &= \sigma^2\tilde{C}\tilde{C}' \\ &= \sigma^2((X'X)^{-1}X' + D)(X(X'X)^{-1} + D') \\ &= V(\mathbf{b}) + \sigma^2 DD'. \end{aligned}$$

By Theorem A.10 we have $\sigma^2 \boldsymbol{D}\boldsymbol{D}' \geq 0$, from which (i) follows.

Note The result (iii) of Theorem 2.4 gives the best linear unbiased estimator of the linear form $\mathbf{d}'\boldsymbol{\beta}$ as

$$\widehat{\mathbf{d}'\boldsymbol{\beta}} = \mathbf{d}'\mathbf{b}_0 \quad \text{with variance } (\mathbf{d}'\mathbf{b}_0) = \mathbf{d}'V(\mathbf{b}_0)\mathbf{d}. \tag{2.1.49}$$

The relation (49) will be used later in predicting the regression process.

2.1.7 Estimating the Variance

To estimate the variance σ^2 of the disturbances ε_t we may follow the concept of the 'sum of squared errors' (6) in principle. In replacing in (6) the vector of residuals $\mathbf{e}$ by the estimated disturbances $\hat{\boldsymbol{\varepsilon}} = \mathbf{y} - \boldsymbol{X}\mathbf{b}_0$ we get

$$S(\mathbf{b}_0) = \hat{\boldsymbol{\varepsilon}}'\hat{\boldsymbol{\varepsilon}} = (\mathbf{y} - \mathbf{X}\mathbf{b}_0)'(\mathbf{y} - \mathbf{X}\mathbf{b}_0). \tag{2.1.50}$$

We now calculate the expectation of $S(\mathbf{b}_0)$. We have

$$\begin{aligned}\hat{\boldsymbol{\varepsilon}} &= \mathbf{y} - \boldsymbol{X}\mathbf{b}_0 = \boldsymbol{\varepsilon} - \boldsymbol{X}(\boldsymbol{X}'\boldsymbol{X})^{-1}\boldsymbol{X}'\boldsymbol{\varepsilon} \\ &= (\boldsymbol{I} - \boldsymbol{X}(\boldsymbol{X}'\boldsymbol{X})^{-1}\boldsymbol{X}')\boldsymbol{\varepsilon} \\ &= \boldsymbol{M}\boldsymbol{\varepsilon}, \text{ say,}\end{aligned}$$

where $\boldsymbol{M}$ is idempotent by Theorem A.28. Thus we get $\boldsymbol{\varepsilon}'\boldsymbol{M}\boldsymbol{M}\boldsymbol{\varepsilon} = \boldsymbol{\varepsilon}'\boldsymbol{M}\boldsymbol{\varepsilon}$ and

$$\begin{aligned}E\hat{\boldsymbol{\varepsilon}}'\hat{\boldsymbol{\varepsilon}} &= E\boldsymbol{\varepsilon}'\boldsymbol{M}\boldsymbol{\varepsilon} \\ &= E\,\mathrm{tr}\,(\boldsymbol{M}\boldsymbol{\varepsilon}\boldsymbol{\varepsilon}') \\ &= \sigma^2\,\mathrm{tr}\,\boldsymbol{M} \\ &= \sigma^2\,\mathrm{tr}\,(\boldsymbol{I}_T - \boldsymbol{I}_K) \quad \text{(see Theorem A.19)} \\ &= \sigma^2\,(T-K).\end{aligned}$$

Based on this, we have as unbiased estimator of σ^2

$$s^2 = \hat{\boldsymbol{\varepsilon}}'\hat{\boldsymbol{\varepsilon}}(T-K)^{-1} = (\mathbf{y} - \boldsymbol{X}\mathbf{b}_0)'(\mathbf{y} - \boldsymbol{X}\mathbf{b}_0)(T-K)^{-1} \tag{2.1.51}$$

and therefore as unbiased estimator of $V(\mathbf{b})$

$$\hat{V}(\mathbf{b}) = s^2(\boldsymbol{X}'\boldsymbol{X})^{-1}. \tag{2.1.52}$$

2.1.8 Univariate Examples

So far we have considered the regression model (48) without any restrictions on the number $K \geq 1$ of included regressors $\boldsymbol{X}_i$. Some of the ideas on regression may become clearer if we consider the case of one exogenous variable. In that context we may have in mind two types of model. Let variable $\mathbf{y}$ be dependent on a regressor $\boldsymbol{X}$ as well as on a so-called *dummy* variable which introduces a constant term. That model is just

$$y_1 = \alpha + \beta x_t + \varepsilon_t \quad (t = 1, \ldots, T). \tag{2.1.53}$$

Let us denote the sample means of $(y_1, \ldots, y_T)$ and $(x_1, \ldots, x_T)$ by

$$\bar{y} = \frac{1}{T}\sum_{t=1}^{T} y_t \quad \text{and} \quad \bar{x} = \frac{1}{T}\sum_{t=1}^{T} x_t,$$

respectively. If we now transform the observed values y_t, x_t as follows:

$$\tilde{y}_t = y_t - \bar{y} \quad \tilde{x}_t = x_t - \bar{x}$$

we shall arrive at the *univariate* regression model

$$\tilde{y}_t = \beta \tilde{x}_t + \varepsilon_t \quad (t = 1, \ldots, T). \tag{2.1.54}$$

Applying formula (15) we get the OLSE of the scaler β in (54) as

$$b_0 = \frac{\sum_t \tilde{x}_t \tilde{y}_t}{\sum \tilde{x}_t^2} = \frac{\sum_t (x_t - \bar{x})(y_t - \bar{y})}{\sum_t (x_t - \bar{x})^2} \tag{2.1.55}$$

which has variance

$$(b_0) = \frac{\sigma^2}{\sum \tilde{x}_t^2}.$$

The estimator s^2 (51) of σ^2 simplifies to

$$s^2 = (T-2)^{-1} \sum_t (\tilde{y}_t - \tilde{x}_t b_0)^2.$$

The estimator of α may be found by re-transformation as

$$\hat{\alpha} = \bar{y} - b_0 \bar{x}.$$

2.1.9 The Classical Normal Regression

The results derived in the above sections hold without any special assumptions on the type of distribution of the disturbances ε_t. To ensure, for example, the truth of the Gauss–Markov Theorem we had to assume only that $\boldsymbol{\varepsilon} \sim (\mathbf{0}, \sigma^2 \boldsymbol{I})$. If we are interested in testing hypotheses on the parameters β_i of $\boldsymbol{\beta}$ we have to specify the common distribution function of $\boldsymbol{\varepsilon}$. As far as this book is concerned we may confine ourselves to the normal distribution, i.e. if necessary we assume that $\boldsymbol{\varepsilon}$ follows the T-variate normal distribution with mean zero and covariance $\sigma^2 \boldsymbol{I}$. A useful abbreviation for this fact is:

$$\boldsymbol{\varepsilon} \sim N(\mathbf{0}, \sigma^2 \boldsymbol{I}) \quad \text{or} \quad \boldsymbol{\varepsilon} \sim N_T(\mathbf{0}, \sigma^2 \boldsymbol{I}). \tag{2.1.56}$$

More explicitly, we may describe this normal distribution by its joint density function (see Definition A.40 with $\boldsymbol{\Sigma} = \boldsymbol{I}$ and $p = T$)

$$f(\boldsymbol{\varepsilon}; \mathbf{0}, \sigma^2 \boldsymbol{I}) = (2\pi\sigma^2)^{-T/2} \exp\left\{ -\frac{1}{2\sigma^2} \sum_{t=1}^{T} \varepsilon_t^2 \right\} \tag{2.1.57}$$

$$= \prod_{t=1}^{T} (2\pi\sigma^2)^{-1/2} \exp\left\{ -\frac{1}{2\sigma^2} \varepsilon_t^2 \right\}.$$

The last term shows the independence of the ε_t $(t = 1, \dots, T)$.

Remark The general T-variate normal distribution $\xi \sim N(\boldsymbol{\mu}, \boldsymbol{\Sigma})$ has the joint density function

$$f(\xi;\boldsymbol{\mu}, \boldsymbol{\Sigma}) = (2\pi)^{-T/2}|\boldsymbol{\Sigma}|^{-1/2}\exp\{-\tfrac{1}{2}(\xi-\boldsymbol{\mu})'\boldsymbol{\Sigma}^{-1}(\xi-\boldsymbol{\mu})\} \tag{2.1.58}$$

where $|\boldsymbol{\Sigma}|$ is the determinant of the covariance matrix $\boldsymbol{\Sigma}$.

2.1.10 The Maximum-Likelihood Principle in Regression

Let $\xi = (\xi_1, \dots, \xi_n)'$ be a random vector with joint density function $f(\xi;\boldsymbol{\theta})$, where $\boldsymbol{\theta} = (\theta_1, \dots, \theta_m)'$ is the vector of unknown parameters varying in the whole parameter set Ω. Then for any fixed sample ξ_0 of ξ

$$L(\boldsymbol{\theta}) = f(\xi_0;\boldsymbol{\theta}) \tag{2.1.59}$$

defines a function of $\boldsymbol{\theta}$ which is called the *likelihood function* of ξ_0.

Definition 2.4 A fixed value $\hat{\boldsymbol{\theta}}$ of $\boldsymbol{\theta}$ is called a *maximum-likelihood estimator* (abbreviated as MLE) of $\boldsymbol{\theta}$ if

$$L(\hat{\boldsymbol{\theta}}) \geq L(\boldsymbol{\theta}) \quad \text{for all } \boldsymbol{\theta} \in \Omega. \tag{2.1.60}$$

Applying this definition to the normal regression model

$$\mathbf{y} = \boldsymbol{X\beta} + \boldsymbol{\varepsilon}, \quad \boldsymbol{\varepsilon} \sim N(\mathbf{0}, \sigma^2 \boldsymbol{I}) \tag{2.1.61}$$

we get the following.

As $\mathbf{y} \sim N(\boldsymbol{X\beta}, \sigma^2\boldsymbol{I})$ (see Theorem A.41), its likelihood function is found to be

$$L(\boldsymbol{\beta}, \sigma^2) = (2\pi\sigma^2)^{-T/2}\exp\left\{-\frac{1}{2\sigma^2}(\mathbf{y}-\boldsymbol{X\beta})'(\mathbf{y}-\boldsymbol{X\beta})\right\}, \tag{2.1.62}$$

It is well known that the logarithmic transformation is a monotonic one. Therefore, maximizing $L(\boldsymbol{\beta}, \sigma^2)$ gives the same solution as maximizing

$$\ln L(\boldsymbol{\beta}, \sigma^2) = -\frac{T}{2}\ln(2\pi\sigma^2) - \frac{1}{2\sigma^2}(\mathbf{y}-\boldsymbol{X\beta})'(\mathbf{y}-\boldsymbol{X\beta}). \tag{2.1.63}$$

If the parameters $\boldsymbol{\beta}$ and σ^2 are not constrained by prior restrictions, the maximum of (62) has to be searched for the whole parameter space $\Omega = \{\boldsymbol{\beta}: \boldsymbol{\beta} \in E^K; \sigma^2: \sigma^2 > 0\}$ where E^K is the Euclidean space of dimension K. Differentiating (63) with respect to $\boldsymbol{\beta}$ and σ^2, respectively, and equating to zero, yields the normal or likelihood equations

(i) $$\sigma^2\frac{\partial \ln L}{\partial \boldsymbol{\beta}} = \boldsymbol{X}'(\mathbf{y}-\boldsymbol{X\beta}) = \mathbf{0} \text{ (see Theorem A.50)}$$

(ii) $$2\sigma^2\frac{\partial \ln L}{\partial \sigma^2} = -T + \frac{1}{\sigma^2}(\mathbf{y}-\boldsymbol{X\beta})'(\mathbf{y}-\boldsymbol{X\beta}) = 0,$$

i.e.

$$\text{(i)} \qquad \boldsymbol{X}'\boldsymbol{X}\hat{\boldsymbol{\beta}} = \boldsymbol{X}'\mathbf{y},$$

$$\text{(ii)} \qquad \hat{\sigma}^2 = \frac{1}{T}(\mathbf{y} - \boldsymbol{X}\hat{\boldsymbol{\beta}})'(\mathbf{y} - \boldsymbol{X}\hat{\boldsymbol{\beta}}). \tag{2.1.64}$$

If rank $\boldsymbol{X} = K$, (i) gives as solution the OLSE $\mathbf{b}_0 = (\boldsymbol{X}'\boldsymbol{X})^{-1}\boldsymbol{X}'\mathbf{y}$, or, in other words the OLSE and the MLE coincide. Comparing the MLE $\hat{\sigma}^2$ (64) and the previous deleted estimator s^2 (51), we have $\hat{\sigma}^2 = s^2(T-K)/T$. Thus the MLE $\hat{\sigma}^2$ is biased for finite samples but unbiased asymptotically.

2.1.11 Testing Linear Hypotheses about the Regression Coefficients

The problem of testing linear hypotheses on some of the β_i's of $\boldsymbol{\beta}' = (\beta_1, \ldots, \beta)$ requires specification of the distribution function of $\boldsymbol{\varepsilon}$. As noted above, we shall confine ourselves to the normal distribution. The so-called joint test of hypotheses may be developed by using the concept of linear restrictions as well as the method of partial estimation. So, for pedagogical reasons, we defer the test of general linear hypotheses to a later section.

Here we shall give only the idea of a separate test of the null hypothesis

$$\beta_i = 0 \quad (i \text{ fixed}) \tag{2.1.65}$$

against the alternative hypothesis that

$$\beta_i \neq 0 \tag{2.1.66}$$

in the model $\mathbf{y} = \boldsymbol{X\beta} + \boldsymbol{\varepsilon}$, $\boldsymbol{\varepsilon} \sim N(\mathbf{0}, \sigma^2\boldsymbol{I})$. It is obvious that a test procedure has to be based on the deviation of the estimate b_i of β_i and the hypothetical value $\beta_i = 0$. b_i as a part of $\mathbf{b}_0$ is normally distributed and has mean zero under the null hypothesis, i.e.

$$b_i \sim N(\mathbf{0}, \operatorname{var}(b_i)) \,|\, H_0$$

where $\operatorname{var}(b_i)$ is the ith diagonal element of $\boldsymbol{V}(\mathbf{b}_0)$,

$$\operatorname{var}(b_i) = \sigma^2(\boldsymbol{X}'\boldsymbol{X})^{-1}_{ii}.$$

If σ^2 can be assumed to be known, the test statistic would be

$$u = \frac{b_i - 0}{\sqrt{\operatorname{var}(b_i)}} \sim N(0, 1) \quad \text{(see Theorem A.41).} \tag{2.1.67}$$

Otherwise, σ^2 has to be estimated by s^2 (51) which has $s^2 \sim \sigma^2\chi^2_{T-K}$ under the assumption of normality of $\boldsymbol{\varepsilon}$. In that case we use the test statistic

$$t = \frac{b_i - 0}{s\sqrt{(\boldsymbol{X}'\boldsymbol{X})^{-1}_{ii}}} \tag{2.1.68}$$

instead of (67). By familiar theorems (see Theorem A.45) this statistic follows a Student's t distribution with $T-K$ degrees of freedom. The null hypothesis (65) may be accepted if the sample value of $|u|$ (67) does not exceed the $\alpha/2$-quantile of the normal distribution (or, if (68) has to be used, $|t|$ must not exceed the $\alpha/2$-quantile of the t_{T-K}-distribution).

2.2 GENERALIZED LINEAR MODEL

2.2.1 The Aitken Estimator

The standard regression model is given by

$$\mathbf{y} = \boldsymbol{X\beta} + \boldsymbol{\varepsilon}, \quad \boldsymbol{\varepsilon} \sim (\mathbf{0}, \boldsymbol{\Sigma}) \tag{2.2.1}$$

(see (2.1.14)) where $\boldsymbol{\Sigma} = E\boldsymbol{\varepsilon\varepsilon}'$ is the dispersion matrix of the disturbances. The optimality of the OLSE $\mathbf{b}_0$ is restrained on the case of uncorrelatedness and identity of the disturbances' distribution, i.e. on the assumption $\boldsymbol{\Sigma} = \sigma^2 \boldsymbol{I}$. In situations where the disturbance terms (elements of $\boldsymbol{\varepsilon}$) do not all have the same distribution, or where they are correlated, a more general estimator has to be used. The generalized linear regression model due to Aitken (1935) is given by (1) where $\boldsymbol{\Sigma} = \sigma^2 \boldsymbol{W}$. That is, the earlier assumption of $\boldsymbol{\Sigma} = \sigma^2 \boldsymbol{I}$ for the dispersion matrix $\boldsymbol{\Sigma}$ is now replaced by the assumption that

$$\boldsymbol{\Sigma} = \sigma^2 \boldsymbol{W} = \sigma^2 \begin{pmatrix} w_{11} & \cdots & w_{1T} \\ \vdots & & \vdots \\ w_{T1} & \cdots & w_{TT} \end{pmatrix}. \tag{2.2.2}$$

If $\boldsymbol{W} = \boldsymbol{I}$, we have the classical model (48). Otherwise, the $T \times T$ symmetric matrix $\boldsymbol{W}$ introduces $\frac{1}{2}T(T+1)$ additional unknown parameters into the estimation problem. As the sample size T is fixed, we cannot hope to estimate sufficiently well all the parameters $\beta_1, \ldots, \beta_K, \sigma^2$, and $w_{ij} (i \leq j)$, simultaneously.

If possible, we may assume that $\boldsymbol{W}$ is known. If not, we have to restrict ourselves to error distributions having a simple structure, as for instance heteroscedasticity or autoregression (see the following sections).

For pedagogical reasons, we first estimate $\boldsymbol{\beta}$ when $\boldsymbol{W}$ is assumed to be fixed (and known). To motivate the discussion of Aitken's generalized least squares, let us show what happens when the OLSE $\mathbf{b}_0$ is taken as an estimate of $\boldsymbol{\beta}$ in the model (1) where $\boldsymbol{\Sigma}$ is specified in (2). The direct application of $\mathbf{b}_0$ yields $E\mathbf{b}_0 = \beta$ (i.e. unbiasedness is not influenced by the dispersion $\boldsymbol{\Sigma}$) but

$$E(\mathbf{b}_0 - \boldsymbol{\beta})(\mathbf{b}_0 - \boldsymbol{\beta})' = \sigma^2 (\boldsymbol{X}'\boldsymbol{X})^{-1} (\boldsymbol{X}'\boldsymbol{W}\boldsymbol{X}) (\boldsymbol{X}'\boldsymbol{X})^{-1} \tag{2.2.3}$$

which differs from the covariance matrix (2.1.47) unless $\boldsymbol{W} = \boldsymbol{I}$.

As will be pointed out more explicitly in Section 2.2.4, the application of $\mathbf{b}_0$ can produce an underestimation of the true variances where 'true' is related to estimators which reflect $\boldsymbol{\Sigma} = \sigma^2 \boldsymbol{W}$ instead of $\boldsymbol{\Sigma} = \sigma^2 \boldsymbol{I}$.

Definition 2.5 The Aitken estimator or, as it is also called, the generalized least squares estimator (GLSE) of $\boldsymbol{\beta}$ based on the dispersion matrix $\boldsymbol{\Sigma} = \sigma^2 \boldsymbol{W}$ is defined as

$$\mathbf{b} = (\boldsymbol{X}'\boldsymbol{W}^{-1}\boldsymbol{X})^{-1} \boldsymbol{X}'\boldsymbol{W}^{-1}\mathbf{y} \tag{2.2.4}$$

Theorem 2.5 (Gauss–Markov–Aitken)

If $\mathbf{y} = X\boldsymbol{\beta} + \boldsymbol{\varepsilon}$ where $\boldsymbol{\varepsilon} \sim (\mathbf{0}, \sigma^2 W)$ then $\mathbf{b}$ is unbiased and, moreover, is the best linear unbiased estimator (BLUE) in the sense that

$$V(\mathbf{b}) = \sigma^2(X'W^{-1}X)^{-1} \leq V(\tilde{\boldsymbol{\beta}}) \tag{2.2.5}$$

for any other linear unbiased estimator $\tilde{\boldsymbol{\beta}}$. (The proof is similar to that of Theorem 2.3).

Theorem 2.6

If $\hat{\boldsymbol{\varepsilon}}_{\mathbf{b}}$ is the residual vector $\mathbf{y} - X\mathbf{b}$, then

$$\hat{\boldsymbol{\varepsilon}}'_{\mathbf{b}} W^{-1} \hat{\boldsymbol{\varepsilon}}_{\mathbf{b}} \leq \hat{\boldsymbol{\varepsilon}}'_{\tilde{\boldsymbol{\beta}}} W^{-1} \hat{\boldsymbol{\varepsilon}}_{\tilde{\boldsymbol{\beta}}} \tag{2.2.6}$$

for any other linear unbiased estimator $\tilde{\boldsymbol{\beta}}$.

A justification of the GLSE $\mathbf{b}$ defined in (4) may be given by the algebraical connection with the OLSE.

As W and W^{-1} are symmetric and positive definite, there exist quadratic and regular matrices M and N such that

$$W = MM' \quad \text{and} \quad W^{-1} = NN' \tag{2.2.7}$$

and, as a trivial conclusion, we have

$$N'MM'N = N'WN = I.$$

Premultiplying the generalized regression model $\mathbf{y} = X\boldsymbol{\beta} + \boldsymbol{\varepsilon}$, $\boldsymbol{\varepsilon} \sim (\mathbf{0}, \sigma^2 W)$ by N' gives

$$N'\mathbf{y} = N'X\boldsymbol{\beta} + N'\boldsymbol{\varepsilon},\ N'\boldsymbol{\varepsilon} \sim (\mathbf{0}, \sigma^2 I) \tag{2.2.8}$$

which is just a classical regression model in the variables $\tilde{\mathbf{y}} = N'\mathbf{y}$, $\tilde{X} = N'X$, $\tilde{\boldsymbol{\varepsilon}} = N'\boldsymbol{\varepsilon}$. Applying Theorem 2.3 to (8) we get the OLSE of $\boldsymbol{\beta}$:

$$\begin{aligned}\hat{\boldsymbol{\beta}} &= (\tilde{X}'\tilde{X})^{-1}\tilde{X}'\tilde{\mathbf{y}} \\ &= (X'NN'X)^{-1}X'NN'\mathbf{y} \\ &= (X'W^{-1}X)^{-1}X'W^{-1}\mathbf{y} = \mathbf{b}.\end{aligned}$$

Thus the GLSE $\mathbf{b}$ (4) in the generalized regression model and the OLSE $\hat{\boldsymbol{\beta}}$ in the model (8) coincide. In other words the introduction of a general disturbance matrix $\boldsymbol{\Sigma} = \sigma^2 W$ where $W \neq I$ does not complicate the estimation of $\boldsymbol{\beta}$ provided W can be assumed to be known.

For the other unknown parameter σ^2 the following unbiased estimator is available:

$$s^2 = (\mathbf{y} - X\mathbf{b})'W^{-1}(\mathbf{y} - X\mathbf{b})(T-K)^{-1} \tag{2.2.9}$$

where $Es^2 = \sigma^2$ may be proved as in Section 2.1.7.

2.2.2 Heteroscedasticity

By a heteroscedastic distribution of $\boldsymbol{\varepsilon}$ it is understood that the disturbances are uncorrelated but not identically distributed, that is

$$E\varepsilon_t\varepsilon_{t'} = \begin{cases} 0 & t \neq t' \\ \sigma^2 k_t & t = t' \quad (k_t > 0) \end{cases}$$

or, in matrix formulation,

$$E\boldsymbol{\varepsilon\varepsilon}' = \sigma^2 \boldsymbol{W} = \sigma^2 \operatorname{diag}(k_1, \ldots, k_T)$$

where the k's can vary with t in the intervals $0 < k_t < \infty$. If $k_t = k$ for $t = 1, \ldots, T$ we have the classical regression model, also called a model with homoscedastic disturbances. Now, $\boldsymbol{W}^{-1} = \operatorname{diag}(k_1^{-1}, \ldots, k_T^{-1})$ and therefore the GLSE (4) is of the special form

$$\mathbf{b} = \left(\sum_{t=1}^{T} \frac{1}{k_t} \boldsymbol{X}_t \boldsymbol{X}_t'\right)^{-1} \left(\sum_{t=1}^{T} \frac{1}{k_t} \boldsymbol{X}_t y_t\right) \tag{2.2.10}$$

where $\boldsymbol{X}_t'$ is the tth row of $\boldsymbol{X}$. It follows that the GLSE $\mathbf{b}$ (10) may be computed by dividing the tth observation on all variables by $\sqrt{k_t}$ and then computing the OLSE on the transformed observations. That procedure assumes knowledge of the k's. A typical situation of heteroscedasticity is described in Huang (1970, p. 146) and also in Goldberger (1964, p. 235). Let us assume that in the model

$$y_t = \alpha + \beta x_t + \varepsilon_t \quad (t = 1, \ldots, T)$$

the variance of ε_t is directly proportional to the square of x_t, that is

$$\operatorname{var}(\varepsilon_t) = \sigma^2 x_t^2.$$

Then we have $\boldsymbol{W} = \operatorname{diag}(x_1^2, \ldots, x_T^2)$, i.e. $k_t = x_t^2$. Applying $\mathbf{b}$ (10) is then equivalent to transforming the data according to

$$\frac{y_t}{x_t} = \alpha\left(\frac{1}{x_t}\right) + \beta + \frac{\varepsilon_t}{x_t}, \quad \operatorname{var}\left(\frac{\varepsilon_t}{x_t}\right) = \sigma^2$$

and calculating the OLSE of $\begin{pmatrix} \alpha \\ \beta \end{pmatrix}$.

Another model of practical importance is that of aggregate data, i.e. we do not have the original samples $\mathbf{y}$ and $\boldsymbol{X}$, but we do have the sample means

$$\bar{y}_t = \frac{1}{n_t} \sum_{j=1}^{n_t} y_j \quad \bar{x}_{ti} = \frac{1}{n_t} \sum_{j=1}^{n_t} x_{ji}$$

which give

$$\bar{y}_t = \sum_{i=1}^{K} \beta_i \bar{x}_{ti} + \bar{\varepsilon}_t \quad (t = 1, \ldots, T)$$

where $\operatorname{var}(\bar{\varepsilon}_t) = \operatorname{var}((\Sigma\varepsilon_t)/n_t) = \sigma^2/n_t$. Thus we have $\boldsymbol{W} = \operatorname{diag}(1/n_1, \ldots, 1/n_T)$ (see Schönfeld, 1969, p. 143).

Blockdiagonality

In many applications we are confronted with the specification of *grouped data* (see, for example, the simple models of analysis of variance). It may be assumed that the regression variables $\mathbf{y}$ and X are observed over m periods and in n situations (individuals, as for example producers, consumers, agricultural objects, etc.). Thus the sample size of each individual is m and, therefore, the global sample size $T = mn$. The model may be written in its matrix form as

$$\mathbf{y} = \begin{pmatrix} \mathbf{y}_1 \\ \vdots \\ \mathbf{y}_n \end{pmatrix} = \begin{pmatrix} X_1 \\ \vdots \\ X_n \end{pmatrix} \boldsymbol{\beta} + \begin{pmatrix} \boldsymbol{\varepsilon}_1 \\ \vdots \\ \boldsymbol{\varepsilon}_n \end{pmatrix} = X\boldsymbol{\beta} + \boldsymbol{\varepsilon} \tag{2.2.11}$$

which is just a multivariate regression model. If we assume that in any group the within-group variances are identical:

$$E\boldsymbol{\varepsilon}_j\boldsymbol{\varepsilon}_j' = \sigma_j^2 I \quad (j = 1, \ldots, n)$$

and that the disturbances are uncorrelated, we have the blockdiagonal dispersion matrix

$$E\boldsymbol{\varepsilon}\boldsymbol{\varepsilon}' = \begin{pmatrix} \sigma_1^2 I & 0 & \ldots & 0 \\ 0 & \sigma_2^2 I & & \cdot \\ \cdot & \cdot & & \cdot \\ \cdot & \cdot & & \cdot \\ 0 & 0 & \ldots & \sigma_n^2 I \end{pmatrix} \tag{2.2.12}$$

$$= \operatorname{diag}(\sigma_1^2 I, \ldots, \sigma_n^2 I)$$

which gives a further example of heteroscedastic disturbances (see Toutenburg, 1975a, p. 65).

More generally, we may assume that the disturbances follow the so-called process of *intra-class correlation* (see Schönfeld, 1969, p. 144). The assumptions on $\boldsymbol{\varepsilon}$ are specified as follows:

$$\varepsilon_{tj} = v_j + u_{tj} \quad t = 1, \ldots, m, j = 1, \ldots, n \tag{2.2.13}$$

where the disturbances v_j are identical for the m realizations of each of the n individuals:

$$Ev_j = 0, \quad \operatorname{var}(v_j) = \sigma_v^2 \quad j = 1, \ldots, n \tag{2.2.14}$$

$$\operatorname{cov}(v_j v_{j'}) = 0 \quad j \neq j'.$$

The disturbance u_{tj} varies over all $T = mn$ realizations and has

$$Eu_{tj} = 0 \quad \operatorname{var}(u_{tj}) = \sigma_u^2 \tag{2.2.15}$$

$$\operatorname{cov}(u_{tj}, u_{t'j'}) = 0 \quad (t, j) \neq (t', j')$$

and, moreover,

$$\operatorname{cov}(u_{tj}, v_{j'}) = 0 \quad \text{for all } t, j, j' \tag{2.2.16}$$

i.e. both processes $\{\mathbf{u}\}$ and $\{\mathbf{v}\}$ are uncorrelated.

The $T \times T$-dispersion matrix of $\boldsymbol{\varepsilon}$ is therefore of the form

$$E\boldsymbol{\varepsilon\varepsilon}' = \operatorname{diag}(\boldsymbol{\phi}, \ldots, \boldsymbol{\phi}) \tag{2.2.17}$$

where $\boldsymbol{\phi}$ is the $m \times m$-matrix of intra-class correlation:

$$\boldsymbol{\phi} = E(\mathbf{u}_j\mathbf{u}_j') = \sigma^2\boldsymbol{\psi} = \sigma^2 \begin{pmatrix} 1 & \gamma & \ldots & \gamma \\ \gamma & 1 & \ldots & \gamma \\ \vdots & \vdots & & \vdots \\ \gamma & \gamma & \ldots & 1 \end{pmatrix} \tag{2.2.18}$$

with

$$\sigma^2 = \sigma_v^2 + \sigma_u^2 \quad \text{and} \quad \gamma = \sigma_v^2/\sigma^2.$$

As pointed out in Schönfeld (1969, p. 147) we may write

$$\boldsymbol{\psi} = (1-\gamma)\left(\boldsymbol{I} + \frac{\gamma}{1-\gamma}\mathbf{ii}'\right) \tag{2.2.19}$$

and

$$\boldsymbol{\psi}^{-1} = \frac{1}{1-\gamma}\left(\boldsymbol{I} - \frac{\gamma}{1+\gamma(m-1)}\mathbf{ii}'\right). \tag{2.2.20}$$

Based on this, we get the GLSE (4) as

$$\mathbf{b} = \left[\sum_{j=1}^{n} D(\mathbf{x}_j, \mathbf{x}_j)\right]^{-1}\left[\sum_{j=1}^{n} d(\mathbf{x}_j, \mathbf{x}_j)\right] \tag{2.2.21}$$

where

$$D(\mathbf{x}_j, \mathbf{x}_j) = \frac{1}{m}\boldsymbol{X}_j'\boldsymbol{X}_j - \frac{\gamma_m}{1+\gamma(m-1)}\bar{\mathbf{x}}_j\bar{\mathbf{x}}_j'$$

and

$$d(\mathbf{x}_j, \mathbf{x}_j) = \frac{1}{m}\boldsymbol{X}_j'\mathbf{y}_j - \frac{\gamma_m}{1+\gamma(m-1)}\bar{\mathbf{x}}_j\bar{y}_j.$$

Remark. Testing for heteroscedasticity is possible if special test statistics for any of the specific models of the above are developed (see Ramsey, 1969, and Huang, 1970, p. 147). As a general test the F-test is available when normality of disturbance can be assumed. On the other hand, the well-known tests for homogeneity of variances may be chosen. A common difficulty is that there is no procedure for determining the optimal grouping of the estimated disturbances $\hat{\varepsilon}_t$, whereas their grouping greatly influences the test procedures.

2.2.3 Autoregressive Disturbances

It is a typical situation in economics that the data are interdependent, with many reasons for interdependence of the successive disturbances. Autocorrelation of first and higher orders in the disturbances can arise, for example, from observation errors in the included variables or from the estimation of missing data by either averaging or extrapolating.

The first order autoregression may be described as

$$\varepsilon_t = \rho\varepsilon_{t-1} + v_t \quad t = \ldots, -2, -1, 0, 1, 2, \ldots \tag{2.2.22}$$

where $\{v_t\}$ is a stochastic process having

$$Ev_t = 0, \quad Ev_t^2 = \sigma_v^2, \quad Ev_t v_{t'} = 0 \quad (t \neq t').$$

Therefore we get

$$\varepsilon_t = \sum_{s=0}^{\infty} \rho^s v_{t-s}$$

and

$$E\varepsilon_t = 0, \quad \operatorname{var}(\varepsilon_t) = \sigma_v^2(1-\rho^2)^{-1} = \sigma^2.$$

$|\rho| < 1$ is the autocorrelation coefficient which has to be estimated. The dispersion matrix of the sample disturbances vector $\boldsymbol{\varepsilon}' = (\varepsilon_1, \ldots, \varepsilon_T)$ is then

$$\boldsymbol{\Sigma} = \sigma^2 \boldsymbol{W} = \sigma^2 \begin{pmatrix} 1 & \rho & \rho^2 & \ldots & \rho^{T-1} \\ \rho & 1 & \rho & \ldots & \rho^{T-2} \\ \vdots & \vdots & \vdots & & \vdots \\ \rho^{T-1} & \rho^{T-2} & \rho^{T-3} & \ldots & 1 \end{pmatrix} \tag{2.2.23}$$

and has

$$\boldsymbol{W}^{-1} = \frac{1}{1-\rho^2} \begin{pmatrix} 1 & -\rho & 0 & \ldots & 0 & 0 \\ -\rho & 1+\rho^2 & -\rho & \ldots & 0 & 0 \\ 0 & -\rho & 1+\rho^2 & \ldots & 0 & 0 \\ \vdots & \vdots & \vdots & & \vdots & \vdots \\ 0 & 0 & 0 & \ldots & 1+\rho^2 & -\rho \\ 0 & 0 & 0 & \ldots & -\rho & 1 \end{pmatrix} \tag{2.2.24}$$

From (23) it follows that the correlation between ε_t and $\varepsilon_{t-\tau}$ is $\sigma^2\rho^\tau$, i.e. the correlation depends on the difference of time $|\tau|$ and decreases for increasing $|\tau|$ as $|\rho| < 1$. The matrix $\boldsymbol{W}$ in (23) contains the correlogram of first order autocorrelated disturbances (see also Figure 2.2.1 for $\rho > 0$). Now, we could insert the dispersion matrix $\boldsymbol{W}$ from (23) into the GLSE (4), but in general ρ and therefore also $\boldsymbol{W}$ is not known and has to be estimated. It is usual to do this as follows. First, the regression parameter $\boldsymbol{\beta}$ of $\mathbf{y} = \boldsymbol{X\beta} + \boldsymbol{\varepsilon}$ is estimated by the OLSE $\mathbf{b}_0 = (\boldsymbol{X}'\boldsymbol{X})^{-1}\boldsymbol{X}'\mathbf{y}$ which gives the estimated residuals $\hat{\boldsymbol{\varepsilon}} = \mathbf{y} - \boldsymbol{X}\mathbf{b}_0$. If we use these residuals in (22):

$$\hat{\varepsilon}_t = \rho\hat{\varepsilon}_{t-1} + v_t \quad (t = 2, \ldots, T)$$

we get as estimator for ρ

$$\hat{\rho} = \frac{\sum_{t=2}^{T} \hat{\varepsilon}_{t-1}\hat{\varepsilon}_t}{\sum_{t=2}^{T} \hat{\varepsilon}_{t-1}^2}. \tag{2.2.25}$$

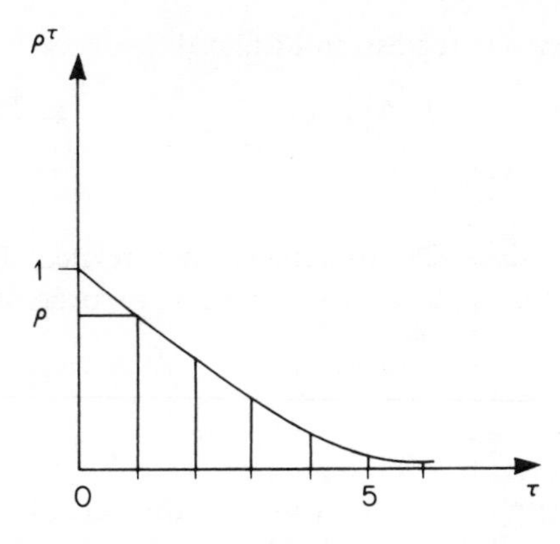

Figure 2.2.1 Correlogram of first order autocorrelated disturbances

Replacing in W^{-1} (24) the true parameter ρ by its estimation $\hat{\rho}$ (25) gives $\hat{W}^{-1}$, and finally we have the *two-stage* Aitken estimator

$$\mathbf{b}^* = (X'\hat{W}^{-1}X)^{-1}X'\hat{W}^{-1}\mathbf{y}. \tag{2.2.26}$$

It may happen that this procedure has to be repeated as an iterative process up to a relative constancy of the estimations $\hat{\rho}$ and $\mathbf{b}^*$ (see Schneeweiss, 1971, p. 183). Schönfeld (1969, p. 210) points out that the two-stage estimator is consistent, if some general conditions are fulfilled.

As an alternative procedure for overcoming autoregression the following data transformation is available. The regression model $\mathbf{y} = X\boldsymbol{\beta} + \boldsymbol{\varepsilon}$, or more explicitly,

$$y_t = \beta_0 + \beta_1 x_{1t} + \ldots + \beta_K x_{Kt} + \varepsilon_t \quad (t = 1, \ldots, T)$$

is transformed to

$$\Delta_\rho y_t = \beta_0(1-\rho) + \beta_1 \Delta_\rho x_{1t} + \ldots + \beta_K \Delta_\rho x_{Kt} + v_t \tag{2.2.27}$$

where the variables are transformed according to the following rules:

$$\Delta_\rho y_t = y_t - \rho y_{t-1}$$
$$\Delta_\rho x_{it} = x_{it} - \rho x_{it-1}$$
$$v_t = \varepsilon_t - \rho\varepsilon_{t-1}$$

where $\mathbf{v}' = (v_1, \ldots, v_T)$ is homoscedastic (see (22)), i.e. $E\mathbf{v}\mathbf{v}' = \sigma_v^2 I$. This means that (27) is a classical (homoscedastic) regression model, which can be estimated by OLSE when ρ is known or has been estimated. In other words, the transformation of variables leads to a two-stage procedure: (i) estimation of ρ by (25), and (ii) estimation of $\boldsymbol{\beta}$ in (27) by the OLSE when ρ is replaced by $\hat{\rho}$.

In practice one can expect that both two-stage procedures of the above will almost coincide.

If ρ is near 1 the so-called first differences

$$\Delta y_t = y_t - y_{t-1},$$
$$\Delta x_{it} = x_{it} - x_{it-1} \tag{2.2.28}$$
$$v_t = \varepsilon_t - \varepsilon_{t-1}$$

are taken. This results in the homogeneous regression of first differences

$$\Delta y_t = \beta_1 \Delta x_{1t} + \ldots + \beta_K \Delta x_{Kt} + v_t \tag{2.2.29}$$

which is estimated by the OLSE.

Note The transformed exogenous variables (28) are almost uncorrelated. The method of first differences is therefore applied as an attempt to overcome multicollinearity (see Section 2.3.1).

Testing for Autoregression

In Section 2.2.4 we will investigate the goodness of the GLSE $\mathbf{b} = (X'W^{-1}X)^{-1}X'W^{-1}\mathbf{y}$ if W is misspecified. Before $\mathbf{b}$ can be applied, however, the assumptions on W have to be checked. As no general test is available for the hypothesis 'ε is spherically distributed', we hàve to test specific hypotheses on W. If an autoregressive distribution is likely, the familiar well-known Durbin–Watson test can be applied (see Durbin and Watson, 1950, 1951) provided the normality of the disturbances distribution can be assumed:

Null hypothesis H_0: $\rho = 0$
against H_1: $\rho > 0$ (if $\hat{\rho} > 0$ was observed).

The test statistic is of the form

$$d = \frac{\sum_{t=2}^{T} (\hat{\varepsilon}_t - \hat{\varepsilon}_{t-1})^2}{\sum_{t=1}^{T} \hat{\varepsilon}_t^2} \tag{2.2.30}$$

where the $\hat{\varepsilon}_t$'s are the components of the OLS residual $\hat{\boldsymbol{\varepsilon}} = \mathbf{y} - X\mathbf{b}_0$. Using the empirical coefficient of autocorrelation (25) we get

$$d \approx 2(1 - \hat{\rho}) \tag{2.2.31}$$

and therefore $0 < d < 4$. If $\hat{\rho} = 0$, $d = 2$ follows. If T and K are fixed, the distribution of the statistic d lies between the distribution of two other statistics d_l

T	$K^* = 1$		$K^* = 2$		$K^* = 3$		$K^* = 4$		$K^* = 5$	
	d_l^*	d_u^*	d_l^*	d_u^*	d_l^*	d_u^*	d_l^*	d_u^*	d_l^*	d_u^*
15	1.08	1.36	0.95	1.54	0.82	1.75	0.69	1.97	0.56	2.21
20	1.20	1.41	1.10	1.54	1.00	1.68	0.90	1.83	0.79	1.99
30	1.35	1.49	1.28	1.57	1.21	1.65	1.14	1.74	1.07	1.83
40	1.44	1.54	1.39	1.60	1.34	1.66	1.29	1.72	1.23	1.79
50	1.50	1.59	1.46	1.63	1.42	1.67	1.38	1.72	1.34	1.77

Figure 2.2.2 5% significance points (Durbin and Watson, 1951). *Note* K^* is the number of exogenous variables when the dummy variable is excluded

and d_u. The critical values d_l^* and d_u^* were tabled by Durbin and Watson (1951) (see Figure 2.2.2). The one-sided test of $H_0: \rho = 0$ against $H_1: \rho > 0$ is as follows:

accept H_0 if $d \geq d_u^*$
reject H_0 if $d \leq d_l^*$
no decision if $d_l^* < d < d_u^*$.

If the sample gives an empirical coefficient of autoregression $\hat{\rho} < 0$, the test statistic is $\tilde{d} = 4 - d$.

Example

We demonstrate the test on autoregression in the following model

$$y_t = \beta_0 + \beta_1 x_{1t} + \varepsilon_t, \quad \varepsilon_t \sim N(0, \sigma^2)$$

or, in matrix formulation,

$$\mathbf{y} = \beta_0 \mathbf{i} + \beta_1 \boldsymbol{X}_1 + \boldsymbol{\varepsilon} = \boldsymbol{X}\boldsymbol{\beta} + \boldsymbol{\varepsilon}, \quad \boldsymbol{\varepsilon} \sim N(\mathbf{0}, \sigma^2 \boldsymbol{I}).$$

Let the following sample of size $T = 6$ be given (see Toutenburg and Rödel, 1978, p. 34):

$$\mathbf{y} = \begin{bmatrix} 1 \\ 3 \\ 2 \\ 3 \\ 0 \\ 2 \end{bmatrix} \qquad \boldsymbol{X} = \begin{bmatrix} 1 & -4 \\ 1 & 3 \\ 1 & 4 \\ 1 & 5 \\ 1 & 3 \\ 1 & 3 \end{bmatrix}.$$

We get

$$\boldsymbol{X}'\boldsymbol{X} = \begin{pmatrix} 6 & 14 \\ 14 & 84 \end{pmatrix} \qquad \boldsymbol{X}'\mathbf{y} = \begin{pmatrix} 11 \\ 34 \end{pmatrix}$$

$$|\boldsymbol{X}'\boldsymbol{X}| = 308$$

$$(\boldsymbol{X}'\boldsymbol{X})^{-1} = \frac{1}{308}\begin{pmatrix} 84 & -14 \\ -14 & 6 \end{pmatrix}$$

$$\mathbf{b}_0 = (\boldsymbol{X}'\boldsymbol{X})^{-1}\boldsymbol{X}'\mathbf{y} = \frac{1}{308}\begin{pmatrix} 448 \\ 50 \end{pmatrix} = \begin{pmatrix} 1.45 \\ 0.16 \end{pmatrix}$$

$$\mathbf{y} = \boldsymbol{X}\mathbf{b}_0 = \begin{bmatrix} 0.81 \\ 1.93 \\ 2.09 \\ 2.25 \\ 1.93 \\ 1.93 \end{bmatrix}, \quad \hat{\boldsymbol{\varepsilon}} = \mathbf{y} - \boldsymbol{X}\mathbf{b}_0 = \begin{bmatrix} 0.19 \\ 1.07 \\ -0.09 \\ 0.75 \\ -1.93 \\ 0.07 \end{bmatrix}$$

$$\hat{\rho} = \frac{-1.54}{5.47} = -0.28, \quad d = 2(1 - \hat{\rho}) = 2.56.$$

As $\hat{\rho} < 0$, the alternative test statistic $\tilde{d} = 4 - d = 1.44$ has to be taken. From Figure 2.2.2 we get for $K^* = 1$ that the critical value corresponding to $T = 6$ is $d_u^* < 1.36$ and therefore $H_0: \rho = 0$ may be accepted.

Note The estimated coefficient of autoregression may be used to improve the prediction of the regressor variable. The sample of the regression model can be described by the index $t = 1, \ldots, T$. For the next period we have the model

$$y_{T+1} = \mathbf{x}'_{T+1}\boldsymbol{\beta} + \varepsilon_{T+1}$$

where y_{T+1} is unobserved and has to be predicted. Following relation (22) we can estimate ε_{T+1} by

$$\hat{\varepsilon}_{T+1} = \hat{\rho}\hat{\varepsilon}_T$$

which gives the predictor

$$\hat{y}_{T+1} = \mathbf{x}'_{T+1}\mathbf{b} + \hat{\rho}\hat{\varepsilon}_T \tag{2.2.32}$$

instead of the usual predictor $\mathbf{x}'_{T+1}\mathbf{b}$ which clearly is the unbiased estimator of the conditional expectation $E(y_{T+1}/\mathbf{x}_{T+1})$ (see (5.1.42)).

For general prediction problems the reader is referred to Bibby and Toutenburg (1978), although some of the main results can be found in Section 5.1 of this book.

2.2.4 Misspecification of the Dispersion Matrix

One of the features of the ordinary least squares estimator $\mathbf{b}_0$ (2.1.15) as given by the Gauss–Markoff theorem is that in the classical model with uncorrelated errors no knowledge of σ^2 is required for point estimation of $\boldsymbol{\beta}$. As a consequence of this an estimate of σ^2 can be obtained using the sum of squares of residuals $\hat{\boldsymbol{\varepsilon}} = \mathbf{y} - \boldsymbol{X}\mathbf{b}_0$. When the residuals are correlated it is necessary for point estimation of $\boldsymbol{\beta}$ to have prior knowledge or assumptions about the covariance matrix $\boldsymbol{W}$, or at least an estimate on it.

Watson (1955) was one of the first to study the effect of a wrong choice of covariance matrix on the properties of the associated estimators of $\boldsymbol{\beta}$. Reasons for this misspecification of the covariance matrix could be one of the following:

(i) The correlation structure of disturbances may have been ignored in order to use OLS estimation and hence simplify calculations. (This is done, for instance, as the first step in model building in order to obtain a rough idea of the underlying relationships.)
(ii) The true matrix $\boldsymbol{W}$ may be unknown and may have to be estimated by $\hat{\boldsymbol{W}}$ (which is stochastic).
(iii) The correlation structure may be better represented by a matrix which is different from $\boldsymbol{W}$.

Assuming the general linear model $\mathbf{y} = \boldsymbol{X}\boldsymbol{\beta} + \boldsymbol{\varepsilon}$, $\boldsymbol{\varepsilon} \sim (\mathbf{0}, \sigma^2 \boldsymbol{W})$, so that $\boldsymbol{W}$ is the true covariance matrix, then the effect of misspecification is to use a covariance matrix $A \neq W$. In any case, the derived estimator will have the form

$$\hat{\beta}(A) = (X'A^{-1}X)^{-1}X'A^{-1}\mathbf{y} \tag{2.2.33}$$

where the existence of A^{-1} and $(X'A^{-1}X)^{-1}$ has to be ensured. (For instance, if $A > \mathbf{0}$, then the above inverses exist.)

Now, the estimator $\hat{\beta}(A)$ is unbiased and has dispersion

$$V[\hat{\beta}(A)] = \sigma^2(X'A^{-1}X)^{-1}X'A^{-1}WA^{-1}X(X'A^{-1}X)^{-1}. \tag{2.2.34}$$

Using the false matrix A results in a *loss in efficiency* in estimating β, which may be measured by the increase of the risk. Using Theorem 2.1 and noting the unbiasedness of $\hat{\beta}(A)$, the loss in efficiency may be expressed by the difference of the dispersion matrices of the GLSE $\mathbf{b} = S^{-1}X'W^{-1}\mathbf{y}$ and $\hat{\beta}(A)$, respectively:

$$\begin{aligned}
&\sigma^{-2}V[\beta(A)] - \sigma^2 V(\mathbf{b}) \\
&= [(X'A^{-1}X)^{-1}X'A^{-1} - S^{-1}X'W^{-1}]W[(X'A^{-1}X)^{-1}X'A^{-1} - S^{-1}X'W^{-1}]' \\
&= CWC', \text{ say.}
\end{aligned} \tag{2.2.35}$$

Using Theorem A.9, we may conclude that $CWC' \geq \mathbf{0}$.

Let us now investigate the most important case, in which the OLSE $\mathbf{b}_0 = (X'X)^{-1}X'\mathbf{y}$ is mistakenly used instead of the true GLSE $\mathbf{b}$. That is, let us assume $A = I$. Letting $U = (X'X)^{-1}$, we get the increase in dispersion due to using the OLSE $\mathbf{b}_0 = \hat{\beta}(I)$ instead of the GLSE as

$$\begin{aligned}
V(\mathbf{b}_0) - V(\mathbf{b}) &= \sigma^2(UX' - S^{-1}X'W^{-1})W(XU - W^{-1}XS^{-1}) \\
&= \sigma^2 UX'WXU - \sigma^2 S^{-1}.
\end{aligned} \tag{2.2.36}$$

From (35) and (36) it is clear that $V(\mathbf{b}_0) - V(\mathbf{b})$ is nonnegative definite and, moreover, that the equality $V(\mathbf{b}_0) = V(\mathbf{b})$ holds if and only if

$$UX' = S^{-1}X'W^{-1},$$

i.e. when $\mathbf{b}_0 = \mathbf{b}$. This fact would imply that

$$X'W^{-1}\mathbf{y} = SUX'\mathbf{y} \tag{2.2.37}$$

is true for all $\mathbf{y}$ and arbitrary X. Denoting by $\mathscr{R}[X] = \{\theta : \theta = X\mathbf{z}\}$ the range space of X spanned by the columns of X, then for any matrix W^{-1} (37) holds for $\mathbf{y} \in \mathscr{R}[X]$, i.e. $\mathbf{y} = X\mathbf{z}$. Therefore, a necessary and sufficient condition for (37) to hold is that this relation is valid for vectors $\mathbf{y}$ orthogonal to X, i.e. we must have

$$X'\mathbf{y} = \mathbf{0} \Rightarrow X'W^{-1}\mathbf{y} = \mathbf{0}. \tag{2.2.38}$$

Assuming that the regressor matrix X includes the dummy variable $X_1 \equiv 1$, then it was proved by McElroy (1967) that (38) holds if and only if the disturbances ε_t have equal variances and equal correlation coefficients.

Theorem 2.7 (McElroy)

Assume the generalized linear regression including a constant term, i.e.

$$\mathbf{y} = X\beta + \varepsilon, \quad X = (\mathbf{1}, \tilde{X}), \quad \varepsilon \sim (\mathbf{0}, \sigma^2 W). \tag{2.2.39}$$

Then OLSE and GLSE coincide if and only if

$$W = (1-\rho)I + \rho \mathbf{i}\mathbf{i}' \tag{2.2.40}$$

where $0 \leq \rho < 1$ and $\mathbf{i}' = (1, \ldots, 1)$.

Note $0 \leq \rho < 1$ ensures that $(1-\rho)I + \rho\mathbf{i}\mathbf{i}'$ is positive definite for all values of the sample size T. For given T it would be replaced by $-1/(T-1) < \rho < 1$.

(For a coordinate-free approach to the above problem the reader is referred to Kruskal, 1968.)

Clearly, an incorrect specification of W will also lead to errors in estimating σ^2. The usual estimator of σ^2 is (see (2.1.51))

$$s^2 = \hat{\boldsymbol{\varepsilon}}'\hat{\boldsymbol{\varepsilon}}(T-K)^{-1}.$$

Suppose that OLSE is used, so that the true dispersion matrix is $\sigma^2 W$. Then the vector of residuals is

$$\hat{\boldsymbol{\varepsilon}} = \mathbf{y} - X\mathbf{b}_0 = [I - XUX']\mathbf{y}$$

where $M = [I - XUX']$ is idempotent and has

$$MX = 0 \quad \text{and} \quad \operatorname{tr} M = T-K.$$

Now the residual sum of squares is

$$\begin{aligned} \hat{\boldsymbol{\varepsilon}}'\hat{\boldsymbol{\varepsilon}} &= \mathbf{y}'M\mathbf{y} = (X\boldsymbol{\beta}+\boldsymbol{\varepsilon})'M(X\boldsymbol{\beta}+\boldsymbol{\varepsilon}) \\ &= \boldsymbol{\varepsilon}'M\boldsymbol{\varepsilon} = \operatorname{tr} M\boldsymbol{\varepsilon}\boldsymbol{\varepsilon}'. \end{aligned} \tag{2.2.41}$$

Therefore

$$E\hat{\boldsymbol{\varepsilon}}'\hat{\boldsymbol{\varepsilon}} = \sigma^2 \operatorname{tr} MW = \sigma^2 \operatorname{tr} W - \operatorname{tr}(X'WX)U. \tag{2.2.42}$$

(This corresponds to equation (4.44) on p. 238 of Goldberger (1964) and to equation (9.19) on p. 145 of Sprent (1969).)

In general, $E\hat{\boldsymbol{\varepsilon}}'\hat{\boldsymbol{\varepsilon}}$ (42) does not equal $\sigma^2(T-K)$ and hence $s^2 = \hat{\boldsymbol{\varepsilon}}'\hat{\boldsymbol{\varepsilon}}(T-K)^{-1}$ is not unbiased. If we write $W = I + (W-I)$ and use the standardization $\operatorname{tr} W = T$, then

$$\begin{aligned} Es^2 = \frac{E\hat{\boldsymbol{\varepsilon}}'\hat{\boldsymbol{\varepsilon}}}{T-K} &= \sigma^2 \frac{\operatorname{tr} M + \operatorname{tr} M(W-I)}{T-K} \\ &= \sigma^2 + \frac{\sigma^2}{T-K} \operatorname{tr} M(W-I) \\ &= \sigma^2 - \frac{\sigma^2}{T-K}[\operatorname{tr}\{(X'WX)U\} - K]. \end{aligned} \tag{2.2.43}$$

The final term represents the bias of s^2 when the OLS is mistakenly used. As far as the estimation of the dispersion matrix of $\mathbf{b}_0$ is concerned, we see that its classical estimator $s^2(X'X)^{-1} = s^2 U$ will be biased on two counts: first $Es^2 \neq \sigma^2$, and secondly $U \neq U(X'WX)U$ (see (36)).

Let us write

$$V(\mathbf{b}_0) = \sigma^2 UX'(I + W - I)XU \qquad (2.2.44)$$
$$= \sigma^2 U + \sigma^2 UX'(W - I)XU.$$

Thus, in estimating the dispersion matrix of $\mathbf{b}_0$ as $s^2 U$, the expectation is

$$Es^2 U = V(\mathbf{b}_0) + \sigma^2 UX'(I - W)XU - \frac{\sigma^2}{T-K}[\mathrm{tr}(X'WXU) - K]U$$
$$= V(\mathbf{b}_0) + \mathrm{bias}[s^2 U], \quad \text{say.} \qquad (2.2.45)$$

Goldberger (1964, p. 239) has investigated the bias of the estimator $s^2 U$ for special heteroscedastic or autoregressive models. As a conclusion he found that there is a tendency to underestimate the variances if $s^2 U$ is mistakenly used. Watson (1955) has obtained bounds for the bias of the estimated variance of the $\mathbf{b}_0$-components, assuming the special case of orthogonal regressors. Swindel (1968) has obtained bounds without the restriction $X'X = U^{-1} = I$.

2.2.5 Estimation of Subvectors

In practice one is often confronted with the problem of comparing the efficiency of estimated subvectors of $\boldsymbol{\beta}$, for example if one has to choose between a single-equation or a system-equation estimator (see Section 6.1.4). Another important example will be given in Section 3.5.2 with the piecewise estimator.

Thus we are led to the problem of giving an explicit formula for the estimator of a subvector of $\boldsymbol{\beta}$ and, moreover, for its variance.

Without loss of generality we can confine ourselves to the situation in which the vector $\boldsymbol{\beta}$ is divided into two subvectors, i.e.

$$\boldsymbol{\beta} = \begin{pmatrix} \boldsymbol{\beta}_1 \\ \boldsymbol{\beta}_2 \end{pmatrix}.$$

If we denote $S_{ij} = X_i' W^{-1} X_j \, (i, j = 1, 2)$ and use the GLSE $\mathbf{b}$ (4) we get

$$\mathbf{b} = \begin{pmatrix} \mathbf{b}_1 \\ \mathbf{b}_2 \end{pmatrix} = \begin{pmatrix} S_{11} & S_{12} \\ S_{21} & S_{22} \end{pmatrix}^{-1} \begin{pmatrix} X_1' W^{-1} \mathbf{y} \\ X_2' W^{-1} \mathbf{y} \end{pmatrix}. \qquad (2.2.46)$$

Applying the formula for partitioned inversion of a matrix (see Theorem A.33) gives

$$S^{-1} = \begin{pmatrix} S_{11} & S_{12} \\ S_{21} & S_{22} \end{pmatrix}^{-1} = \begin{pmatrix} S_{11}^{-1} + S_{11}^{-1} S_{12} D^{-1} S_{21} S_{11}^{-1} & -S_{11}^{-1} S_{12} D^{-1} \\ -D^{-1} S_{21} S_{11}^{-1} & D^{-1} \end{pmatrix} \qquad (2.2.47)$$

where $D = S_{22} - S_{21} S_{11}^{-1} S_{12}$.

If we denote

$$E = X_2' - S_{21} S_{11}^{-1} X_1' \qquad (2.2.48)$$

we have

$$\left. \begin{aligned} D &= EW^{-1}X_2, \quad EW^{-1}X_1 = 0 \\ D &= EW^{-1}E' \end{aligned} \right\}. \qquad (2.2.49)$$

Therefore we get the components $\mathbf{b}_1$ and $\mathbf{b}_2$ of $\mathbf{b}$ which correspond to the subvectors $\boldsymbol{\beta}_1$ and $\boldsymbol{\beta}_2$ of $\boldsymbol{\beta}$ as

$$\mathbf{b}_2 = \boldsymbol{D}^{-1}\boldsymbol{E}\boldsymbol{W}^{-1}\mathbf{y} \tag{2.2.50}$$

and

$$\mathbf{b}_1 = \boldsymbol{S}_{11}^{-1}\boldsymbol{X}_1'\boldsymbol{W}^{-1}\mathbf{y} - \boldsymbol{S}_{11}^{-1}\boldsymbol{S}_{12}\mathbf{b}_2. \tag{2.2.51}$$

Now we calculate the components of the dispersion matrix

$$\boldsymbol{V}(\mathbf{b}) = \sigma^2\boldsymbol{S}^{-1} = \begin{pmatrix} \boldsymbol{V}_{11} & \boldsymbol{V}_{12} \\ \boldsymbol{V}_{21} & \boldsymbol{V}_{22} \end{pmatrix}. \tag{2.2.52}$$

Using (48)–(51), the following relationships hold:

$$\mathbf{b}_2 - \boldsymbol{\beta}_2 = \boldsymbol{D}^{-1}\boldsymbol{E}\boldsymbol{W}^{-1}\boldsymbol{\varepsilon}, \tag{2.2.53}$$

$$\boldsymbol{V}(\mathbf{b}_2) = \boldsymbol{V}_{22} = \sigma^2\boldsymbol{D}^{-1}, \tag{2.2.54}$$

$$\mathbf{b}_1 - \boldsymbol{\beta}_1 = \boldsymbol{S}_{11}^{-1}[\boldsymbol{X}_1' - \boldsymbol{S}_{12}\boldsymbol{D}^{-1}\boldsymbol{E}]\boldsymbol{W}^{-1}\boldsymbol{\varepsilon}, \tag{2.2.55}$$

$$\boldsymbol{V}(\mathbf{b}_1) = \boldsymbol{V}_{11} = \sigma^2\boldsymbol{S}_{11}^{-1} + \boldsymbol{V}_{12}\boldsymbol{V}_{22}^{-1}\boldsymbol{V}_{21}, \tag{2.2.56}$$

$$\boldsymbol{V}_{21}' = \boldsymbol{V}_{12} = \boldsymbol{E}(\mathbf{b}_1 - \boldsymbol{\beta}_1)(\mathbf{b}_2 - \boldsymbol{\beta}_2)' = -\sigma^2\boldsymbol{S}_{11}^{-1}\boldsymbol{S}_{12}\boldsymbol{D}^{-1}. \tag{2.2.57}$$

2.3 LINEAR RESTRICTIONS AND RELATED PROBLEMS

2.3.1 Multicollinearity

In practical applications of linear models one is often confronted with independent variables $\boldsymbol{X}_1, \ldots, \boldsymbol{X}_K$ which are correlated. A high-order correlation will lead to an *ill-conditioned* matrix $\boldsymbol{X}'\boldsymbol{X}$ of second moments. If $\det(\boldsymbol{X}'\boldsymbol{X})$ is close to zero we say that the model is *weak multicollinear*. If ordinary least squares are used, then the components $\boldsymbol{\beta}_i$ are estimated with ever greater uncertainty as the elements of the dispersion matrix $\boldsymbol{V}(\mathbf{b}_0) = \sigma^2(\boldsymbol{X}'\boldsymbol{X})^{-1}$ become large. That is, multicollinearity may produce large standard errors of the estimated parameters. A *strong* (or extreme) *multicollinearity* holds if there are exact linear relations among the observed values of the regressors $\boldsymbol{X}_1, \ldots, \boldsymbol{X}_K$ so that rank $\boldsymbol{X} = p < K$. As a consequence we have the fact that the vector $\boldsymbol{\beta}$ is not unbiased estimable: there exist only estimators of subvectors (see Section 2.1.1). This becomes clearer if we observe that we cannot realize the condition (2.1.24) for lack of bias: as rank $\boldsymbol{X} = p < K$, the rank of the product $\boldsymbol{C}'\boldsymbol{X}$ does not exceed p whereas $\boldsymbol{I}_K$ is of rank K. Thus $\boldsymbol{C}'\boldsymbol{X} \neq \boldsymbol{I}$ in general.

Example (Goldberger, 1964, p. 192)

Let the simple two-regressor model

$$y_t = \beta_1 + \beta_2 x_{1t} + \varepsilon_t \quad (t = 1, \ldots, T)$$

be given. Rank $(\mathbf{i}, \mathbf{x}_1) = 1$ implies that $x_{11} = \ldots = x_{1T} = a$ (a real scalar). Then we have $\bar{x}_1 = a$, $\sum_t (x_{1t} - \bar{x}_1)^2 = 0$ and the estimator $\mathbf{b}_0$ (2.1.55) is not defined. Let

$$\begin{pmatrix} \hat{\beta}_1 \\ \hat{\beta}_2 \end{pmatrix} = \boldsymbol{C}\mathbf{y}$$

be a linear estimator. The unbiasedness condition $\boldsymbol{CX} = \boldsymbol{I}$ specializes to

$$\begin{pmatrix} \sum c_{1t} & a\sum c_{1t} \\ \sum c_{2t} & a\sum c_{2t} \end{pmatrix} = \begin{pmatrix} 1 & 0 \\ 0 & 1 \end{pmatrix}$$

which has no solution. From $x_t = a$ (for all t) it follows that

$$y_t = (\beta_1 + \beta_2 a) + \varepsilon_t$$

holds, so that β_1 and β_2 are not separately estimable. We can only estimate the linear combination $(\beta_1 + \beta_2 a)$ by $\bar{y}$.

Let us go back to the problem cited above and assume, for example, that the extreme multicollinearity is caused by linear dependence of $\mathbf{x}_1$ on the remaining observed regressors, that is

$$\mathbf{x}_1 = \sum_{i=2}^{K} \alpha_i \mathbf{x}_i.$$

Then for any scalar $\lambda \neq 0$

$$\begin{aligned} \boldsymbol{X\beta} = \sum_{i=1}^{K} \mathbf{x}_i \beta_i &= (1-\lambda)\beta_1 \mathbf{x}_1 + \sum_{i=2}^{K} (\beta_i + \lambda_i \beta_i)\mathbf{x}_i \\ &= \tilde{\beta}_1 \mathbf{x}_1 + \sum_{i=2}^{K} \tilde{\beta}_i \mathbf{x}_i, \text{ say} \\ &= \boldsymbol{X\tilde{\beta}}. \end{aligned}$$

Although $\boldsymbol{\beta} \neq \boldsymbol{\tilde{\beta}}$ (so far $\lambda \neq 0$), both vectors yield the same hyperplane (response surface) $\boldsymbol{X\beta} = \boldsymbol{X\tilde{\beta}}$ so that, from the vector of observations $\mathbf{y}$, we cannot distinguish between $\boldsymbol{\beta}$ and $\boldsymbol{\tilde{\beta}}$. The parameter vector $\boldsymbol{\beta}$ is said to be *not identifiable*.

To overcome multicollinearity we can go various ways:

(i) apply methods of experimental design to decrease the correlation between the regressor's values,
(ii) transform the variables (use, for example, first differences, etc.),
(iii) change the list of regressors (see Goldberger, 1964, p. 194),
(iv) use prior information, especially exact linear restrictions.

2.3.2 Restricted Least Squares

We will confine ourselves to method (iv), since this is related to the subject of prior information. Furthermore, in the following, we will concentrate on algebraical aspects of linear restrictions. The statistical background of restrictions is discussed more widely in Chapter III.

Let us begin with a review of some geometrical properties of the least squares estimator (see, for example, Seber, 1966).

First, we note the fact (see Section 2.1.1) that the OLSE $X\mathbf{b}_0$ of $X\boldsymbol{\beta}$ minimizes the sum of squares

$$S(\boldsymbol{\beta}) = (y - X\boldsymbol{\beta})'(\mathbf{y} - X\boldsymbol{\beta})$$

and, moreover, $X\mathbf{b}_0$ is the orthogonal projection of $\mathbf{y}$ on the *range space* $\mathscr{R}[X]$. Here $\mathscr{R}[X]$ is spanned by the columns of X and we may write

$$\mathscr{R}[X] = \{\boldsymbol{\theta} : \boldsymbol{\theta} = X\boldsymbol{\beta}\}. \tag{2.3.1}$$

Conversely, let a symmetric idempotent matrix $\boldsymbol{P}$ of rank $\boldsymbol{P} = p \leq K$ be given, where $\boldsymbol{P}$ represents the orthogonal projection of the Euclidean space E^K on $\mathscr{R}[X]$. Then $X\mathbf{b}_0 = \boldsymbol{P}\mathbf{y}$.

Now return to the problem of estimating the vector $\boldsymbol{\beta}$ in the case where the regressor matrix is not of full rank (that is, the model is underlying an extreme multicollinearity). Let rank $X = p < K$. To get a unique determined estimate of $\boldsymbol{\beta}$ we will introduce *linear restrictions* on $\boldsymbol{\beta}$, namely

$$\mathbf{r} = \boldsymbol{R}\boldsymbol{\beta}. \tag{2.3.2}$$

Here $\mathbf{r}$ is a given $(K-p) \times 1$-vector and $\boldsymbol{R}$ a $(K-p) \times K$ (known) matrix of rank $\boldsymbol{R} = K - p$. Furthermore, we assume that $\boldsymbol{R}$ is such that

$$\operatorname{rank}\begin{pmatrix} X \\ R \end{pmatrix} = K \tag{2.3.3}$$

holds.

Notation It is usual to say that X and $\boldsymbol{R}$ are *complementary matrices* if condition (3) holds.

Then the following theorem can be proved.

Theorem 2.8

Assume the generalized linear regression model

$$\mathbf{y} = X\boldsymbol{\beta} + \boldsymbol{\varepsilon}, \quad \boldsymbol{\varepsilon} \sim (\mathbf{0}, \sigma^2 W), \quad \text{rank } X = p < K$$

under exact linear constraints

$$\mathbf{r} = \boldsymbol{R}\boldsymbol{\beta}, \quad \text{rank } \boldsymbol{R} = K - p, \quad \text{rank}\begin{pmatrix} X \\ R \end{pmatrix} = K.$$

Then:

(i) The projection $\boldsymbol{P}$ of the Euclidean space E^K on $\boldsymbol{R}[\tilde{X}]$ has

$$\boldsymbol{P} = X(X'W^{-1}X + R'R)^{-1}X'W^{-1}. \tag{2.3.4}$$

(ii) The GLSE under constraint $\mathbf{r} = \boldsymbol{R\beta}$ is of the form

$$\mathbf{b}_R(\mathbf{r}) = (X'W^{-1}X + R'R)^{-1}(X'W^{-1}\mathbf{y} + R'\mathbf{r}). \tag{2.3.5}$$

Proof (i) See Toutenburg, 1975a, pp. 31–33.

(ii) The minimization of $S(\boldsymbol{\beta})$ under constraint $\mathbf{r} = \boldsymbol{R\beta}$ is equivalent to the minimization of the function

$$S(\boldsymbol{\beta}) + 2\lambda'(\boldsymbol{R\beta} - \mathbf{r}) \tag{2.3.6}$$

where λ is a $(K-p)\times 1$-vector of Lagrangian multipliers. In generalized regression $S(\boldsymbol{\beta})$ is of the form

$$S(\boldsymbol{\beta}) = (\mathbf{y} - X\boldsymbol{\beta})'W^{-1}(\mathbf{y} - X\boldsymbol{\beta}). \tag{2.3.7}$$

Differentiating (6) with respect to $\boldsymbol{\beta}$ and $\boldsymbol{\lambda}$, respectively, leads to the normal equations

$$\tilde{X}'\tilde{X}\boldsymbol{\beta} - \tilde{X}'\mathbf{y} + R'\boldsymbol{\lambda} = \mathbf{0} \tag{2.3.8}$$

$$\boldsymbol{R\beta} - \mathbf{r} = \mathbf{0} \tag{2.3.9}$$

where $\tilde{X} = W^{-1/2}X$.

If we note that

$$X(\tilde{X}'\tilde{X} + R'R)^{-1}R' = \mathbf{0} \tag{2.3.10}$$

(see Tan, 1971, and Toutenburg, 1975a, p. 32) we may conclude that $R(\tilde{X}'\tilde{X} + R'R)^{-1}R'$ is idempotent:

$$\begin{aligned} & R(\tilde{X}'\tilde{X} + R'R)^{-1}R'R(\tilde{X}'\tilde{X} + R'R)^{-1}R' \\ &= R(\tilde{X}'\tilde{X} + R'R)^{-1}(R'R + \tilde{X}'\tilde{X} - \tilde{X}'\tilde{X})(\tilde{X}'\tilde{X} + R'R)^{-1}R' \\ &= R(\tilde{X}'\tilde{X} + R'R)^{-1}R'. \end{aligned}$$

The matrix $(\tilde{X}'\tilde{X} + R'R)$ and its inverse are positive definite (Theorems A.11, A.6). As rank $R = K - p$, it follows that $R(\tilde{X}'\tilde{X} + R'R)^{-1}R' > \mathbf{0}$ (Theorem A.5) and therefore this matrix is regular (Theorem A.4). A regular idempotent matrix is just the identity matrix (Theorem A.23):

$$R(\tilde{X}'\tilde{X} + R'R)^{-1}R' = I. \tag{2.3.11}$$

Now we will solve the normal equations. From $\boldsymbol{R\beta} = r$ it follows that $R'\boldsymbol{R\beta} = R'\mathbf{r}$. Using that identity, the first normal equation (8) becomes

$$(\tilde{X}'\tilde{X} + R'R)\boldsymbol{\beta} = (\tilde{X}'\mathbf{y} + R'\mathbf{r}) - R'\lambda. \tag{2.3.12}$$

Premultiplying by $R(\tilde{X}'\tilde{X} + R'R)^{-1}$ gives

$$\begin{aligned} \boldsymbol{R\beta} &= R(\tilde{X}'\tilde{X} + R'R)^{-1}(\tilde{X}'\mathbf{y} + R'\mathbf{r}) - R(\tilde{X}'\tilde{X} + R'R)^{-1}R'\lambda \\ &= \mathbf{r} - \lambda \text{ (see (10) and (11))} \end{aligned}$$

so that $\lambda = \mathbf{0}$. Solving (12) for $\boldsymbol{\beta}$ gives the estimator $\mathbf{b}_R(\mathbf{r})$ (5). We note that $\mathbf{b}_R(\mathbf{r})$ is *conditionally unbiased*. Under this notion we understand that $E[\mathbf{b}_R(\mathbf{r})/\mathbf{r} = \boldsymbol{R\beta}]$

$= \boldsymbol{\beta}$ holds (we use again $\boldsymbol{S} = \boldsymbol{X}'\boldsymbol{W}^{-1}\boldsymbol{X}$):

$$E[\mathbf{b}_R(\mathbf{r})] = (\boldsymbol{S}+\boldsymbol{R}'\boldsymbol{R})^{-1}(\boldsymbol{X}'\boldsymbol{W}^{-1}\boldsymbol{\beta}+\boldsymbol{R}'\mathbf{r}) \qquad (2.3.13)$$
$$= \boldsymbol{\beta} \quad \text{if} \quad \mathbf{r} = \boldsymbol{R}\boldsymbol{\beta}.$$

Then we get the dispersion matrix

$$V[\mathbf{b}_R(\mathbf{r})] = \sigma^2(\boldsymbol{S}+\boldsymbol{R}'\boldsymbol{R})^{-1}\boldsymbol{S}(\boldsymbol{S}+\boldsymbol{R}'\boldsymbol{R})^{-1}. \qquad (2.3.14)$$

In the context of biased estimation, as for example mixed or ridge estimation, a special case of $\mathbf{b}_R(\mathbf{r})$ becomes interesting. Let $\mathbf{r} = \mathbf{0}$. Then we are led to the so-called *restricted least-squares estimator* (RLSE)

$$\mathbf{b}_R(\mathbf{0}) = \mathbf{b}_R = (\boldsymbol{S}+\boldsymbol{R}'\boldsymbol{R})^{-1}\boldsymbol{X}'\boldsymbol{W}^{-1}\mathbf{y}. \qquad (2.3.15)$$

This estimator has

$$E\mathbf{b}_R = (\boldsymbol{S}+\boldsymbol{R}'\boldsymbol{R})^{-1}\boldsymbol{S}\boldsymbol{\beta} \quad \text{if } \mathbf{0} \neq \boldsymbol{R}\boldsymbol{\beta} \qquad (2.3.16)$$
$$= \boldsymbol{\beta} \qquad \text{if } \mathbf{0} = \boldsymbol{R}\boldsymbol{\beta}.$$

Furthermore, the following theorem holds (Leamer, 1979, p. 127).

Theorem 2.9

The restricted LSE $\mathbf{b}_R$ which is subject to the constraint $\boldsymbol{R}\boldsymbol{\beta} = \mathbf{0}$ lies on the ellipsoid

$$\left(\boldsymbol{\beta}-\frac{\mathbf{b}}{2}\right)'\boldsymbol{S}\left(\boldsymbol{\beta}-\frac{\mathbf{b}}{2}\right) = \frac{\mathbf{b}'\boldsymbol{S}\mathbf{b}}{4}. \qquad (2.3.17)$$

Furthermore, any point on this ellipsoid is a restricted LSE of $\boldsymbol{\beta}$ for some $\boldsymbol{R}$ (see Figure 2.3.1).

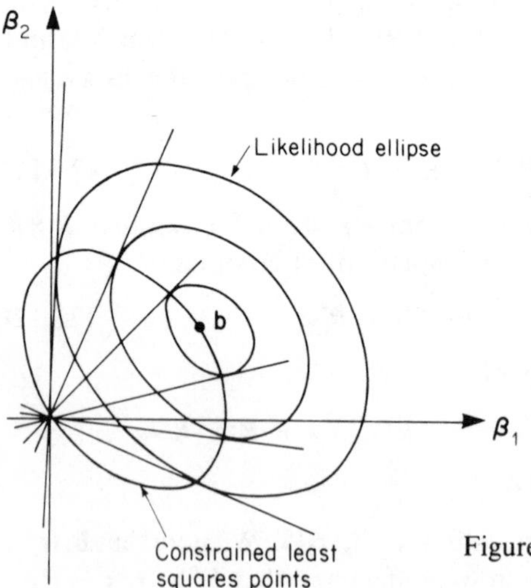

Figure 2.3.1 The ellipse of constrained least squares (Leamer, 1979, p. 128)

2.3.3 Ridge Regression

As pointed out in Section 2.1.6. (Gauss–Markoff Theorem), the OLSE $\mathbf{b}_0$ has minimum variance in the class of linear unbiased estimators of $\boldsymbol{\beta}$. This property is satisfactory as long as the numerical results are accepted by the user. If the matrix $X'X$ comes to a *near-degeneracy*, the estimator $\mathbf{b}_0$ becomes unacceptable because of large variances of the estimates of the $\boldsymbol{\beta}$-components (see the discussion on multicollinearity). Various attempts to overcome this 'variance inflation' (Hocking *et al.*, 1976) have been made, resulting in biased estimators as alternatives to the OLSE. The family of ridge estimators proposed by Hoerl and Kennard (1970a, b) is perhaps the first biased procedure to have found widespread practical application.

Note Nothing is lost by considering only the classical regression model, i.e. $W = I$.

The ridge estimator of $\boldsymbol{\beta}$ as proposed by Hoerl and Kennard (1970a) then is defined by

$$\mathbf{b}(k) = (X'X + kI)^{-1}X'\mathbf{y}. \tag{2.3.18}$$

Clearly $\mathbf{b}(0) = \mathbf{b}_0$, the OLS estimator.

As a reason for proposing this estimator (18), Hoerl and Kennard (1970a) investigate the scalar MSE of the unbiased estimator $\mathbf{b}_0$ in the light of the eigenvalues $\lambda_1 \geq \ldots \geq \lambda_K$ of $X'X$. The scalar MSE may be interpreted as the average squared distance from $\mathbf{b}_0$ to $\boldsymbol{\beta}$:

$$\begin{aligned} \text{SMSE } \mathbf{b}_0 &= E(\mathbf{b}_0 - \boldsymbol{\beta})'(\mathbf{b}_0 - \boldsymbol{\beta}) \\ &= \operatorname{tr} \sigma^2 (X'X)^{-1} \\ &= \sigma^2 \sum_{i=1}^{K} (1/\lambda_i). \end{aligned} \tag{2.3.19}$$

Hence if the data are such that they result in a matrix $X'X$ with one or more small eigenvalues, the distance from $\mathbf{b}_0$ to $\boldsymbol{\beta}$ will tend to be large. This fact will be observed, for example, if the column vectors $\mathbf{x}_i$ of X move from orthogonality, i.e. if $X'X$ moves from the identity matrix to an ill-conditioned one. Thus a reasonable heuristic interpretation for the set-up of $\mathbf{b}(k)$ is given. Another, more analytic reasoning was found by the 'fathers' of the ridge estimator (see Hoerl and Kennard, 1970a) who introduced the so-called ridge-trace. Let $\hat{\boldsymbol{\beta}}$ be any estimator of the unknown $\boldsymbol{\beta}$. Then the sum of squared errors $S(\hat{\boldsymbol{\beta}}) = (y - X\hat{\boldsymbol{\beta}})$ $(y - X\hat{\boldsymbol{\beta}})$ may be written as

$$\begin{aligned} S(\hat{\boldsymbol{\beta}}) &= (\mathbf{y} - X\mathbf{b}_0)'(\mathbf{y} - X\mathbf{b}_0) + (\mathbf{b}_0 - \hat{\boldsymbol{\beta}})'X'X(\mathbf{b}_0 - \hat{\boldsymbol{\beta}}) \\ &= S_{\min} + \phi(\hat{\boldsymbol{\beta}}). \end{aligned} \tag{2.3.20}$$

There is a manifold of estimators $\boldsymbol{\beta}$ that will satisfy the relationship

$$S(\hat{\boldsymbol{\beta}}) = S_{\min} + \phi_0 \tag{2.3.21}$$

where $\phi_0 > 0$ is any fixed increment. Minimizing the length of $\hat{\boldsymbol{\beta}}$ subject to a given increment ϕ_0 leads to

$$\min_{\hat{\boldsymbol{\beta}}} \{\hat{\boldsymbol{\beta}}'\hat{\boldsymbol{\beta}} + (1/k)[(\mathbf{b}_0 - \hat{\boldsymbol{\beta}})'\boldsymbol{X}'\boldsymbol{X}(\mathbf{b}_0 - \hat{\boldsymbol{\beta}}) - \phi_0]\} \tag{2.3.22}$$

where $1/k$ is the Lagrangian multiplier. The solution of this problem is just the ridge estimator $\mathbf{b}(k)$ where k is chosen to satisfy the restraint (21) for a given ϕ_0.

Let us now investigate the properties of $\mathbf{b}(k)$. We may write $\mathbf{b}(k)$ as

$$\mathbf{b}(k) = \boldsymbol{G}_k\boldsymbol{X}'\mathbf{y} \tag{2.3.23}$$

where

$$\boldsymbol{G}_k = (\boldsymbol{X}'\boldsymbol{X} + k\boldsymbol{I})^{-1},$$

and

$$\boldsymbol{G}_k\boldsymbol{X}'\boldsymbol{X} = \boldsymbol{I} - k\boldsymbol{G}_k. \tag{2.3.24}$$

Using this we may write the expectation of the ridge estimator as

$$E\mathbf{b}(k) = \boldsymbol{G}_k\boldsymbol{X}'\boldsymbol{X}\boldsymbol{\beta} = \boldsymbol{\beta} - k\boldsymbol{G}_k\boldsymbol{\beta} \tag{2.3.25}$$

so that

$$\text{bias } \mathbf{b}(k) = -k\,\boldsymbol{G}_k\boldsymbol{\beta}. \tag{2.3.26}$$

The dispersion matrix is

$$\boldsymbol{V}[\mathbf{b}(k)] = \sigma^2\boldsymbol{G}_k\boldsymbol{X}'\boldsymbol{X}\boldsymbol{G}_k. \tag{2.3.27}$$

This results in the MSE-matrix of $\mathbf{b}(k)$:

$$\boldsymbol{M} = \text{MSE}[\mathbf{b}(k)] = \boldsymbol{G}_k(\sigma^2\boldsymbol{X}'\boldsymbol{X} + k^2\boldsymbol{\beta}\boldsymbol{\beta}')\boldsymbol{G}_k. \tag{2.3.28}$$

Without loss of generality (see Goldstein and Smith, 1974) we may confine ourselves to the case where $\boldsymbol{X}'\boldsymbol{X}$ is diagonal, i.e. we assume

$$\boldsymbol{X} = \begin{bmatrix} \boldsymbol{\Lambda} \\ \boldsymbol{0} \end{bmatrix} \tag{2.3.29}$$

where $\boldsymbol{\Lambda}$ is the matrix of the eigenvalues $\lambda_1 \geq \ldots \geq \lambda_K > 0$ of $\boldsymbol{X}'\boldsymbol{X}$. Inserting (29) in (28), we find that the mean-square error matrix $\boldsymbol{M}$ has elements

$$m_{ii} = \frac{\sigma^2\lambda_i + k^2\beta_i^2}{(\lambda_i + k)^2} \tag{2.3.30}$$

and

$$m_{ij} = \frac{k^2\beta_i\beta_j}{(\lambda_i + k)(\lambda_j + k)} \qquad (i \neq j). \tag{2.3.31}$$

Note that the scalar MSE of $\mathbf{b}(k)$

$$\text{tr}\,\boldsymbol{M} = \sum_{i=1}^{K} m_{ii} = \sum_{i=1}^{K} \frac{\sigma^2\lambda_i + k^2\beta_i^2}{(\lambda_i + k)^2} \tag{2.3.32}$$

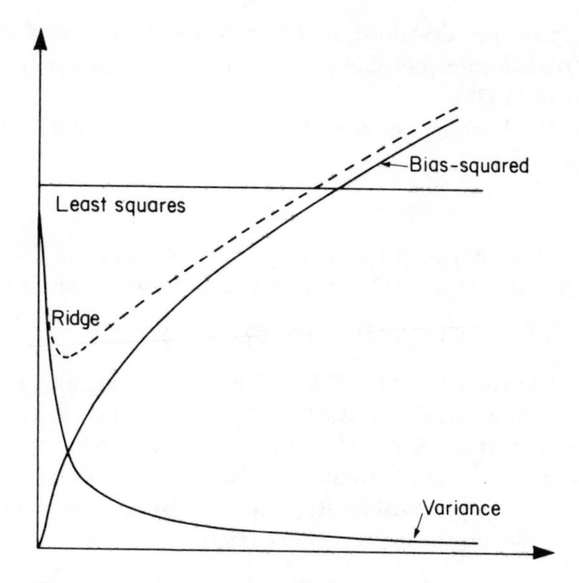

Figure 2.3.2 Mean square error functions of OLSE and ridge regression

is a function of k which starts from $\Sigma\,\sigma^2/\lambda_i$ when $k = 0$, then tends to a minimum for a positive k_{opt} and later starts increasing as far as $k_{\text{opt}} < \infty$ (see Figure 2.3.2). $k = 0$ corresponds to the OLSE $\mathbf{b}_0$. In other words, for a range of positive k the OLSE can be improved using the ridge technique, i.e.

$$\operatorname{tr}\{\operatorname{MSE}(\mathbf{b}_0) - \operatorname{MSE}[\mathbf{b}(k)]\} \geq 0 \quad \text{if } 0 < k < k^* \tag{2.3.33}$$

where k^* is unknown.

Denoting $(\boldsymbol{X}'\boldsymbol{X})^{-1} = \boldsymbol{U}$ and using

$$\begin{aligned}\operatorname{MSE}(\mathbf{b}_0) = \sigma^2 \boldsymbol{U} &= \sigma^2 \boldsymbol{G}_k(\boldsymbol{G}_k^{-1}\boldsymbol{U}\boldsymbol{G}_k^{-1})\boldsymbol{G}_k \\ &= \sigma^2 \boldsymbol{G}_k(\boldsymbol{U}^{-1} + k^2\boldsymbol{U} + 2k\boldsymbol{I})\boldsymbol{G}_k\end{aligned}$$

then Theorem 2.1 (with $\boldsymbol{A} = \boldsymbol{I}$) leads to the conclusion that (33) holds if and only if

$$\operatorname{MSE}(\mathbf{b}_0) - \operatorname{MSE}[\mathbf{b}(k)] = k\boldsymbol{G}_k[\sigma^2(2\boldsymbol{I} + k\boldsymbol{U}) - k\boldsymbol{\beta}\boldsymbol{\beta}']\boldsymbol{G}_k \geq 0. \tag{2.3.34}$$

Thus

$$2\sigma^2\boldsymbol{I} - k\boldsymbol{\beta}\boldsymbol{\beta}' \geq 0 \tag{2.3.35}$$

is sufficient for (34) to hold. Using Theorem A.17 this is equivalent to

$$k \leq 2\sigma^2/\boldsymbol{\beta}'\boldsymbol{\beta} \tag{2.3.36}$$

(see Theobald, 1974). Interpreting the ridge estimator as a special mixed estimator leads to a similar condition (see Section 3.3.4).

If prior information on the length of $\boldsymbol{\beta}$ as $\boldsymbol{\beta}'\boldsymbol{\beta} \leq r^2$ is available, this will lead to the minimax principle described in Chapter IV. In other words, the ridge estimator $\mathbf{b}(k)$ can be interpreted as a minimax linear estimator (see Section 4.2.4 and Swamy *et al.*, 1978).

The value of $\boldsymbol{\beta}$ which minimizes $\sigma^{-2}(\mathbf{y}-\boldsymbol{X\beta})'(\mathbf{y}-\boldsymbol{X\beta})$ subject to the constraint $\boldsymbol{\beta}'\boldsymbol{\beta} \leq r^2$ is

$$\hat{\boldsymbol{\beta}}(\mu) = (\boldsymbol{X}'\boldsymbol{X} + \sigma^2\mu\boldsymbol{I})^{-1}\boldsymbol{X}'\mathbf{y} \tag{2.3.37}$$

where μ is a value satisfying $\hat{\boldsymbol{\beta}}'(\mu)\hat{\boldsymbol{\beta}}(\mu) = r^2$ (see Swamy *et al.*, 1978). As may be shown (Swamy and Mehta, 1977), $\hat{\boldsymbol{\beta}}(\mu)$ dominates $\mathbf{b}_0$ with respect to the MSE, iff

$$\boldsymbol{\beta}'[(2/\mu)\boldsymbol{I} + \sigma^2(\boldsymbol{X}'\boldsymbol{X})^{-1}]^{-1}\boldsymbol{\beta} \leq 1. \tag{2.3.38}$$

If r^2 is large, then the constraint $\boldsymbol{\beta}'\boldsymbol{\beta} \leq r^2$ is less binding and μ will be small. Conversely, choosing a small μ and setting up the estimator $\hat{\boldsymbol{\beta}}(\mu)$ corresponds to using prior information $\boldsymbol{\beta}'\boldsymbol{\beta} \leq r^2$ with a large r. Such less-binding prior information seems to be almost always true. However, due to the unknown σ^2 the estimator $\hat{\boldsymbol{\beta}}(\mu)$ is not practicable. Replacing σ^2 by its unbiased estimator s^2 (2.1.51) leads to the approximate ridge estimator

$$\hat{\boldsymbol{\beta}}^*(\mu) = (\boldsymbol{X}'\boldsymbol{X} + s^2\mu\boldsymbol{I})^{-1}\boldsymbol{X}'\mathbf{y} \tag{2.3.39}$$

(see the related problems for mixed estimation and minimax estimation, respectively).

Notes The problem of ridge regression has given rise to a series of generalizations. Bibby (1972) considered the estimator

$$\hat{\boldsymbol{\beta}}(\boldsymbol{K}) = (\boldsymbol{X}'\boldsymbol{X} + \boldsymbol{K})^{-1}\boldsymbol{X}'\mathbf{y}$$

where $\boldsymbol{K}$ is any matrix, and also the estimator

$$\hat{\boldsymbol{\beta}}(k, l) = [\boldsymbol{X}'\boldsymbol{X} + k\boldsymbol{I} + l(\boldsymbol{X}'\boldsymbol{X} + k\boldsymbol{I})^{-1}]\boldsymbol{X}'\mathbf{y}$$

(k, l arbitrary scalars).

Furthermore, numerous simulation experiments have been carried out in order to clear up the improvement region of the ridge estimator (see, for example, Hocking *et al.*, 1976; Hemmerle and Brantle, 1978; Marquardt and Snee, 1975; Mayer and Willke, 1973; Gunst and Mason, 1977). Baldwin and Hoerl (1978) have given bounds on the MSE of the ridge estimator. A fairly extensive review of modern literature of ridge regression may be found in the paper of Vinod (1978).

2.3.4 The Shrunken Estimator

As a special biased estimator which is a function of the GLSE $\mathbf{b}$ the following linear estimator may be defined (see Mayer and Willke, 1973)

$$\hat{\boldsymbol{\beta}}(\rho) = (1+\rho)^{-1}\mathbf{b}, \quad \rho \geq 0 \text{ (known)}. \tag{2.3.40}$$

We now derive

$$\text{bias } \hat{\boldsymbol{\beta}}(\rho) = -\rho(1+\rho)^{-1}\boldsymbol{\beta} \tag{2.3.41}$$

$$V[\hat{\boldsymbol{\beta}}(\rho)] = \sigma^2(1+\rho)^{-2}\boldsymbol{S}^{-1} = (1+\rho)^{-2}V(\mathbf{b}) \tag{2.3.42}$$

which gives the risk

$$\begin{aligned} R(\hat{\boldsymbol{\beta}}(\rho), A) &= \operatorname{tr} A \operatorname{MSE}[\hat{\boldsymbol{\beta}}(\rho)] \\ &= \operatorname{tr} A(1+\rho)^{-2}[\rho^2\boldsymbol{\beta}\boldsymbol{\beta}' + \sigma^2\boldsymbol{S}^{-1}]. \end{aligned} \tag{2.3.43}$$

Comparing with the GLSE $\mathbf{b}$ we have

$$\begin{aligned} &R(\mathbf{b}, A) - \mathbf{R}(\hat{\boldsymbol{\beta}}(\rho), A) \\ &\quad = (1+\rho)^{-2}\rho\sigma^2 \operatorname{tr} A[(2+\rho)\boldsymbol{S}^{-1} - \sigma^{-2}\rho\boldsymbol{\beta}\boldsymbol{\beta}'] \geq 0 \text{ iff} \\ &\quad\quad \rho \leq 2\{(\sigma^{-2}\boldsymbol{\beta}'A\boldsymbol{\beta})(\operatorname{tr}\boldsymbol{S}^{-1}A)^{-1} - 1\}^{-1}. \end{aligned} \tag{2.3.44}$$

This has to be fulfilled for all $\boldsymbol{\beta}$. Using Theorem A.13 we get the following theorem, which gives a sufficient condition.

Theorem 2.10

The shrunken estimator $\hat{\boldsymbol{\beta}}(\rho)$ has a smaller risk $R(\cdot, A)$ than the GLSE $\mathbf{b}$, if

$$\rho \leq 2\{\sigma^{-2}\boldsymbol{\beta}'\boldsymbol{\beta}\lambda_{\max}(A)(\operatorname{tr}\boldsymbol{S}^{-1}A)^{-1} - 1\}^{-1}. \tag{2.3.45}$$

(see further results in Section 4.5).

Confining himself to the classical regression model, Farebrother (1978b) has proposed generalizations of the shrinkage technique using the canonical form (see (2.1.29))

$$\mathbf{y} = \tilde{\boldsymbol{X}}\tilde{\boldsymbol{\beta}} + \varepsilon, \quad \varepsilon \sim (\mathbf{0}, \sigma^2\boldsymbol{I})$$

where $\tilde{\boldsymbol{X}} = \boldsymbol{XP}$, $\tilde{\boldsymbol{\beta}} = \boldsymbol{P}'\boldsymbol{\beta}$ and $\tilde{\boldsymbol{X}}'\tilde{\boldsymbol{X}} = \boldsymbol{\Lambda}$ (the matrix of eigenvalues of $\boldsymbol{X}'\boldsymbol{X}$).

Given arbitrary matrices $\underset{K,K}{C}$ and $\underset{K,1}{\mathbf{c}}$, we may define the general shrunken (or shrinkage) estimator of $\boldsymbol{\beta}$ by

$$\begin{aligned} \tilde{\boldsymbol{\beta}}_* &= \mathbf{b}_0 + C'(\mathbf{c} - \mathbf{b}_0) \\ &= \mathbf{c} + (\boldsymbol{I} - C)'(\mathbf{b}_0 - \mathbf{c}). \end{aligned} \tag{2.3.46}$$

As

$$\operatorname{bias} \tilde{\boldsymbol{\beta}}_* = -C'(\tilde{\boldsymbol{\beta}} - \mathbf{c})$$

and

$$V(\tilde{\boldsymbol{\beta}}_*) = \sigma^2(\boldsymbol{I} - C)'\boldsymbol{\Lambda}^{-1}(\boldsymbol{I} - C)$$

we get

$$\operatorname{MSE}(\tilde{\boldsymbol{\beta}}_*) = \sigma^2(\boldsymbol{I} - C)'\boldsymbol{\Lambda}^{-1}(\boldsymbol{I} - C) + C'(\tilde{\boldsymbol{\beta}} - \mathbf{c})(\tilde{\boldsymbol{\beta}} - \mathbf{c})'C. \tag{2.3.47}$$

As $\operatorname{MSE}(\mathbf{b}_0) = \sigma^2\boldsymbol{\Lambda}^{-1}$ (see (2.1.33)) we derive the difference

$$\begin{aligned} \Delta(\mathbf{b}_0, \tilde{\boldsymbol{\beta}}_*) &= \operatorname{MSE}(\mathbf{b}_0) - \operatorname{MSE}(\tilde{\boldsymbol{\beta}}_*) = \sigma^2 C'\boldsymbol{\Lambda}^{-1}C \\ &\quad + \sigma^2 C'\boldsymbol{\Lambda}^{-1}(\boldsymbol{I} - C) + \sigma^2(\boldsymbol{I} - C)'\boldsymbol{\Lambda}^{-1}C - C'(\tilde{\boldsymbol{\beta}} - \mathbf{c})(\tilde{\boldsymbol{\beta}} - \mathbf{c})'C. \end{aligned} \tag{2.3.48}$$

Theorem 2.11 (Farebrother, 1976, 1978b)

Assume that

$$C'\Lambda^{-1}(I-C)+(I-C)'\Lambda^{-1}C \geq \mathbf{0}.$$

Then $\Delta(\mathbf{b}_0, \tilde{\beta}_*) \geq \mathbf{0}$ holds if

$$(\tilde{\beta}-\mathbf{c})'\Lambda(\tilde{\beta}-\mathbf{c}) \leq \sigma^2 \tag{2.3.49}$$

or

$$(\beta-P\mathbf{c})'(X'X)(\beta-P\mathbf{c}) \leq \sigma^2. \tag{2.3.50}$$

In the context of minimax linear estimation (Chapter IV) we will use just this information to derive improved estimators.

As an important example of the general shrunken estimator we note that the R-optimal biased estimator $\hat{\beta}_2$ (2.4.6) with $W = I$, that is

$$\hat{\beta}_2 = \beta\beta' X'(X\beta\beta' X' + \sigma^2 I)^{-1}\mathbf{y}$$

may be written

$$P\hat{\beta}_2 = \tilde{\beta}\tilde{\beta}'(\tilde{\beta}\tilde{\beta}' + \sigma^2\Lambda^{-1})^{-1}\Lambda^{-1}\tilde{X}'\mathbf{y}. \tag{2.3.51}$$

This is just of the form (46) with $\mathbf{c} = \mathbf{0}$ and $C' = I - \tilde{\beta}\tilde{\beta}'(\tilde{\beta}\tilde{\beta}' + \sigma^2\Lambda^{-1})^{-1}$.

2.3.5 Testing Linear Hypotheses

Assume the general linear model with normally distributed disturbances

$$\mathbf{y} = X\beta + \varepsilon, \quad \varepsilon \sim N(\mathbf{0}, \sigma^2 W). \tag{2.3.52}$$

In the following we will summarize the standard procedure for testing the general linear hypothesis

$$H_0: R\beta = \mathbf{r} \tag{2.3.53}$$

against the alternative hypothesis

$$H_1: R\beta \neq \mathbf{r} \tag{2.3.54}$$

where it is assumed:

> R is a $(K-s)\times K$-matrix, rank $R = K-s$,
> $\mathbf{r}$ is a $(K-s)\times 1$-vector,
> $s = 0, 1, \ldots, K-1$.

The null hypothesis implies two essential special cases:

(i) $s = 0$. Then R is regular and we get

$$H_0: \beta = R^{-1}\mathbf{r} = \beta^*, \text{ say}, \qquad H_1: \beta \neq \beta^*. \tag{2.3.55}$$

That is, the whole parameter vector β is underlying the hypothesis.

(ii) $s > 0$. Choosing a $s \times K$-matrix $\boldsymbol{B}$ which is complementary to $\boldsymbol{R}$ and denoting

$$\boldsymbol{X}\begin{pmatrix}\boldsymbol{B}\\ \boldsymbol{R}\end{pmatrix}^{-1} = \boldsymbol{X}^* = (\boldsymbol{X}_1^*, \boldsymbol{X}_2^*),$$

$$\boldsymbol{\beta}_1^* = \boldsymbol{B}\boldsymbol{\beta}, \quad \boldsymbol{\beta}_2^* = \boldsymbol{R}\boldsymbol{\beta}$$

gives

$$\begin{aligned}\mathbf{y} &= \boldsymbol{X}\begin{pmatrix}\boldsymbol{B}\\ \boldsymbol{R}\end{pmatrix}^{-1}\begin{pmatrix}\boldsymbol{B}\\ \boldsymbol{R}\end{pmatrix}\boldsymbol{\beta} + \varepsilon \\ &= \boldsymbol{X}_1^*\boldsymbol{\beta}_1^* + \boldsymbol{X}_2^*\boldsymbol{\beta}_2^* + \varepsilon.\end{aligned}$$

Thus the null-hypothesis (53) becomes

$$H_0 : \boldsymbol{\beta}_2^* = \mathbf{r}. \tag{2.3.56}$$

That is, if $s > 0$, only a subvector of $\boldsymbol{\beta}$ will be tested. Without loss of generality we will now assume that the hypotheses are of the form either (53) or (56). With other words, the asterisk may be omitted in (56). Denoting the unrestricted parameter space by

$$\Omega = \{\boldsymbol{\beta}; \sigma^2 : \boldsymbol{\beta} \in E^K, \sigma^2 > 0\}$$

and, analogously, the restricted parameter space by $\omega = \{\boldsymbol{\beta}; \sigma^2 : \boldsymbol{\beta} \in E^K$ and $\mathbf{r} = \boldsymbol{R}\boldsymbol{\beta}; \sigma^2 > 0\}$ we may use the likelihood-ratio

$$\lambda(\mathbf{y}) = \frac{\max\limits_{\omega} L(\boldsymbol{\beta}, \sigma^2)}{\max\limits_{\Omega} L(\boldsymbol{\beta}, \sigma^2)} \tag{2.3.57}$$

to arrive at a test statistic. As shown in Section 2.1.10, the likelihood function $L(\boldsymbol{\beta}, \sigma^2)$ has its maximum for the ML-estimates of $\boldsymbol{\beta}$ and σ^2, i.e.

$$\begin{aligned}\max L(\boldsymbol{\beta}, \sigma^2) &= L(\hat{\boldsymbol{\beta}}, \hat{\sigma}^2) \\ &= (2\pi\hat{\sigma}^2)^{-T/2}\exp\{-\tfrac{1}{2}2(\mathbf{y} - \boldsymbol{X}\hat{\boldsymbol{\beta}})'\boldsymbol{W}^{-1}(\mathbf{y} - \boldsymbol{X}\hat{\boldsymbol{\beta}})\} \\ &= (2\pi\hat{\sigma}^2)^{-T/2}\exp(-T/2).\end{aligned}$$

(Clearly, analogous to (2.1.64) and (2.2.9), the ML estimator of σ^2 in generalized regression is $\hat{\sigma}^2 = (\mathbf{y} - \boldsymbol{X}\hat{\boldsymbol{\beta}})'\boldsymbol{W}^{-1}(\mathbf{y} - \boldsymbol{X}\hat{\boldsymbol{\beta}})/T$.) Thus the likelihood ratio (57) becomes

$$\lambda(\mathbf{y}) = \left(\frac{\hat{\sigma}_\omega^2}{\hat{\sigma}_\Omega^2}\right)^{-T/2}.$$

Using a monotone transform of this we arrive at the usual test statistic

$$\begin{aligned}F &= \{(\lambda(\mathbf{y}))^{-2/T} - 1\}(T - K)(K - s)^{-1} \\ &= \frac{\hat{\sigma}_\omega^2 - \hat{\sigma}_\Omega^2}{\hat{\sigma}_\Omega^2}\,\frac{T - K}{K - s}.\end{aligned} \tag{2.3.58}$$

If $s = 0$, this statistic has the form

$$F = \frac{(\mathbf{b}-\boldsymbol{\beta}^*)' S(\mathbf{b}-\boldsymbol{\beta}^*)}{(\mathbf{y}-X\mathbf{b})' W^{-1}(\mathbf{y}-X\mathbf{b})} \frac{T-K}{K} \tag{2.3.59}$$

and, under the normality assumption, we have

$$F \sim F_{K,T-K}(\sigma^{-2}(\boldsymbol{\beta}-\boldsymbol{\beta}^*)' S(\boldsymbol{\beta}-\boldsymbol{\beta}^*)) \quad \text{if } \boldsymbol{\beta} \neq \boldsymbol{\beta}^* \tag{2.3.60}$$

and

$$F \sim F_{K,T-K} \qquad \text{under } H_0. \tag{2.3.61}$$

In the other case $s > 0$ we get

$$F = \frac{(\mathbf{b}_2-\mathbf{r}) D(\mathbf{b}_2-\mathbf{r})}{(\mathbf{y}-X\mathbf{b})' W^{-1}(\mathbf{y}-X\mathbf{b})} \frac{T-K}{K-s} \tag{2.3.62}$$

where $\mathbf{b}_2$ is the subvector of the GLSE $\mathbf{b}$ corresponding to the vector $\boldsymbol{\beta}_2$ of the null-hypothesis (56) and $\sigma^2 D^{-1}$ is the dispersion matrix of $\mathbf{b}_2$ (see (2.2.46) and (2.2.49)). Furthermore, we have the result that F (62) follows an F-distribution:

$$F \sim F_{K-s,T-K}[\sigma^{-2}(\boldsymbol{\beta}_2-\mathbf{r})' D(\boldsymbol{\beta}_2-\mathbf{r})] \quad \text{if } \boldsymbol{\beta}_2 \neq \mathbf{r} \tag{2.3.63}$$

and

$$F \sim F_{K-s,T-K} \qquad \text{under } H_0 \text{ (56).} \tag{2.3.64}$$

Choosing a critical value $F^{\alpha}_{K-s,T-K}$ of the central F-distribution, the null-hypotheses (55) or (56), respectively, will be accepted if

$$0 \leq F < F^{\alpha}_{K-s,T-K} \qquad (s = 0, 1, \ldots, K-1). \tag{2.3.65}$$

(For a detailed proof of the above results the reader is referred to Toutenburg, 1975a, pp. 47–54.)

2.3.6 Exact Linear Restrictions in the Full Rank Case

To overcome extreme multicollinearity, exact linear restrictions $\mathbf{r} = R\boldsymbol{\beta}$ (2) were introduced to ensure a geometrical solution of the estimation problem. As a necessary condition, R was restricted to have rank $K-p$ and, furthermore, to be complementary to X. If the matrix X, although perhaps ill-conditioned and so far weak multicollinear, is of full rank K, then $\boldsymbol{\beta}$ is estimable. Then the auxiliary knowledge of a relation such as

$$\mathbf{r} = R\boldsymbol{\beta}, \quad \text{rank} \underset{J,K}{R} = J \tag{2.3.66}$$

may be interpreted as prior information on $\boldsymbol{\beta}$. Its use should help to improve the precision of the estimation procedure. The well-known method for taking into account such exact linear restrictions in minimizing the quadratic function $S(\boldsymbol{\beta}) = (\mathbf{y}-X\boldsymbol{\beta})' W^{-1}(\mathbf{y}-X\boldsymbol{\beta})$ is to solve the problem

$$\min_{\boldsymbol{\beta},\lambda} \{S(\boldsymbol{\beta}) - 2\lambda'(R\boldsymbol{\beta}-\mathbf{r})\} = \min_{\boldsymbol{\beta},\lambda} \tilde{S}(\boldsymbol{\beta}) \tag{2.3.67}$$

where λ is a $K \times 1$-vector of Lagrangian multipliers. Differentiating with respect to $\boldsymbol{\beta}$ and λ gives (Theorems A.50, A.53)

$$\frac{1}{2}\frac{\partial \tilde{S}(\boldsymbol{\beta})}{\partial \boldsymbol{\beta}} = -X'W^{-1}\mathbf{y} + X'W^{-1}X\boldsymbol{\beta} - R'\lambda, \tag{2.3.68}$$

$$\frac{1}{2}\frac{\partial S(\boldsymbol{\beta})}{\partial \lambda} = \mathbf{r} - R\boldsymbol{\beta}. \tag{2.3.69}$$

Equating (68) to zero leads to

$$\hat{\boldsymbol{\beta}} = \mathbf{b} + S^{-1}R'\hat{\lambda}. \tag{2.3.70}$$

Premultiplying (70) by R gives

$$R\hat{\boldsymbol{\beta}} = R\mathbf{b} + RS^{-1}R'\hat{\lambda} = \mathbf{r} \quad \text{(using (66))}$$

whence

$$\hat{\lambda} = [RS^{-1}R']^{-1}(\mathbf{r} - R\mathbf{b}).$$

Note that $RS^{-1}R' > 0$ (Theorem A.5) and therefore its inverse exists (Theorem A.4). Thus the resulting estimator becomes (see Goldberger, 1964, p. 257)

$$\hat{\boldsymbol{\beta}} = \tilde{\mathbf{b}}_R = \mathbf{b} + S^{-1}R'[RS^{-1}R']^{-1}(\mathbf{r} - R\mathbf{b}). \tag{2.3.71}$$

We can show that conditional unbiasedness holds, that is

$$E[\tilde{\mathbf{b}}_R | \mathbf{r} = R\boldsymbol{\beta}] = \boldsymbol{\beta} \tag{2.3.72}$$

and, moreover, that $\tilde{\mathbf{b}}_R$ has dispersion

$$V(\tilde{\mathbf{b}}_R) = \sigma^2 S^{-1} - \sigma^2 S^{-1}R'[RS^{-1}R']^{-1}RS^{-1}. \tag{2.3.73}$$

Comparing $\tilde{\mathbf{b}}_R$ with the unrestricted GLSE $\mathbf{b}$ gives

$$V(\mathbf{b}) - V(\tilde{\mathbf{b}}_R) = \sigma^2 S^{-1}R'[RS^{-1}R']^{-1}RS^{-1} \geq 0. \tag{2.3.74}$$

We will call $\tilde{\mathbf{b}}_R$ the *conditional restricted least squares estimator* (conditional RLSE).

It may be proved (Toutenburg, 1975a, p. 99–100) that the estimator $\tilde{\mathbf{b}}_R$ is BLUE of $\boldsymbol{\beta}$ in the class $\{\hat{\boldsymbol{\beta}} = D_1\mathbf{y} + D_2\mathbf{r}\}$ of heterogeneous estimators which are conditional unbiased in the sense $E\{\hat{\boldsymbol{\beta}} - \boldsymbol{\beta} | \mathbf{r} = R\boldsymbol{\beta}\} = \mathbf{0}$.

Later we will use this estimator in the context of hypothesis testing (see Section 3.4.2).

2.4 OPTIMAL LINEAR ESTIMATORS

2.4.1 Heterogeneous Set-Up

Given the quadratic risk (2.1.27) of an estimator $\hat{\boldsymbol{\beta}}$, i.e.

$$R(\hat{\boldsymbol{\beta}}, A) = E(\hat{\boldsymbol{\beta}} - \boldsymbol{\beta})'A(\hat{\boldsymbol{\beta}} - \boldsymbol{\beta}) = \operatorname{tr} \operatorname{AMSE}(\hat{\boldsymbol{\beta}}) \tag{2.4.1}$$

where $A > 0$ is assumed, and using Theorem 2.1 about the equivalence concerning the risk (for $A \geq 0$) and the mean square error, we will now derive estimators which minimize the 'distance' to the unknown vector β.

We call an estimator R-optimal if its expected quadratic loss is less than that of any other estimator, that is, R-optimality of an estimator $\hat{\beta}$ requires

$$R(\hat{\beta}, A) \leq R(\tilde{\beta}, A) \tag{2.4.2}$$

for any other estimator $\tilde{\beta}$.

Clearly, R-optimality will be constrained to a fixed class of estimators which are under consideration. We will use the linear set-up

$$\hat{\beta} = Cy + c \tag{2.4.3}$$

(see (2.1.16)) which leads to the risk

$$R(\hat{\beta}, A) = \sigma^2 \operatorname{tr} ACWC' + [(CX - I)\beta + c]'A[(CX - I)\beta + c] \tag{2.4.4}$$

(see also equation (2.1.21) which defines MSE $\hat{\beta}$ in the case $W = I$). Minimizing (4) with respect to C and c simplifies by noting that the first term in (4) is free of c. Thus $c_{\text{opt}} = \hat{c}$ is that value of c which equates the second term to zero:

$$\hat{c} = -(\hat{C}X - I)\beta$$

where $\hat{C}$ is the solution of

$$\frac{\partial}{\partial C} \operatorname{tr} ACWC' = 2WCA = 0 \text{ (see Theorem A.54).}$$

Thus $\hat{C} = 0$ and $\hat{c} = \beta$. This gives the R-optimal heterogeneous 'estimator'

$$\hat{\beta}_1 = \beta. \tag{2.4.5}$$

This trivial estimator has zero bias, zero risk—and (Bibby and Toutenburg, 1978, p. 76) 'zero usefulness'. Nevertheless, the above results demonstrate that nothing is lost in confining ourselves to homogeneous estimators as far as R-optimality is concerned.

2.4.2 Homogeneous Estimation

Putting $c = 0$ in (3) gives the homogeneous set-up $\hat{\beta} = Cy$. Substituting $c = 0$ in (4) and minimizing the risk with respect to C gives

$$\frac{\partial}{\partial C} R(\hat{\beta}, A) = 2A[C(X\beta\beta'X' + \sigma^2 W) - \beta\beta'X'].$$

Equating to zero gives the optimal $C_{\text{opt}} = \hat{C}$ which leads to the R-optimal estimator $\hat{\beta}_2$, say:

$$\hat{\beta}_2 = \hat{C}y = \beta\beta'X'(X\beta\beta'X' + \sigma^2 W)^{-1} y. \tag{2.4.6}$$

(This expression was derived independently by Toutenburg (1968), Rao (1971,

p. 389), Bibby (1972), and Theil (1971).) Using the inversion formula (Theorem A. 35) we have

$$(X\beta\beta'X' + \sigma^2 W)^{-1} = \sigma^{-2}W^{-1} - \frac{\sigma^{-4}W^{-1}X\beta\beta'X'W^{-1}}{1+\sigma^{-2}\beta'X'W^{-1}X\beta} \tag{2.4.7}$$

and therefore $\hat{\beta}_2$ simplifies to

$$\hat{\beta}_2 = \frac{\beta'X'W^{-1}y}{\sigma^2 + \beta'S\beta}\beta \tag{2.4.8}$$

(see Bibby (1972) and Farebrother (1975)). From (8) it follows that

$$E(\hat{\beta}_2) = \beta\left[\frac{\beta'S\beta}{\sigma^2 + \beta'S\beta}\right] = \beta\alpha, \text{ say} \tag{2.4.9}$$

where $0 < \alpha < 1$. Hence, on average, $\hat{\beta}_2$ leads to an underestimate of β. Clearly, since $\alpha \neq 0$, $\hat{\beta}_2$ is a biased estimator:

$$\text{bias } \hat{\beta}_2 = (\alpha - 1)\beta = -\frac{\sigma^2}{\sigma^2 + \beta'S\beta}\beta. \tag{2.4.10}$$

Its dispersion matrix becomes

$$V(\hat{\beta}_2) = \frac{\beta'S\beta}{\sigma^2 + \beta'S\beta}\sigma^2\beta\beta' \tag{2.4.11}$$

so that the MSE of $\hat{\beta}_2$ has the form

$$\text{MSE}(\hat{\beta}_2) = \frac{\sigma^2\beta\beta'}{\sigma^2 + \beta'S\beta}. \tag{2.4.12}$$

Writing $\hat{\beta}_2$ (8) as

$$\begin{aligned}\hat{\beta}_2 &= \beta\left[\beta'\left(\frac{\sigma^2}{\beta'\beta}I + S\right)\beta\right]^{-1}\beta'X'W^{-1}y \\ &= \beta(\beta'V\beta)^{-1}\beta'V\beta^{+}, \text{ say}\end{aligned} \tag{2.4.13}$$

where

$$V = \frac{\sigma^2}{\beta'\beta}I + S \tag{2.4.14}$$

and

$$\beta^{+} = V^{-1}X'W^{-1}y \tag{2.4.15}$$

leads to the interpretation of $\hat{\beta}_2$ as a two-stage estimator. The first stage calculates β^{+}, which is of ridge type. In the second stage β^{+} is regressed on β with variance V. Hence $\hat{\beta}_2$ may be interpreted as a combination between ridge and generalized regression (Bibby and Toutenburg, 1978, p. 78).

Moreover, in the case $K = 1$, $\boldsymbol{\beta}$ is scalar and $\boldsymbol{X} = \mathbf{x}$ is a vector so that (8) simplifies to

$$\hat{\boldsymbol{\beta}}_2 = \frac{\mathbf{x}'\mathbf{y}}{\mathbf{x}'\mathbf{x} + \sigma^2/\boldsymbol{\beta}'\boldsymbol{\beta}} \tag{2.4.16}$$

which clearly gives a shrinkage of the OLSE $\mathbf{x}'\mathbf{y}/\mathbf{x}'\mathbf{x}$.

Now, $\hat{\boldsymbol{\beta}}_2$ is a function of the unknown 'coefficient of variation' $\sigma^{-1}\boldsymbol{\beta}$ and, therefore, it is not practicable. To overcome this difficulty one has to use prior information about $\boldsymbol{\beta}$ such that $\boldsymbol{\beta}$ is bounded both in direction and size (see Toutenburg, 1968, and Bibby and Toutenburg, 1978, p. 90).

Finally, let us compare $\hat{\boldsymbol{\beta}}_1$ and $\hat{\boldsymbol{\beta}}_2$. Clearly

$$R(\hat{\boldsymbol{\beta}}_1, \boldsymbol{A}) \leq R(\hat{\boldsymbol{\beta}}_2, \boldsymbol{A}) \tag{2.4.17}$$

holds, as $R(\hat{\boldsymbol{\beta}}_2, \boldsymbol{A})$ cannot be negative.

2.4.3 Homogeneous Unbiased Estimation

Inserting condition $\boldsymbol{CX} - \boldsymbol{I} = \boldsymbol{0}$ (2.1.24) for lack of bias and $\mathbf{c} = \mathbf{0}$ in the risk (4) gives

$$R(\hat{\boldsymbol{\beta}}, \boldsymbol{A}) = \sigma^2 \operatorname{tr} \boldsymbol{ACWC}'. \tag{2.4.18}$$

Therefore the optimal $\boldsymbol{C}$ is the solution to the constrained optimization problem:

$$\min_{C} \operatorname{tr} \boldsymbol{ACWC}' \quad \text{subject to } \boldsymbol{CX} = \boldsymbol{I}.$$

The solution to this is

$$\hat{\boldsymbol{C}} = \boldsymbol{S}^{-1}\boldsymbol{X}'\boldsymbol{W}^{-1}.$$

Thus the R-optimal homogeneous unbiased estimator of $\boldsymbol{\beta}$ is

$$\hat{\boldsymbol{\beta}}_3 = \hat{\boldsymbol{C}}\mathbf{y} = \boldsymbol{S}^{-1}\boldsymbol{X}'\boldsymbol{W}^{-1}\mathbf{y} = \mathbf{b}, \tag{2.4.19}$$

the well-known GLSE.

Using the fact that $\hat{\boldsymbol{\beta}}_3$ is obtained by adding extra constraints (namely $\boldsymbol{CX} = \boldsymbol{I}$) to the R-risk, gives

$$R(\hat{\boldsymbol{\beta}}_1, \boldsymbol{A}) \leq R(\hat{\boldsymbol{\beta}}_2, \boldsymbol{A}) \leq R(\hat{\boldsymbol{\beta}}_3, \boldsymbol{A}). \tag{2.4.20}$$

Although possessing the greatest risk of the three-estimators $\hat{\boldsymbol{\beta}}_1$, $\hat{\boldsymbol{\beta}}_2$, $\hat{\boldsymbol{\beta}}_3$, the GLSE $\hat{\boldsymbol{\beta}}_3 = \mathbf{b}$ has the advantage of being the only practicable solution among the three R-optimal set-ups. This fact underlines the statement that $\mathbf{b}$ *is best if no other than the sample and model information is available.*

III Mixed Estimation

3.1 LINEAR STOCHASTIC RESTRICTIONS

3.1.1 Sample and Prior Information

As a starting point which was the basis of the standard regression procedures described in Chapter II, we take a T-dimensional sample of the interrelated variables $\mathbf{y}$ and $X_1, \ldots, X_k$. If the linear regression model $\mathbf{y} = X\boldsymbol{\beta} + \boldsymbol{\varepsilon}$ may be assumed to be a realistic picture of the underlying (e.g. economic) relationship, then the 'best' estimator $\boldsymbol{\beta}$ of the parameter vector $\boldsymbol{\beta}$ is known (compare the corresponding theorems for the various types of linear models as given in Chapter II). The property of a fixed estimator to be 'best' clearly is related to a given class of possible estimator functions (linear, homogeneous or inhomogeneous, biased or unbiased) and a given concept of comparing two estimators (see Section 2.1.4 for the definition of measures of goodness by a risk function). But, on the other hand, the property 'best' fully depends on the information contained in the sample of the model

$$\mathbf{y} = X\boldsymbol{\beta} + \boldsymbol{\varepsilon}, \quad \boldsymbol{\varepsilon} \sim (\mathbf{0}, \sigma^2 W). \tag{3.1.1}$$

In other words, if no other information than that contained in the sample matrix $(\mathbf{y}, X)$ and given with the model (1) is available, then the GLSE $\mathbf{b} = S^{-1}X'W^{-1}\mathbf{y}$ is the most favourable linear unbiased estimator of $\boldsymbol{\beta}$. This has been known for a long time, and many attempts have been made in statistical research to overcome its *restrained usefulness*:

(i) by experimental design which allows the decrease of the risk $R(\mathbf{b}, A) = \sigma^2 \operatorname{tr} A(X'W^{-1}X)^{-1}$ if X is chosen in a corresponding manner,
(ii) by the admission of biased estimators (see Section 2.3 or Bibby and Toutenburg, 1978, for a fuller discussion),
(iii) by the use of prior (or outside) information on the parameters of the model,
(iv) by the methods of simultaneous (multivariate) estimation, if the model of interest may be connected with a system of other linear equations (see Chapter VI).

In this chapter we shall look at (iii).

3.1.2 Examples of Prior Information

In addition to having observations on the endogenous and exogenous variables (what is called the sample) we now assume that we have *auxiliary information* on the vector of regression coefficients. When this takes the form of inequalities, the minimax principle (see Chapter IV) or simplex algorithms (see, for example, Mantel, 1969) can be used to find estimators, or at least numerical solutions, which involve the proposed restrictions of $\boldsymbol{\beta}$. If the auxiliary information is such that it can be written as linear equalities in the form

$$\mathbf{r} = \boldsymbol{R\beta} + \boldsymbol{\phi}, \tag{3.1.2}$$

then *restrictive estimation techniques* are available based on an idea of Theil and Goldberger (1961).

In (2) $\mathbf{r}$ is a $(J \times 1)$ stochastic vector which is assumed to be realized (i.e. known), $\boldsymbol{R}$ is a $(J \times K)$ fixed matrix with rank $J \leq K$, and it is also assumed that the disturbance term has

$$\boldsymbol{\phi} \sim (\mathbf{0}, \boldsymbol{V}), \quad \boldsymbol{V} > \boldsymbol{0}. \tag{3.1.3}$$

The auxiliary information is assumed to be independent of the sample and the model (1), i.e. we have

$$E\boldsymbol{\phi}\boldsymbol{\varepsilon}' = \boldsymbol{0}. \tag{3.1.4}$$

The covariance matrix $\boldsymbol{V}$ of $\boldsymbol{\phi}$ has to be assumed known. (In Section 4.5.2 an idea for realizing this assumption for a general case will be given.) Often it is possible to restrict the prior information in such a way that

$$\boldsymbol{V} = \sigma^2/k\,\boldsymbol{W} \tag{3.1.5}$$

can be assumed. Here k is a constant of proportionality relating the variances of $\boldsymbol{\varepsilon}$ and $\boldsymbol{\phi}$. According to (2.2.7) there is no loss in generality from restricting ourselves to equivariance and independence of disturbances, i.e. in assuming $\boldsymbol{W} = \boldsymbol{I}$. We shall also use this model to get notational simplicity if necessary. Moreover, if test procedures come to be of interest, we shall assume normality for $\boldsymbol{\varepsilon}$ and $\boldsymbol{\phi}$.

Now, let us give some practical examples of the situation described by (2).

(i) A separate estimate of $\boldsymbol{\beta}$ might exist, say from an earlier sample of the underlying model or from comparable models. The second situation occurs if for instance an economic model of a country is estimated for regions (districts) of the country separately. If this separate estimate of $\boldsymbol{\beta}$ is called $\mathbf{b}^*$ then we may write

$$\mathbf{b}^* = \boldsymbol{\beta} + \boldsymbol{\phi}$$

or, in our introduced notation (2),

$$\mathbf{b}^* = \mathbf{r}, \quad \boldsymbol{R} = \boldsymbol{I}.$$

(ii) If only a part of $\boldsymbol{\beta}$, say $\boldsymbol{\beta}_1$, with $\boldsymbol{\beta} = (\boldsymbol{\beta}_1, \boldsymbol{\beta}_2)'$ is prior estimated, then we might have

$$\mathbf{b}_1^* = \boldsymbol{\beta}_1 + \boldsymbol{\phi} \quad \text{or} \quad \mathbf{r} = \boldsymbol{R\beta} + \boldsymbol{\phi}$$

where $\mathbf{r} = \mathbf{b}_1^*$ and $\boldsymbol{R} = (\boldsymbol{I}, \boldsymbol{0})$. (Here the dimension of $\boldsymbol{I}$ in $\boldsymbol{R}$ corresponds to the dimension of $\boldsymbol{\beta}_1$.)

(iii) Another special case occurs when $\boldsymbol{V}$, the dispersion matrix of $\boldsymbol{\phi}$, is zero. This corresponds to knowing that $\mathbf{r}$ equals $\boldsymbol{R\beta}$ with certainty. For instance, we might know exactly some coefficients of $\boldsymbol{\beta}$. This happens in the multivariate regression as well as in the econometric model when restrictions which identify the partial equations are available (see Chapter VI). The exact prior knowledge of particular elements of $\boldsymbol{\beta}$ leads to (ii), where $\boldsymbol{\phi} = 0$.

(iv) Alternatively, there may be information concerning the ratios between certain coefficients. For instance, if the ratio $\boldsymbol{\beta}_1 : \boldsymbol{\beta}_2 : \boldsymbol{\beta}_3$ is $ab : b : 1$, this may be written

$$\mathbf{0} = \begin{bmatrix} 1 & -a & 0 \\ 0 & 1 & -b \end{bmatrix} \begin{bmatrix} \boldsymbol{\beta}_1 \\ \boldsymbol{\beta}_2 \\ \boldsymbol{\beta}_3 \end{bmatrix}$$

If there were some uncertainty about the ratios, the left-hand side of the above equation would not be zero (i.e. we introduce the disturbance $\boldsymbol{\phi}$).

(v) Another example concerns prior knowledge of a linear combination (e.g. the sum) of the coefficients. This occurs in the analysis of variance where $\Sigma \boldsymbol{\beta}_i = c$ is the condition of reparametrization. This gives $\mathbf{r} = \mathbf{c}$, $\boldsymbol{R} = (1, \ldots, 1)$, $J = 1$. Another example is yielded by the well-known Cobb–Douglas production function with constant returns to scales, for which $\boldsymbol{\beta}_1 + \boldsymbol{\beta}_2 = 1$.

(vi) If we have reason to believe that particular coefficients lie in a certain region, we may write this prior information $a_i \leq \beta_i \leq b_i$ (for certain i) as $\beta_i = 1/2\,(a_i + b_i) + \phi_i$. (For further discussion see Section 4.5.2 and Hartung, 1978.)

(vii) Alternatively, an equivalent Bayesian formulation of this prior information (2) is possible (see, for example, Shiller, 1973, and Teräsvirta, 1979a). In the following, we shall confine ourselves to the classical approach.

3.1.3 The Mixed Estimation Procedure

Durbin (1953) was one of the first who used sample and auxiliary information, simultaneously, in generating a stepwise estimator of the parameters. In this fundamental paper Durbin estimated one β_i of a two-regressor model using a prior estimate of the other β_j. Theil and Goldberger (1961) and Theil (1963) founded the mixed estimation technique for the general case (2) of linear stochastic restrictions by 'mixing' both the sample and the prior information.

The mixed estimator is derived as follows. Let the sample model (1) and the model of restriction (2) be given. Then combine these by enlarging the model (1) with (2) giving the *mixed model*

$$\begin{pmatrix} \mathbf{y} \\ \mathbf{r} \end{pmatrix} = \begin{pmatrix} \boldsymbol{X} \\ \boldsymbol{R} \end{pmatrix} \boldsymbol{\beta} + \begin{pmatrix} \boldsymbol{\varepsilon} \\ \boldsymbol{\phi} \end{pmatrix}. \tag{3.1.6}$$

Calling the augmented variables $\tilde{\mathbf{y}}$, $\tilde{X}$, and $\tilde{\varepsilon}$ this can be written

$$\tilde{\mathbf{y}} = \tilde{X}\boldsymbol{\beta} + \tilde{\varepsilon} \quad (\text{where rank } \tilde{X} = K). \tag{3.1.7}$$

Using assumption (4) on the uncorrelatedness of the disturbances $\boldsymbol{\varepsilon}$ and $\boldsymbol{\phi}$, we know that $\tilde{\varepsilon} \sim (0, \sigma^2 \tilde{W})$ where

$$\tilde{W} = \begin{pmatrix} W & 0 \\ 0 & \sigma^{-2} V \end{pmatrix}. \tag{3.1.8}$$

Since W and V are positive definite we may conclude that $\tilde{W}$ is also.

By this idea, stochastic constraints like (2) may easily be incorporated within the linear regression model (1). (As noted in Bibby and Toutenburg, 1978, a change to correlated disturbances $\boldsymbol{\varepsilon}$ and $\boldsymbol{\phi}$ would change the off-diagonal submatrices of $\tilde{W}$.)

Now, the mixed model (7) along with (8) forms a generalized linear regression model (see (2.2.1)). Thus, applying Theorem 2.5 (Gauss–Markoff–Aitken) immediately yields the BLUE of $\boldsymbol{\beta}$ in the mixed model as

$$\hat{\boldsymbol{\beta}} = (\tilde{X}'\tilde{W}^{-1}\tilde{X})^{-1}\tilde{X}'\tilde{W}^{-1}\tilde{\mathbf{y}}.$$

Inserting the original variables/matrices gives the following theorem.

Theorem 3.1

The optimal unbiased linear estimator in the model (7) is the mixed estimator

$$\hat{\boldsymbol{\beta}}(\sigma^2) = (\sigma^{-2}S + R'V^{-1}R)^{-1}(\sigma^{-2}X'W^{-1}\mathbf{y} + R'V^{-1}\mathbf{r}). \tag{3.1.9}$$

The dispersion matrix of $\hat{\boldsymbol{\beta}}(\sigma^2)$ is

$$V[\hat{\boldsymbol{\beta}}(\sigma^2)] = (\sigma^{-2}S + R'V^{-1}R)^{-1}. \tag{3.1.10}$$

Gain in Efficiency

One of the main effects of the auxiliary information (2) is to reduce the dispersion matrix of the optimal estimator. Comparing the unrestricted Aitken-estimator $\mathbf{b} = S^{-1}X'W^{-1}\mathbf{y}$ having dispersion matrix $V(\mathbf{b}) = \sigma^2 S^{-1}$ and the mixed estimator $\hat{\boldsymbol{\beta}}(\sigma^2)$ gives

$$V(b) - V[\hat{\boldsymbol{\beta}}(\sigma^2)] = \sigma^2 S^{-1} - (\sigma^{-2}S + R'V^{-1}R)^{-1}. \tag{3.1.11}$$

This difference of dispersion matrices is nonnegative definite as $V^{-1}[\hat{\boldsymbol{\beta}}(\sigma^2)] - V^{-1}(\mathbf{b}) = \sigma^2 R'V^{-1}R \geq 0$ (see Theorem A.9). By Theorem 2.1, it follows that $\hat{\boldsymbol{\beta}}(\sigma^2)$ has smaller mean square error and smaller risk $R(.,A)$ than the GLSE $\mathbf{b}$.

Note Due to the unknown σ^2, $\hat{\boldsymbol{\beta}}(\sigma^2)$ is unknown, too. This problem does not arise if the variances and covariances of $\boldsymbol{\phi}$ are proportional to those of $\boldsymbol{\varepsilon}$, i.e. if $V = (\sigma^2/k)W$ can be assumed (see (5)). In that case the mixed estimator becomes

$$\hat{\boldsymbol{\beta}}(k) = (S + kR'V^{-1}R)^{-1}(X'W^{-1}\mathbf{y} + kR'V^{-1}\mathbf{r}) \tag{3.1.12}$$

and, moreover, $\hat{\boldsymbol{\beta}}(k)$ is known when k is known. Before continuing this, let us now look at the case of a general dispersion matrix $\boldsymbol{V}$. This seems to be more realistic, for, at first sight, it seems to need less prior information than assumed with (5).

3.2 AUXILIARY INFORMATION ON σ^2

3.2.1 Using the Sample Information

To ensure practicability of the mixed estimator $\hat{\boldsymbol{\beta}}(\sigma^2)$, the unknown σ^2 has to be replaced by a known scalar value. One possibility in this area is to estimate σ^2 by s^2(2.2.9), as was proposed by Theil (1963). Defining $\varphi = 1/\sigma^2$ as the 'precision' of the regression process and replacing σ^2 by its unbiased estimator $\hat{\sigma}^2 = s^2$ gives $\hat{\varphi} = 1/s^2$ and therefore

$$\hat{\boldsymbol{\beta}}(s^2) = (\hat{\varphi}\boldsymbol{S} + \boldsymbol{R}'\boldsymbol{V}^{-1}\boldsymbol{R})^{-1}(\hat{\varphi}\boldsymbol{X}'\boldsymbol{W}^{-1}\mathbf{y} + \boldsymbol{R}'\boldsymbol{V}^{-1}\mathbf{r}). \tag{3.2.1}$$

It is clear that the replacement of φ by its unbiased estimate $\hat{\varphi}$ has influence on the behaviour of the modified estimator $\hat{\boldsymbol{\beta}}(s^2)$ with respect to the GLSE $\mathbf{b}$, i.e. we cannot expect that the matrix difference analogous to (11) is nonnegative definite in general. Let us assume that the difference $\hat{\varphi} - \varphi$ is $O(T^{-1/2})$ in probability (where T is the sample size). This assumption is justified for instance, in the normal regression case. Then it can be proved (Theil, 1963) that

$$\hat{\boldsymbol{\beta}}(s^2) = \hat{\boldsymbol{\beta}}(\sigma^2) + O(T^{-1}). \tag{3.2.2}$$

It follows immediately that $\hat{\boldsymbol{\beta}}(s^2)$ is asymptotically unbiased and that its bias is $O(T^{-1})$. Furthermore, both estimators $\hat{\boldsymbol{\beta}}(s^2)$ and $\hat{\boldsymbol{\beta}}(\sigma^2)$ have the same asymptotic covariance matrix $\boldsymbol{V}[\hat{\boldsymbol{\beta}}(\sigma^2)]$ of (3.1.10). This result shows that if $\hat{\varphi} - \varphi$ is $O(T^{-1/2})$ in probability, it is impossible to get a gain in asymptotic efficiency in estimating $\boldsymbol{\beta}$. As shown by Theil (1963), this assertion does not depend on the chosen estimator s^2 which is based on the 'unrestricted' residuals $\mathbf{y} - \boldsymbol{X}\mathbf{b}$, where $\mathbf{b}$ is the GLSE. If σ^2 is estimated by a $\hat{\sigma}^2$ which is based on residuals reflecting to the restriction, then only negligible terms are influenced by this procedure. Nevertheless, one can use the 'restricted' variance estimator

$$\hat{\sigma}^2 = \frac{(\mathbf{y} - \boldsymbol{X}\hat{\boldsymbol{\beta}}(s^2))'\boldsymbol{W}^{-1}(\mathbf{y} - \boldsymbol{X}\hat{\boldsymbol{\beta}}(s^2))}{T - \operatorname{tr} s^{-2}\boldsymbol{S}(s^{-2}\boldsymbol{S} + \boldsymbol{R}'\boldsymbol{V}^{-1}\boldsymbol{R})^{-1}} \tag{3.2.3}$$

where the bias of $\hat{\sigma}^2$ is of higher order of smallness than T^{-1} (for further results, especially for formulae for bias and moment matrix of $\hat{\boldsymbol{\beta}}(s^2)$, see Nagar and Kakwani, 1964). Summarizing, one could state that replacing φ by a (stochastic) estimate $\hat{\varphi}$, e.g. $\hat{\varphi} = 1/s^2$, gives an estimator $\hat{\boldsymbol{\beta}}(s^2)$ of $\boldsymbol{\beta}$ which is 'nearly' the optimal estimator $\hat{\boldsymbol{\beta}}(\sigma^2)$, where 'nearly' reflects the order of smallness of the differences of the estimators and their moment matrices.

Thus one would expect that a sufficient sample size guarantees $\hat{\boldsymbol{\beta}}(s^2)$ to be a good approximation of $\hat{\boldsymbol{\beta}}(\sigma^2)$ such that

$$\boldsymbol{V}(\mathbf{b}) - \boldsymbol{V}[\hat{\boldsymbol{\beta}}(s^2)] \geq \boldsymbol{0}. \tag{3.2.4}$$

3.2.2 Using a Constant

We may reformulate Theorem 3.1 so that the mixed estimator $\hat{\boldsymbol{\beta}}(\sigma^2)$ (3.1.9) is a solution of the normal equation

$$(\sigma^{-2}\boldsymbol{X}'\boldsymbol{W}^{-1}\mathbf{y}+\boldsymbol{R}'\boldsymbol{V}^{-1}\mathbf{r})=(\sigma^{-2}\boldsymbol{S}+\boldsymbol{R}'\boldsymbol{V}^{-1}\boldsymbol{R})\hat{\boldsymbol{\beta}}. \tag{3.2.5}$$

We now replace in this equation the unknown σ^{-2} by a nonstochastic constant $c \geq 0$. In this way we get the family of normal equations

$$c\boldsymbol{X}'\boldsymbol{W}^{-1}\mathbf{y}+\boldsymbol{R}'\boldsymbol{V}^{-1}\mathbf{r}=(c\boldsymbol{S}+\boldsymbol{R}'\boldsymbol{V}^{-1}\boldsymbol{R})\,\hat{\boldsymbol{\beta}}(c) \tag{3.2.6}$$

which defines a family $\mathscr{F}_c=\{\hat{\boldsymbol{\beta}}_c\}$ of estimators which do not contain unknown parameters and therefore are practicable. The members of $\mathscr{F}_c$ are indexed by the value of the chosen scalar c.

As rank $\boldsymbol{R}=J$ was assumed, $\boldsymbol{R}'\boldsymbol{V}^{-1}\boldsymbol{R}>\boldsymbol{0}$ and therefore the matrix

$$\boldsymbol{M}_c=(c\boldsymbol{S}+\boldsymbol{R}'\boldsymbol{V}^{-1}\boldsymbol{R}) \tag{3.2.7}$$

is positive definite for all $c \geq 0$. Since

$$\boldsymbol{X}'\boldsymbol{W}^{-1}\mathbf{y}=\boldsymbol{X}'\boldsymbol{W}^{-1}(\boldsymbol{X}\boldsymbol{\beta}+\boldsymbol{\varepsilon})=\boldsymbol{S}\boldsymbol{\beta}+\boldsymbol{X}'\boldsymbol{W}^{-1}\boldsymbol{\varepsilon}$$

and

$$\boldsymbol{R}'\boldsymbol{V}^{-1}\mathbf{r}=\boldsymbol{R}'\boldsymbol{V}^{-1}(\boldsymbol{R}\boldsymbol{\beta}+\boldsymbol{\phi})=\boldsymbol{R}'\boldsymbol{V}^{-1}\boldsymbol{R}\boldsymbol{\beta}+\boldsymbol{R}'\boldsymbol{V}^{-1}\boldsymbol{\phi},$$

the left-hand term of (6) equals $\boldsymbol{M}_c\boldsymbol{\beta}+\boldsymbol{X}'\boldsymbol{W}^{-1}\boldsymbol{\varepsilon}+\boldsymbol{R}'\boldsymbol{V}^{-1}\boldsymbol{\phi}$. Hence on solving (6) for $\hat{\boldsymbol{\beta}}(c)$ we get

$$\hat{\boldsymbol{\beta}}(c)=\boldsymbol{M}_c^{-1}(c\boldsymbol{X}'\boldsymbol{W}^{-1}\mathbf{y}+\boldsymbol{R}'\boldsymbol{V}^{-1}\mathbf{r})=\boldsymbol{\beta}+\boldsymbol{M}_c^{-1}(c\boldsymbol{X}'\boldsymbol{W}^{-1}\boldsymbol{\varepsilon}+\boldsymbol{R}'\boldsymbol{V}^{-1}\boldsymbol{\phi}) \tag{3.2.8}$$

(see Theil, 1963, for a corresponding class of estimators, and Toutenburg, 1975b). Considering the expectation, it may be seen that $\hat{\boldsymbol{\beta}}(c)$ is unbiased (for all c) and has dispersion matrix

$$V[\hat{\boldsymbol{\beta}}(c)]=\boldsymbol{M}_c^{-1}(c^2\sigma^2\boldsymbol{S}+\boldsymbol{R}'\boldsymbol{V}^{-1}\boldsymbol{R})\boldsymbol{M}_c^{-1}. \tag{3.2.9}$$

Since $\hat{\boldsymbol{\beta}}(\sigma^2)$ was the best estimator within the family of linear unbiased estimators of the mixed model (3.1.6), it follows immediately that

$$V[\hat{\boldsymbol{\beta}}(c)]-V[\hat{\boldsymbol{\beta}}(\sigma^2)]\geq \boldsymbol{0} \tag{3.2.10}$$

must hold. This equation gives the loss in efficiency which results from the replacement of the unknown σ^2 by the constant c.

In this context the limits $c \to 0$ and $c \to \infty$ are of special interest (see Teräsvirta and Toutenburg, 1980; Hartung, 1978).

Applying the well-known formula for matrix inversion (Theorem A.34) on $c^{-1}\boldsymbol{M}_c=\boldsymbol{S}+\boldsymbol{R}'(c\boldsymbol{V})^{-1}\boldsymbol{R}$ gives

$$c\boldsymbol{M}_c^{-1}=\boldsymbol{S}^{-1}-\boldsymbol{S}^{-1}\boldsymbol{R}'(\boldsymbol{R}\boldsymbol{S}^{-1}\boldsymbol{R}'+c\boldsymbol{V})^{-1}\boldsymbol{R}\boldsymbol{S}^{-1}. \tag{3.2.11}$$

Writing

$$\hat{\boldsymbol{\beta}}(c)=\{c\boldsymbol{M}_c^{-1}\}\boldsymbol{X}'\boldsymbol{W}^{-1}\mathbf{y}+\{c\boldsymbol{M}_c^{-1}\}(c^{-1}\boldsymbol{R}'\boldsymbol{V}^{-1}\mathbf{r}) \tag{3.2.12}$$

gives the following:

(i) As $c \to \infty$, $cM_c^{-1} \to S^{-1}$. For both terms of $\hat{\beta}(c)$ (12) we have then

$$\{cM_c^{-1}\}X'W^{-1}\mathbf{y} \underset{c \to \infty}{\to} S^{-1}X'W^{-1}\mathbf{y}$$

and

$$\{cM_c^{-1}\}(c^{-1}R'V^{-1}\mathbf{r}) \underset{c \to \infty}{\to} \mathbf{0}.$$

The limit $c \to \infty$ means that the sample information is the dominating one, for c was chosen as a prior 'estimate' of σ^{-2}. Now, combining both limits from above gives

$$\lim_{c \to \infty} \hat{\beta}(c) = \mathbf{b} = S^{-1}X'W^{-1}\mathbf{y} \tag{3.2.13}$$

which is the optimal unbiased estimator based only on the sample information.

(ii) As $c \to 0$,

$$cM_c^{-1} \to S^{-1} - S^{-1}R'(RS^{-1}R')^{-1}RS^{-1}.$$

The first part of $\hat{\beta}(c)$ (12) has the limit

$$\{cM_c^{-1}\}X'W^{-1}\mathbf{y} \underset{c \to 0}{\to} \mathbf{b} - S^{-1}R'(RS^{-1}R')R\mathbf{b}.$$

The second part of $\hat{\beta}(c)$ gives

$$\{cM_c^{-1}\}(c^{-1}R'V^{-1}\mathbf{r}) = c^{-1}[S^{-1} - S^{-1}R'(RS^{-1}R' + cV)^{-1}RS^{-1}]R'V^{-1}\mathbf{r}$$

(using (11))

$$\begin{aligned} &= c^{-1}S^{-1}R'V^{-1}\mathbf{r} \\ &\quad - c^{-1}S^{-1}R'(RS^{-1}R' + cV)^{-1}(RS^{-1}R' + cV)^{-1}\mathbf{r} \\ &\quad + S^{-1}R'(RS^{-1}R' + cV)^{-1}\mathbf{r} \\ &\underset{c \to 0}{\to} S^{-1}R'(RS^{-1}R')^{-1}\mathbf{r}. \end{aligned}$$

Thus we have the result

$$\hat{\beta}(c) \underset{c \to 0}{\to} \tilde{\mathbf{b}}_R = \mathbf{b} + S^{-1}R'(RS^{-1}R')^{-1}(\mathbf{r} - R\mathbf{b}) \tag{3.2.14}$$

which is the well-known conditional restricted estimator under the exact prior restriction $\mathbf{r} = R\beta$ (see (2.3.71))

Although $c \to 0$ seems to indicate a high degree of uncertainty in the sample ($c \to 0$ stands for $\sigma^2 \to \infty$), the limiting estimator (14) is not based on the prior information alone. However, the result (14) does show that the uncertainty of the prior information (expressed by the disturbance ϕ) is negligible if $c \to 0$.

The same result (14) is obtained if we take the limit $V \to \mathbf{0}$, i.e. if the stochastic restriction $\mathbf{r} = R\beta + \phi$ tends to the exact linear restriction $\mathbf{r} = R\beta$, that is (see (3.5.16))

$$\hat{\beta}(c) \to \tilde{\mathbf{b}}_R \text{ as } V \to \mathbf{0}. \tag{3.2.15}$$

3.2.3 Optimal Choice of c

To find a constant c which should be best in some sense, let us take the risk of $\hat{\boldsymbol{\beta}}(c)$ (see (2.1.28))

$$R[\hat{\boldsymbol{\beta}}(c)] = \operatorname{tr} \boldsymbol{A}\boldsymbol{V}[\hat{\boldsymbol{\beta}}(c)] \tag{3.2.16}$$

(which is a scalar function of c) and calculate the first derivative with respect to c (assume $c > 0$). Now we have

$$\frac{\partial}{\partial c} R[\hat{\boldsymbol{\beta}}(c)] = \operatorname{tr} \boldsymbol{A} \frac{\partial}{\partial c} \boldsymbol{V}[\hat{\boldsymbol{\beta}}(c)]$$

(see Theorem A.53). Writing (9) in an equivalent form

$$\boldsymbol{V}[\hat{\boldsymbol{\beta}}(c)] = \boldsymbol{M}_c^{-1} + (c^2\sigma^2 - c)\boldsymbol{M}_c^{-1}\boldsymbol{S}\boldsymbol{M}_c^{-1}$$

and differentiating gives (see equation B.3.2)

$$\begin{aligned}\frac{\partial}{\partial c} \boldsymbol{V}[\hat{\boldsymbol{\beta}}(c)] &= 2(c\sigma^2 - 1)\boldsymbol{M}_c^{-1}\boldsymbol{S}\boldsymbol{M}_c^{-1} - 2(c^2\sigma^2 - c)\boldsymbol{M}_c^{-1}\boldsymbol{S}\boldsymbol{M}_c^{-1}\boldsymbol{S}\boldsymbol{M}_c^{-1} \\ &= 2(c\sigma^2 - 1)\boldsymbol{M}_c^{-1}\boldsymbol{S}[\boldsymbol{S}^{-1} - c\boldsymbol{M}_c^{-1}]\boldsymbol{S}\boldsymbol{M}_c^{-1}.\end{aligned}$$

Since $c^{-1}\boldsymbol{M}_c - \boldsymbol{S} = c^{-1}\boldsymbol{R}'\boldsymbol{V}^{-1}\boldsymbol{R} \geq \boldsymbol{0}$ it follows that $[\boldsymbol{S}^{-1} - c\boldsymbol{M}_c^{-1}] \geq \boldsymbol{0}$ (see Theorem A.12ii). Moreover, we may conclude that

$$\operatorname{tr} \boldsymbol{A}\boldsymbol{M}_c^{-1}\boldsymbol{S}[\boldsymbol{S}^{-1} - c\boldsymbol{M}_c^{-1}]\boldsymbol{S}\boldsymbol{M}_c^{-1} = \alpha \geq 0. \tag{3.2.17}$$

So we have

$$\frac{\partial}{\partial c} R[\hat{\boldsymbol{\beta}}(c)] = 2(c\sigma^2 - 1)\alpha$$

which gives the minimum $c = \sigma^{-2}$ in accordance with Theorem 3.1.

Now α (17) is nonnegative and so we get the relations

(i) $R[\hat{\boldsymbol{\beta}}(c)]$ is monotonically increasing in c when $c > \sigma^{-2}$,
(ii) $R[\boldsymbol{\beta}(c)]$ is monotonically decreasing in c when $c < \sigma^{-2}$.

Based on these relations we may assume the following prior information on σ^2 (Toutenburg, 1977a). Let an upper and a lower bound of σ^2 be known such that

$$\sigma_1^2 < \sigma^2 < \sigma_2^2. \tag{3.2.18}$$

It is evident that the smaller the difference $\sigma_2^2 - \sigma_1^2$, the higher is the amount of information given by (18). If we choose now a value c^* such that

$$c^*(\lambda) = \sigma_2^{-2} + \lambda(\sigma_1^{-2} - \sigma_2^{-2}) \quad (0 < \lambda < 1) \tag{3.2.19}$$

we may conclude from the shown monotonicity of the risk $R[\hat{\boldsymbol{\beta}}(c)]$ that the estimator

$$\hat{\boldsymbol{\beta}}(c(\lambda)) \quad \text{with } 0 < \lambda < 1 \tag{3.2.20}$$

has a smaller risk than all other estimators $\hat{\boldsymbol{\beta}}(c(\lambda))$ for $\lambda < 0$ or $\lambda > 1$. As the optimal value $c = \sigma^{-2}$ is unknown, one would choose the compromise $\lambda = 0.5$.

Note Another estimator of $\boldsymbol{\beta}$ in the mixed model (3.1.7) is the 'mixed' OLSE

$$\mathbf{b}_R^* = (X'X + R'R)^{-1}(X'\mathbf{y} + R'\mathbf{r}) \tag{3.2.21}$$

which may be interpreted as a misspecified estimator, ignoring the true dispersion (see also Section 2.2.4).

The dispersion matrix of $\mathbf{b}_R^*$ is

$$V(\mathbf{b}_R^*) = (X'X + R'R)^{-1}(\sigma^2 X'WX + R'VR)(X'X + R'R)^{-1}. \tag{3.2.22}$$

Since $\hat{\boldsymbol{\beta}}(\sigma^2)$ is the BLUE of $\boldsymbol{\beta}$ in the mixed model (3.1.7), the difference $V(\mathbf{b}_R^*) - V[\hat{\boldsymbol{\beta}}(\sigma^2)]$ is nonnegative definite. But, what is the relation between $\mathbf{b}_R^*$ and $\hat{\boldsymbol{\beta}}(c)$? Corresponding to similar qualitative results of Swamy (1971) we may conclude the following. Let L be a $(T+J) \times (T \times J - K)$-matrix of full rank $(T+J-K)$ such that

$$L'\begin{pmatrix} X \\ R \end{pmatrix} = L'\tilde{X} = \mathbf{0}.$$

If $\tilde{X}'\tilde{W}L$ is close to zero, then in small samples the relation $V[\hat{\boldsymbol{\beta}}(c)] - V(\mathbf{b}_R^*) \geq 0$ holds. The relation $\tilde{X}'\tilde{W}L \approx \mathbf{0}$ is valid, for instance, when the variances of $\boldsymbol{\varepsilon}$ are close to the variances of $\boldsymbol{\phi}$ and both are small enough.

3.2.4 Comparison of $\hat{\boldsymbol{\beta}}(c)$ and the GLSE

Following the previous arguments we intend to choose the value of c in such a way that the modified and so far practicable mixed estimator $\hat{\boldsymbol{\beta}}(c)$ is better than the unrestricted GLSE $\mathbf{b}$ in the sense that

$$\boldsymbol{\Delta}(c) = V(\mathbf{b}) - V[\hat{\boldsymbol{\beta}}(c)] \geq \mathbf{0}. \tag{3.2.23}$$

Using the relation

$$M_c S^{-1} M_c = c^2 S + (R'V^{-1}R)S^{-1}(R'V^{-1}R) + 2c(R'V^{-1}R)$$

and $V(\mathbf{b}) = \sigma^2 S^{-1}$, $V[\hat{\boldsymbol{\beta}}(c)]$ from (9), we may write (23) as

$$\begin{aligned}\boldsymbol{\Delta}(c) &= M_c^{-1}\{\sigma^2 M_c S^{-1} M_c - c^2\sigma^2 S - (R'V^{-1}R)\}M_c^{-1} \\ &= M_c^{-1}(R'V^{-1}R)B(c;\sigma^2)(R'V^{-1}R)M_c^{-1}\end{aligned} \tag{3.2.24}$$

where $B(c;\sigma^2)$ is the following matrix:

$$B(c;\sigma^2) = \sigma^2 S^{-1} + (2c\sigma^2 - 1)(R'V^{-1}R)^{-1}. \tag{3.2.25}$$

Since our intention is to choose c such that $\boldsymbol{\Delta}(c) \geq \mathbf{0}$, *the family of (in this context) admissible estimators* is

$$\mathscr{F}_c = \{\hat{\boldsymbol{\beta}}(c);\ c \text{ such that } B(c;\sigma^2) \geq \mathbf{0}\} \tag{3.2.26}$$

for $B(c;\sigma^2) \geq \mathbf{0}$ ensures that $\boldsymbol{\Delta}(c) \geq \mathbf{0}$.

Case (i) $B(0;\sigma^2) = \sigma^2 S^{-1} - (R'V^{-1}R)^{-1}$ is negative definite.

With $\boldsymbol{B}(0;\sigma^2)=\boldsymbol{\Delta}(0)$ negative definite and $\boldsymbol{B}(\frac{1}{2}\sigma^{-2};\sigma^2)=\sigma^2\boldsymbol{S}^{-1}$ positive definite and $\mathbf{a}'\mathbf{B}(c;\sigma^2)\mathbf{a}$ ($\mathbf{a}\neq\mathbf{0}$ a fixed $K\times 1$-vector) being a function which is continuous in c, there must exist a critical value $c_0(\mathbf{a})$ such that

$$\mathbf{a}'\boldsymbol{B}(c_0;\sigma^2)\mathbf{a}=0,\quad 0<c_0(\mathbf{a})<\tfrac{1}{2}\sigma^{-2},\quad \mathbf{a}'\boldsymbol{B}(c;\sigma^2)\mathbf{a}\geq 0\quad \text{for } c>c_0(\mathbf{a}). \tag{3.2.27}$$

Solving $\mathbf{a}'\boldsymbol{B}(c_0;\sigma^2)\mathbf{a}=0$ gives the critical value $c_0(\mathbf{a})$ as

$$c_0(\mathbf{a})=(2\sigma^2)^{-1}-\frac{\mathbf{a}'\boldsymbol{S}^{-1}\mathbf{a}}{2\mathbf{a}'(\boldsymbol{R}'\boldsymbol{V}^{-1}\boldsymbol{R})^{-1}\mathbf{a}} \tag{3.2.28}$$

which clearly is unknown due to the unknown σ^2.

Using the prior information (18) on σ^2 helps to overcome this difficulty.

Theorem 3.2

Let constants σ_1^2, σ_2^2 be prior known such that

(i) $0<\sigma_1^2<\sigma^2<\sigma_2^2<\infty$, and

(ii) $\boldsymbol{B}(0;\sigma_2^2)$ is negative definite.

Then the family of estimators $\hat{\boldsymbol{\beta}}(c)$ which have smaller risk than the GLSE $\mathbf{b}$ is $\mathscr{F}_c=\{\hat{\boldsymbol{\beta}}(c);\,c\geq\sigma_1^{-2}\}$.

Proof From $\boldsymbol{B}(0;\sigma_2^2)$ negative definite it follows that $\boldsymbol{B}(0;\sigma^2)$ is also negative definite. Now, $\sigma_1^{-2}>\frac{1}{2}\sigma^{-2}$ and so (27) is fulfilled, i.e. $\boldsymbol{\Delta}(c)\geq\boldsymbol{0}$ for $c\geq\sigma_1^{-2}$.

Note According to the monotonicity of $R[\hat{\boldsymbol{\beta}}(c)]$ with respect to c, $c=\sigma_1^{-2}$ is the best choice within the family $\mathscr{F}_c=\{\hat{\boldsymbol{\beta}}(c);\,c\geq\sigma_1^{-2}\}$.

Case (ii) $\boldsymbol{B}(0;\sigma^2)$ is nonnegative definite.

Then $\boldsymbol{B}(c;\sigma^2)\geq\boldsymbol{0}$ for all $c>0$ and therefore we have $\mathscr{F}_c=\{\hat{\boldsymbol{\beta}}(c);\,c\geq 0\}$. Since $\boldsymbol{B}(0;\sigma_1^2)\geq 0$ implies the same for $\boldsymbol{B}(0;\sigma^2)$ we have proved the following theorem.

Theorem 3.3

Let a constant σ_1^2 be prior known such that

(i) $0<\sigma_1^2<\sigma^2$, and

(ii) $\boldsymbol{B}(0;\sigma_1^2)\geq\boldsymbol{0}$.

Then the class of estimators $\hat{\boldsymbol{\beta}}(c)$ which ensure a smaller risk than the GLSE $\mathbf{b}$ is $\mathscr{F}_c=\{\hat{\boldsymbol{\beta}}(c);\,c\geq 0\}$ and

$$\begin{aligned}\max_{c\geq 0}\{R(\mathbf{b})-R[\hat{\boldsymbol{\beta}}(c)]\}&=\max_{\lambda}\operatorname{tr}\boldsymbol{A}\boldsymbol{\Delta}(c^*(\lambda))\\&=\operatorname{tr}\boldsymbol{A}\boldsymbol{\Delta}(c^*(\lambda))\quad\text{for}\quad 0<\lambda<1\end{aligned}$$

(see (19)).

Note If the type of distribution of ε is known, then a reasonable value of σ_1 may be determined as a function of the α-quantiles of this distribution (see (4.3.40) for the determination of σ_1 if ε follows a normal distribution).

3.2.5 Measuring the Gain in Efficiency

The influence of the used prior information $\mathbf{r} = \boldsymbol{R\beta} + \boldsymbol{\phi}$ on the efficiency of estimation may be measured, for instance, by the scalar

$$\delta(c) = \frac{R(\mathbf{b}) - R(\hat{\boldsymbol{\beta}}(c))}{R(\mathbf{b})} \tag{3.2.29}$$

which gives the decrease of risk relative to the unrestricted estimation by $\mathbf{b}$. To make δ practicable, we have to replace σ^2 by an estimator, say c^{-1}. Using (23) to (25) and the risk function $R(\hat{\boldsymbol{\beta}}) = \operatorname{tr} \boldsymbol{V}(\hat{\boldsymbol{\beta}})$, and abbreviating $(\boldsymbol{R}' \boldsymbol{V}^{-1} \boldsymbol{R}) = \boldsymbol{A}$, we get

$$\hat{\delta}(c) = \frac{\operatorname{tr} \boldsymbol{M}_c^{-1} \boldsymbol{A}\{c^{-1}\boldsymbol{S}^{-1} + \boldsymbol{A}^{-1}\}\boldsymbol{A}\boldsymbol{M}_c^{-1}}{\operatorname{tr} c^{-1}\boldsymbol{S}^{-1}} \tag{3.2.30}$$

where $0 \leq \hat{\delta}(c) \leq 1$ so far c is chosen according to Theorem 3.2 or Theorem 3.3.

If we use the form of $\boldsymbol{V}[\hat{\boldsymbol{\beta}}(c)]$ derived above, namely

$$\boldsymbol{V}[\hat{\boldsymbol{\beta}}(c)] = \boldsymbol{M}_c^{-1} - c(c\sigma^2 - 1)\boldsymbol{M}_c^{-1}\boldsymbol{S}\boldsymbol{M}_c^{-1}$$

and, furthermore, if we have found a good 'estimator' c of σ^{-2} such that $c\sigma^2 - 1 \approx 0$, we could use the approximation $\boldsymbol{V}[\hat{\boldsymbol{\beta}}(c)] \approx \boldsymbol{M}_c^{-1}$. Then (30) becomes

$$\hat{\delta}(c) \approx \frac{\operatorname{tr}\{c^{-1}\boldsymbol{S}^{-1} - \boldsymbol{M}_c^{-1}\}}{\operatorname{tr} c^{-1}\boldsymbol{S}^{-1}} \tag{3.2.31}$$

$$= \frac{\operatorname{tr}\{\boldsymbol{S}^{-1} - (\boldsymbol{S} + c^{-1}\boldsymbol{R}'\boldsymbol{V}^{-1}\boldsymbol{R})^{-1}\}}{\operatorname{tr} \boldsymbol{S}^{-1}}$$

$$= 1 - \frac{\operatorname{tr}(\boldsymbol{S} + c^{-1}\boldsymbol{R}'\boldsymbol{V}^{-1}\boldsymbol{R})^{-1}}{\operatorname{tr}\boldsymbol{S}^{-1}} = \hat{\delta}_0(c), \text{ say.} \tag{3.2.32}$$

As $\boldsymbol{S}^{-1} - (\boldsymbol{S} + c^{-1}\boldsymbol{R}'\boldsymbol{V}^{-1}\boldsymbol{R})^{-1} \geq \boldsymbol{0}$, $\operatorname{tr}\boldsymbol{S}^{-1} \geq \operatorname{tr}(\boldsymbol{S} + c^{-1}\boldsymbol{R}'\boldsymbol{V}^{-1}\boldsymbol{R})^{-1}$ and therefore $0 \leq \hat{\delta}_0(c) \leq 1$. The nearer $\hat{\delta}_0(c)$ is to one, the higher is the influence of the prior information on the efficiency of estimation. If $\hat{\delta}_0(c)$ tends to zero, this may be caused for instance by $c \to \infty$. $\hat{\delta}_0(c) \approx 0$ indicates that the sample information is the dominating one (see (13)).

As a further aspect of prior information we have the measure of the *shares* of prior and sample information in the so-called posterior precision (this was introduced by Theil, 1963). The posterior knowledge is measured by the approximated moment matrix of the mixed estimator, i.e. $\boldsymbol{V}[\hat{\boldsymbol{\beta}}(c)] \approx \boldsymbol{M}_c^{-1}$. Then

$$\lambda(c; \text{sample}) = \frac{1}{K} \operatorname{tr} \boldsymbol{S}(\boldsymbol{S} + c^{-1}\boldsymbol{R}'\boldsymbol{V}^{-1}\boldsymbol{R})^{-1} \tag{3.2.33}$$

and

$$\lambda(c; \text{prior}) = \frac{1}{K} \operatorname{tr} c^{-1} \boldsymbol{R}' \boldsymbol{V}^{-1} \boldsymbol{R} (\boldsymbol{S} + c^{-1} \boldsymbol{R}' \boldsymbol{V}^{-1} \boldsymbol{R})^{-1} \qquad (3.2.34)$$

are the estimated shares of sample and prior information, respectively, in our posterior knowledge which is expressed by $\boldsymbol{M}_c^{-1}$. Clearly, both measures add up to 1:

$$\lambda(c; \text{sample}) + \lambda(c; \text{prior}) = \frac{1}{K} \operatorname{tr} \boldsymbol{I}_K = 1.$$

3.3 MIXED ESTIMATION UNDER INCORRECT PRIOR INFORMATION

3.3.1 Introduction

In the previous sections of this chapter we have introduced auxiliary information on the parameter vector $\boldsymbol{\beta}$ which in general allowed the presentation in the form of linear equalities $\mathbf{r} = \boldsymbol{R\beta}$. To give this a more sensitive and realistic field of validity, we assumed that $\mathbf{r} = \boldsymbol{R\beta}$ holds in the mean, i.e. we assumed the prior information to be of the form $\mathbf{r} = \boldsymbol{R\beta} + \boldsymbol{\phi}$ (see (3.1.2)) where $\boldsymbol{\phi} \sim (\mathbf{0}, \boldsymbol{V})$ should be a disturbance independent of the sample disturbance $\boldsymbol{\varepsilon}$. Now, there are many applications where the assumption $E\mathbf{r} = \boldsymbol{R\beta}$ does not hold, so that the true but unknown form of the auxiliary information is in fact

$$\mathbf{r} = \boldsymbol{R\beta} + \mathbf{g} + \boldsymbol{\phi} \qquad (3.3.1)$$

where $\mathbf{g} \neq \mathbf{0}$.

Model	Estimator	Gains in efficiency
(i) $\mathbf{y} = \boldsymbol{X\beta} + \boldsymbol{\varepsilon}$, $\boldsymbol{\varepsilon} \sim (\mathbf{0}, \sigma^2 \boldsymbol{I})$ $\mathbf{r} = \boldsymbol{R\beta} + \mathbf{g} + \boldsymbol{\phi}$ $\mathbf{g} \neq \mathbf{0}$, $\boldsymbol{\phi} \sim (\mathbf{0}, (\sigma^2/k)\boldsymbol{I})$	Mixed estimator (biased) $\mathbf{b}_R(k) = (\boldsymbol{X}'\boldsymbol{X} + k\boldsymbol{R}'\boldsymbol{R})^{-1}(\boldsymbol{X}'\mathbf{y} + k\boldsymbol{R}'\mathbf{r})$ $R[\mathbf{b}_R(k)]$ in (3.3.13); $\mathbf{b}_R(k) \underset{k \to \infty}{\longrightarrow} \mathbf{b}_R$	$R[\mathbf{b}_R(k)] \leq R(\mathbf{b}_0)$ iff $\sigma^{-2}\mathbf{g}'\boldsymbol{S}_k\mathbf{g} \leq 1$ or (sufficient condition) if $0 < k \leq k^*$ (3.3.22)
(ii) as in (i), but $\mathbf{r} = \mathbf{0}$, $\boldsymbol{R} = \boldsymbol{I}$	Ridge estimator (biased) $\mathbf{b}(k) = (\boldsymbol{X}'\boldsymbol{X} + k\boldsymbol{I})^{-1}\boldsymbol{X}'\mathbf{y}$ $R[\mathbf{b}(k)]$ in (3.3.27)	$R(\mathbf{b}(k)) \leq R(\mathbf{b}_0)$ iff $\sigma^{-2}\boldsymbol{\beta}'(k^{-1}\boldsymbol{I} + (\boldsymbol{X}'\boldsymbol{X})^{-1})\boldsymbol{\beta} \leq 1$ or (sufficient condition) if $0 < k < \{\sigma^{-2}\boldsymbol{\beta}'\boldsymbol{\beta} - (\lambda_{\max}(\boldsymbol{X}'\boldsymbol{X}))^{-1}\}$

Figure 3.3.1 Biased mixed estimators (incorrect prior information)

Example (Teräsvirta, 1979b)
Let a one-input distributed lag model be given:

$$y_t = \sum_{i=0}^{K-1} \beta_i x_{t-i} + \varepsilon_t \qquad (t = 1, \ldots, T) \qquad (3.3.2)$$

or, in its matrix version,

$$\mathbf{y} = \boldsymbol{X\beta} + \boldsymbol{\varepsilon}, \qquad \boldsymbol{\varepsilon} \sim N(\mathbf{0}, \sigma^2 \boldsymbol{I}).$$

Furthermore, if we assume that the components of $\boldsymbol{\beta}$ lie on a polynomial of degree $p-1 < K$, then

$$\beta_i = \sum_{j=0}^{p-1} \alpha_j i^j \qquad (i = 0, \ldots, K-1) \tag{3.3.3}$$

(see Almon, 1965).

Now, the pth differences of a polynomial of degree $p-1$ are zero, i.e. we may rewrite (3) as $\boldsymbol{R\beta} = \mathbf{0}$. Shiller (1973) replaced the exact restriction (3) or its equivalent $\boldsymbol{R\beta} = \mathbf{0}$ by stochastic restrictions '$\boldsymbol{R\beta}$ is close to zero'. This was achieved by introducing a prior distribution

$$\gamma_1 = \boldsymbol{R\beta} \sim N(\mathbf{0}, (\sigma^2/k)\boldsymbol{I}). \tag{3.3.4}$$

Reformulating this in the sampling theoretic framework gives the form

$$\mathbf{r} = \boldsymbol{R\beta} + \boldsymbol{\phi}, \qquad \boldsymbol{\phi} \sim N(0, (\sigma^2/k)\boldsymbol{I}). \tag{3.3.5}$$

Adopting the polynomial approach (3) and choosing a nontrivial degree $p-1 < K$, then despite the assumptions (4) and (5), in practice $E\gamma_1 \neq \mathbf{0}$ or $E\mathbf{r} \neq \mathbf{0}$. Thus the correct form of (5) would be

$$\mathbf{r} = \boldsymbol{R\beta} + \mathbf{g} + \boldsymbol{\phi} \qquad \text{with } \mathbf{g} \neq \mathbf{0}. \tag{3.3.6}$$

Now, as in previous investigations, *the question arises how incorrect can the prior information be so that the resulting estimator is still advantageous compared with unrestricted estimators*? That question becomes more interesting than before, as prior information like (5) which is incorrect in the mean leads to biased estimators. Therefore the variance of estimators cannot be used for comparisons of goodness of estimation. Instead of the dispersion matrix we have to use the mean square error to find out the price to be paid for reducing the variances in order to lead to a reduced mean square error.

3.3.2 The Biased Mixed Estimator

For notational simplicity let us concentrate on the classical regression case

$$\mathbf{y} = \boldsymbol{X\beta} + \boldsymbol{\varepsilon}, \qquad \boldsymbol{\varepsilon} \sim (\mathbf{0}, \sigma^2 \boldsymbol{I}). \tag{3.3.7}$$

Moreover, we assume that we have incorrect prior information of type (6) where the variances of the components of $\boldsymbol{\phi}$ are proportional to those of the ε_t's (see (3.1.5)), i.e. we assume that we have auxiliary information of the type

$$\mathbf{r} = \boldsymbol{R\beta} + \mathbf{g} + \boldsymbol{\phi}, \qquad \mathbf{g} \neq 0, \qquad \boldsymbol{\phi} \sim (\mathbf{0}, (\sigma^2/k)\boldsymbol{I}). \tag{3.3.8}$$

Then the mixed estimator $\hat{\boldsymbol{\beta}}(\sigma^2)$ (3.1.9) specializes to

$$\mathbf{b}_R(k) = (\boldsymbol{X}'\boldsymbol{X} + k\boldsymbol{R}'\boldsymbol{R})^{-1}(\boldsymbol{X}'\mathbf{y} + k\boldsymbol{R}'\mathbf{r}). \tag{3.3.9}$$

The scalar k can be interpreted as a *precision parameter* of the stochastic information (8). When $k \to 0+$ in (9) we have the ordinary least squares estimator $\mathbf{b}_0 = (X'X)^{-1}X'\mathbf{y}$, and $k \to \infty$ leads to the restricted estimator $\mathbf{b}_R$ (2.3.15).

In the following we abbreviate the inverse in (9) as

$$Z_k = (X'X + kR'R)^{-1}. \tag{3.3.10}$$

With the restriction (8), it is clear that the mixed estimator is no longer unbiased:

$$\begin{aligned} E\mathbf{b}_R(k) &= EZ_k(X'X\boldsymbol{\beta} + X'\boldsymbol{\varepsilon} + kR'R\boldsymbol{\beta} + kR'\mathbf{g} + kR'\boldsymbol{\phi}) \\ &= \boldsymbol{\beta} + kZ_kR'\mathbf{g}, \end{aligned}$$

that is, we have in general the bias

$$\text{bias}\, b_R(k) = kZ_kR'\mathbf{g}. \tag{3.3.11}$$

The dispersion matrix of $b_R(k)$ is just as before (see (3.1.10));

$$V[\mathbf{b}_R(k)] = \sigma^2 Z_k. \tag{3.3.12}$$

From (11) and (12) we get the risk of the estimator immediately:

$$\begin{aligned} R[\mathbf{b}_R(k), \boldsymbol{\beta}, A] = R[\mathbf{b}_R(k)] &= \operatorname{tr} A \,\mathrm{MSE}\, \mathbf{b}_R(k) \\ &= \operatorname{tr} A\,(\sigma^2 Z_k + k^2 Z_k R'\mathbf{g}\mathbf{g}'RZ_k). \end{aligned} \tag{3.3.13}$$

3.3.3 Comparing the Mixed Estimator and the OLSE

To investigate the conditions which make the biased mixed estimator better than the OLSE $\mathbf{b}_0$ we use the following representation (see Teräsvirta, 1979a). Let us denote

$$(X'X)^{-1} = U \quad \text{and} \quad (k^{-1}I + RUR')^{-1} = S_k \tag{3.3.14}$$

and apply the matrix inversion formula of Theorem A.34 to the matrix Z_k

$$\begin{aligned} Z_k &= (U^{-1} + kR'R)^{-1} \\ &= U - UR'(k^{-1}I + RUR')^{-1}RU. \end{aligned}$$

This gives the risk (13) as

$$R[\mathbf{b}_R(k)] = \operatorname{tr} A\,(\sigma^2 U - \sigma^2 UR'S_kRU + UR'S_k\mathbf{g}\mathbf{g}'S_kRU). \tag{3.3.15}$$

Since $R(\mathbf{b}_0) = \sigma^2 \operatorname{tr} AU$ we have for the difference of the risks of both estimators:

$$R(\mathbf{b}_0) - R[\mathbf{b}_R(k)] = \operatorname{tr} A\{UR'S_k[\sigma^2 S_k^{-1} - \mathbf{g}\mathbf{g}']S_kRU\}. \tag{3.3.16}$$

Assuming $A \geq 0$, then (16) is nonnegative for all $A \geq 0$ iff the matrix $\{\;\} \geq 0$ (see Theorem A.15). Now this matrix $\{\;\} \geq 0$ iff the inner matrix $\sigma^2 S_k^{-1} - \mathbf{g}\mathbf{g}' \geq 0$. Now, using Theorem A.17 we get as a necessary and sufficient condition for (16) to be nonnegative that

$$\sigma^{-2}\mathbf{g}'S_k\mathbf{g} \leq 1 \tag{3.3.17}$$

(see Yancey *et al.*, 1974).

Differentiating this quadratic form with respect to k gives (Theorem A.53, equation B.3.2)

$$\sigma^{-2}\mathbf{g}'\boldsymbol{S}_k(k^{-2}\boldsymbol{I}+\boldsymbol{RUR}')\boldsymbol{S}_k\mathbf{g} > 0.$$

Thus (17) is an increasing continuous function of the precision constant k. As $\lim_{k\to 0+} \sigma^{-2}\mathbf{g}'\boldsymbol{S}_k\mathbf{g} = 0$, it follows that there must exist a critical value k^* such that $\sigma^{-2}\mathbf{g}'\boldsymbol{S}_{k*}\mathbf{g} = 1$. Clearly, k^* is a function of σ^2 and therefore it is unknown. So we have proved the following theorem.

Theorem 3.4

Assume the classical regression model (7) and auxiliary information on $\boldsymbol{\beta}$ of type (8). Then for all k with $0 < k < k^*$ the mixed estimator $\mathbf{b}_{\boldsymbol{R}}(k)$ as in (9) is better than the OLSE $\mathbf{b}_0$, i.e.

$$R(\mathbf{b}_0) - R[\mathbf{b}_{\boldsymbol{R}}(k)] \geq 0 \qquad (0 < k < k^*). \tag{3.3.18}$$

Note If we write $\hat{\boldsymbol{\beta}}(c)$ from (3.2.8) in an equivalent form (assumed $\boldsymbol{W} = \boldsymbol{I}, \boldsymbol{V} = \boldsymbol{I}$) we have

$$\hat{\boldsymbol{\beta}}(c) = (\boldsymbol{X}'\boldsymbol{X} + c^{-1}\boldsymbol{R}'\boldsymbol{R})^{-1}(\boldsymbol{X}'\mathbf{y} + c^{-1}\boldsymbol{R}'\mathbf{r}).$$

Comparing this with $\mathbf{b}_{\boldsymbol{R}}(k)$ (9) gives a formal correspondence of c^{-1} and k. Thus, using (3.2.14), we may conclude that

$$\lim_{k\to\infty} \mathbf{b}_{\boldsymbol{R}}(k) = \mathbf{b}_{\boldsymbol{R}}(\infty) = \tilde{\mathbf{b}}_{\boldsymbol{R}}, \tag{3.3.19}$$

the restricted least squares estimator. So, if we let $k \to \infty$ in (17) we arrive at the necessary and sufficient condition

$$\sigma^{-2}\mathbf{g}'(\boldsymbol{RUR}')^{-1}\mathbf{g} \leq 1 \tag{3.3.20}$$

for the restricted LSE to be superior to the OLSE $\mathbf{b}_0$ (or to be at least equivalent to $\mathbf{b}_0$, as far as the risk is concerned).

Teräsvirta (1979a) has given the critical condition (17) a more illustrative form. Applying the spectral decomposition (Theorem A.2) on $\boldsymbol{RUR}'$ gives

$$\boldsymbol{P\Lambda P}' = \boldsymbol{RUR}'$$

where $\boldsymbol{\Lambda} = \operatorname{diag}(\lambda_1, \ldots, \lambda_J), \lambda_1 \geq \ldots \geq \lambda_J$ is the matrix of eigenvalues of $\boldsymbol{RUR}'$. $\boldsymbol{P}$, the matrix of eigenvectors, is orthogonal: $\boldsymbol{P}'\boldsymbol{P} = \boldsymbol{PP}' = \boldsymbol{I}$. Let $\boldsymbol{P}'\mathbf{g} = \tilde{\mathbf{g}}$. Then (17) becomes

$$\sigma^{-2}\tilde{\mathbf{g}}'(k^{-1}\boldsymbol{I} + \boldsymbol{\Lambda})^{-1}\tilde{\mathbf{g}} = \sigma^{-2}\sum_{j=1}^{J}\frac{\tilde{g}_j^2}{k^{-1}+\lambda_j} \leq 1. \tag{3.3.21}$$

Now, $\lambda_j \geq \lambda_J$ (all j) and so

$$\sum_{j=1}^{J}\frac{\tilde{g}_j^2}{k^{-1}+\lambda_j} \leq \frac{\tilde{\mathbf{g}}'\tilde{\mathbf{g}}}{k^{-1}+\lambda_J} = \frac{\mathbf{g}'\mathbf{g}}{k^{-1}+\lambda_J}.$$

So we have found as a sufficient condition for (17) to hold:

$$0 < k \leq (\sigma^{-2}\mathbf{g}'\mathbf{g} - \lambda_J)^{-1} \tag{3.3.22}$$

(if $\sigma^{-2}\mathbf{g}'\mathbf{g} - \lambda_J > 0$). The smaller $\sigma^{-2}\mathbf{g}\mathbf{g}' - \lambda_J$, the greater are the chances of finding mixed estimators $\mathbf{b}_R(k)$ which dominate the unrestricted OLSE $\mathbf{b}_0$. Assume the regressor matrix $\boldsymbol{X}$ and the prior matrix $\boldsymbol{R}$ fixed; then these chances grow if the incorrectness of the prior information is of an inconsiderable degree and/or if the sample has a great variance σ^2 so that $\sigma^{-2}\mathbf{g}'\mathbf{g}$ becomes small.

3.3.4 Ridge and Mixed Estimation

Our intention is to investigate the connection between two biased estimators: the mixed estimator $\mathbf{b}_R(k)$ (9) and the ridge estimator $\mathbf{b}(k)$ defined in Section 2.3.3:

$$\mathbf{b}(k) = (\boldsymbol{X}'\boldsymbol{X} + k\boldsymbol{I})^{-1}\boldsymbol{X}'\mathbf{y}. \tag{3.3.23}$$

This means that the ridge estimator $\mathbf{b}(k)$ has the form of a mixed estimator. Indeed, letting $\boldsymbol{R} = \boldsymbol{I}$ and $\mathbf{r} = \mathbf{0}$ in (9), we see that

$$\mathbf{b}_I(k) = \mathbf{b}(k). \tag{3.3.24}$$

So we have the interesting interpretation that the ridge estimator is a special mixed estimator derived under the incorrect prior information

$$\mathbf{0} = \boldsymbol{\beta} + \mathbf{g} + \boldsymbol{\phi}, \quad \boldsymbol{\phi} \sim (\mathbf{0}, (\sigma^2/k)\boldsymbol{I}). \tag{3.3.25}$$

Moreover, we can use the relation between the biased mixed estimator and the OLSE $\mathbf{b}_0$ to check the *improvement region* of $\mathbf{b}(k)$ derived in Section 2.3.3.

With $\boldsymbol{R} = \boldsymbol{I}$ and $\mathbf{r} = \mathbf{0}$ we get

$$\boldsymbol{S}_k = (k^{-1}\boldsymbol{I} + \boldsymbol{U})^{-1} \tag{3.3.26}$$

and (see 15) the risk of the ridge estimator is

$$R[\mathbf{b}(k)] = \operatorname{tr} \boldsymbol{A}(\sigma^2\boldsymbol{U} - \sigma^2\boldsymbol{U}\boldsymbol{S}_k\boldsymbol{U} + \boldsymbol{U}\boldsymbol{S}_k\boldsymbol{\beta}\boldsymbol{\beta}'\boldsymbol{S}_k\boldsymbol{U}). \tag{3.3.27}$$

Then the ridge estimator dominates the ordinary least squares if and only if (see (17))

$$\sigma^{-2}\boldsymbol{\beta}'(k^{-1}\boldsymbol{I} + \boldsymbol{U})^{-1}\boldsymbol{\beta} \leq 1. \tag{3.3.28}$$

Applying spectral decomposition (Theorem A.2) on $\boldsymbol{U}$ gives

$$\boldsymbol{P}_U\boldsymbol{\Lambda}_U\boldsymbol{P}'_U = \boldsymbol{U}$$

where $\boldsymbol{\Lambda}_U = \operatorname{diag}(\lambda_1(\boldsymbol{U}), \ldots, \lambda_K(\boldsymbol{U}))$; $\lambda_1(\boldsymbol{U}) \geq \ldots \geq \lambda_K(\boldsymbol{U}) > \boldsymbol{0}$ is the matrix of eigenvalues of $\boldsymbol{U} = (\boldsymbol{X}'\boldsymbol{X})^{-1}$ and $\boldsymbol{P}_U$ the matrix of the corresponding eigenvectors. Let $\boldsymbol{P}'_U\boldsymbol{\beta} = \tilde{\boldsymbol{\beta}}$; then (28) becomes

$$\sigma^{-2}\tilde{\boldsymbol{\beta}}'(k^{-1}\boldsymbol{I} + \boldsymbol{\Lambda}_U)^{-1}\tilde{\boldsymbol{\beta}} = \sum_{i=1}^{K} \frac{\tilde{\beta}_i^2\sigma^{-2}}{k^{-1} + \lambda_i(\boldsymbol{U})} \leq 1. \tag{3.3.29}$$

Thus, corresponding to (22) we have as a sufficient condition for the biased ridge estimator $\mathbf{b}(k)$ to dominate the unbiased OLSE $\mathbf{b}_0$ that k is in the interval

$$0 < k \le (\sigma^{-2}\boldsymbol{\beta}'\boldsymbol{\beta} - \lambda_K(U))^{-1}. \quad (3.3.30)$$

Now, from the representation $U = P_U \Lambda_U P_U'$ it follows that

$$U^{-1} = P_U \Lambda_U^{-1} P_U' = P_U[\operatorname{diag}(\lambda_1^{-1}(U), \ldots, \lambda_K^{-1}(U))]P_U' \quad (3.3.31)$$

must hold. Thus $\Lambda_U^{-1} = (\Lambda_U)^{-1}$ is the matrix of eigenvalues of $U^{-1} = X'X$ where $\lambda_K^{-1}(U) \ge \ldots \ge \lambda_1^{-1}(U) > 0$. Set $\lambda_i^{-1}(U) = \tilde{\lambda}_i(U^{-1}) = \tilde{\lambda}_i(X'X)$. Then $\lambda_K^{-1}(U) = \lambda_{\max}(X'X)$ and we may write (30) in terms of $(X'X)$:

$$0 < k < \{\sigma^{-2}\boldsymbol{\beta}'\boldsymbol{\beta} - (\lambda_{\max}(X'X))^{-1}\}^{-1}. \quad (3.3.32)$$

Let us use the canonical representation of the model $\mathbf{y} = X\boldsymbol{\beta} + \boldsymbol{\varepsilon}$, i.e.

$$\mathbf{y} = XP_U P_U'\boldsymbol{\beta} + \boldsymbol{\varepsilon} = \tilde{X}\tilde{\boldsymbol{\beta}} + \boldsymbol{\varepsilon} \quad (3.3.33)$$

where $\tilde{X} = XP_U$. Then situations where ridge estimation can be expected to dominate OLS estimation with respect to the risk may now be expressed as follows, see (32):

(i) the ratio $\boldsymbol{\beta}'\boldsymbol{\beta}/\sigma^2$ should not be too large. (Sometimes $\boldsymbol{\beta}'\boldsymbol{\beta}/\sigma^2$ is given the name 'signal-to-noise ratio'.)

(ii) The maximal eigenvalue $\lambda_{\max}(X'X)$ is rather small and therefore all other eigenvalues of $X'X$ are rather small. This means that the matrix $X'X$ should be relatively ill-conditioned, in which case the OLSE $\mathbf{b}_0$ would have high variance.

3.3.5 Optimal Choice of k

So far we have found condition (17) to be necessary and sufficient for the mixed estimator $\mathbf{b}_R(k)$ to have smaller risk than $\mathbf{b}_0$ for all $A \ge 0$. A sufficient condition for (17) to hold was given in (22). Following the approach given in Teräsvirta (1979a), we will now handle the problem of minimizing the risk $R[\mathbf{b}_R(k)]$ as a function of k. Differentiating (15) with respect to k gives (see equation B.3.2 and Theorem A.53)

$$\frac{\partial}{\partial k}R[\mathbf{b}_R(k)] = -k^{-2}\sigma^2 \operatorname{tr} AUR'S_k^2RU + 2k^{-2}\mathbf{g}'S_k^2RUAUR'S_k\mathbf{g}$$

$$= -u + v, \text{ say} \quad (3.3.34)$$

where $u > 0$, and u is independent of g, the mean error of prior information.

Thus we may evaluate

$$\frac{\partial}{\partial k}R[\mathbf{b}_R(k)] \underset{k \to 0}{\rightarrow} -\sigma^2 \operatorname{tr} RUAUR' < 0. \quad (3.3.35)$$

Differentiating (34) further gives the result that

$$\frac{\partial^2}{\partial k^2} R[\mathbf{b}_R(k)] \underset{k \to 0}{\to} 2(\sigma^2 \operatorname{tr} \boldsymbol{AUR'RUR'RU} + \mathbf{g}'\boldsymbol{RUAUR}'\mathbf{g}) \geq 0. \quad (3.3.36)$$

Clearly, the risk $R[\mathbf{b}_R(k)]$ as a function of k first decreases, then reaches the minimum at $k = \mathrm{k}_0$, say, and then increases so far as $k_0 < \infty$. Some information about the possible decrease of the risk when $\mathbf{b}_R(k)$ is used instead of $\mathbf{b}_0$ can be obtained by evaluating the size of $\partial R[\mathbf{b}_R(k)]/\partial k$ at zero. This derivative (35) is negative and, moreover, the second derivative (36) is positive and depends on the mean error g. If this error is large the second derivative tends to be large, also.

Equating (34) to zero in general would not result in an explicit formula for the k-value which gives the minimum of the risk. Thus we continue to investigate the location of the minimum of $R[\mathbf{b}_R(k)]$. From (35) and the above arguments we may deduce that we must be interested in a condition upon k which ensures that we are on the descending part of the risk function. That is, as long as $\partial R[\mathbf{b}_R(k)]/\partial k$ stays nonpositive, the corresponding value of k lies on the descending part of the risk. Let us give the first derivative (34) the equivalent form

$$\frac{\partial}{\partial k} R[\mathbf{b}_R(k)] = -k^{-2} \operatorname{tr} \boldsymbol{AUR'S_kQS_kRU}. \quad (3.3.37)$$

Then this is nonpositive for all $\boldsymbol{A} \geq \boldsymbol{0}$ if and only if

$$\boldsymbol{Q} = \sigma^2 \boldsymbol{I} - \mathbf{gg}'\boldsymbol{S}_k - \boldsymbol{S}_k\mathbf{gg}' \geq \boldsymbol{0}. \quad (3.3.38)$$

The eigenvalues of a nonnegative definite matrix must be nonnegative, i.e. $\boldsymbol{Q} \geq \boldsymbol{0}$ if the largest eigenvalue of $\mathbf{gg}'\boldsymbol{S}_k + \boldsymbol{S}_k\mathbf{gg}'$ is not greater than σ^2. Now, the nonzero eigenvalues of this matrix are

$$\mathbf{g}'\boldsymbol{S}_k\mathbf{g} \pm (\mathbf{g}'\boldsymbol{S}_k^2\mathbf{gg}'\mathbf{g})^{1/2}. \quad (3.3.39)$$

Moreover, we get by the Cauchy–Schwarz inequality that

$$\mathbf{g}'\boldsymbol{S}_k\mathbf{g} \leq (\mathbf{g}'\boldsymbol{S}_k^2\mathbf{gg}'\mathbf{g})^{1/2}.$$

Thus $\boldsymbol{Q} \geq \boldsymbol{0}$ if

$$\sigma^{-2}\{\mathbf{g}'\boldsymbol{S}_k\mathbf{g} + (\mathbf{g}'\boldsymbol{S}_k^2\mathbf{gg}'\mathbf{g})^{1/2}\} \leq 1. \quad (3.3.40)$$

Comparing this result with condition (17) we may conclude that being on the descending part of the risk $R[\mathbf{b}_R(k)]$ is sufficient for the mixed estimator to dominate $\mathbf{b}_0$.

Teräsvirta (1979a) proposed the following procedure for testing the hypothesis that k lies on the descending part of the risk, i.e.

$$H_0\colon \quad \frac{\partial}{\partial k} R[\mathbf{b}_R(k)] \leq 0.$$

Let $\lambda_1(\boldsymbol{S}_k) \geq \ldots \geq \lambda_J(\boldsymbol{S}_k)$ denote the eigenvalues of $\boldsymbol{S}_k$. Clearly,

$$\lambda_1(\boldsymbol{S}_k) = (k^{-1} + \lambda_J)^{-1}, \quad \lambda_J(\boldsymbol{S}_k) = (k^{-1} + \lambda_1)^{-1} \quad (3.3.41)$$

where $\lambda_1 \geq \ldots \geq \lambda_J$ are the eigenvalues of $\boldsymbol{RUR}'$ (see (21)). By Theorem A.13 we have

$$\frac{(\mathbf{g}'\boldsymbol{S}_k^{1/2})\,\boldsymbol{S}_k(\boldsymbol{S}_k^{1/2}\mathbf{g})}{(\mathbf{g}'\boldsymbol{S}_k^{1/2})\,(\boldsymbol{S}_k^{1/2}\mathbf{g})} \leq \lambda_1(\boldsymbol{S}_k)$$

and

$$\lambda_J(\boldsymbol{S}_k) \leq \frac{\mathbf{g}'\boldsymbol{S}_k\mathbf{g}}{\mathbf{g}'\mathbf{g}}.$$

This gives the relation

$$\mathbf{g}'\boldsymbol{S}_k\mathbf{g} + (\mathbf{g}'\boldsymbol{S}_k^2\mathbf{g}\mathbf{g}'\mathbf{g})^{1/2} \leq [1 + \{\lambda_1(\boldsymbol{S}_k)\}^{1/2}\{\lambda_J(\boldsymbol{S}_k)\}^{-1/2}]\mathbf{g}'\boldsymbol{S}_k\mathbf{g}. \tag{3.3.42}$$

If we use (42) and (41), we can deduce that

$$\sigma^{-2}\{1 + (k^{-1} + \lambda_1)^{1/2}(k^{-1} + \lambda_J)^{-1/2}\}\,\mathbf{g}'\boldsymbol{S}_k\mathbf{g} \leq 1 \tag{3.3.43}$$

is sufficient to realize condition (40).

Let us go back to the models (7) and (8) and assume further that normality conditions hold, i.e.

$$\boldsymbol{\varepsilon} \sim N_T(\mathbf{0}, \sigma^2\boldsymbol{I}) \quad \text{and} \quad \boldsymbol{\phi} \sim N_J(\mathbf{0}, (\sigma^2/k)\boldsymbol{I}).$$

Then we have the result

$$\begin{aligned} \mathbf{r} - \boldsymbol{R}\mathbf{b}_0 &= \mathbf{r} - \boldsymbol{R\beta} - \boldsymbol{RUX}'\boldsymbol{\varepsilon} + \boldsymbol{\phi} \\ &= \mathbf{g} - \boldsymbol{RUX}'\boldsymbol{\varepsilon} + \boldsymbol{\phi} \\ &\sim N_J(\mathbf{g}, \sigma^2\boldsymbol{S}_k^{-1}). \end{aligned} \tag{3.3.44}$$

Using Theorem A.44 we may then conclude that

$$(\mathbf{r} - \boldsymbol{R}\mathbf{b}_0)'\boldsymbol{S}_k(\mathbf{r} - \boldsymbol{R}\mathbf{b}_0) \sim \sigma^2\chi_J^2(\lambda) \tag{3.3.45}$$

with the noncentrality parameter

$$\lambda = \sigma^{-2}\mathbf{g}'\boldsymbol{S}_k\mathbf{g}. \tag{3.3.46}$$

As

$$\begin{aligned} s^2 &= (\mathbf{y} - \boldsymbol{X}\mathbf{b}_0)'(\mathbf{y} - \boldsymbol{X}\mathbf{b}_0)(T - K)^{-1} \\ &= \mathbf{y}'(\boldsymbol{I} - \boldsymbol{XUX}')\mathbf{y}(T - K)^{-1} \sim \sigma^2\chi_{T-K}^2 \end{aligned}$$

and s^2 is independent of $(\mathbf{r} - \boldsymbol{R}\mathbf{b}_0)'\boldsymbol{S}_k(\mathbf{r} - \boldsymbol{R}\mathbf{b}_0)$ (the proof of this is left to the reader), we deduce from Theorem A.45 that the statistic

$$\gamma(k) = \frac{(\mathbf{r} - \boldsymbol{R}\mathbf{b}_0)'\boldsymbol{S}_k(\mathbf{r} - \boldsymbol{R}\mathbf{b}_0)}{\mathbf{y}'(\boldsymbol{I} - \boldsymbol{XUX}')\mathbf{y}}\,\frac{T - K}{J} \tag{3.3.47}$$

follows a noncentral $F_{J,T-K}(\lambda)$ distribution with λ defined in (46). If we now wish to test whether a fixed k is such that we are still on the left of the minimum of the risk function we have to show that (see (43))

$$\lambda \leq \{1 + (k^{-1} + \lambda_1)^{1/2}(k^{-1} + \lambda_J)^{-1/2}\}^{-1} = \lambda_0, \text{ say.} \tag{3.3.48}$$

This is done in the well-known way: if $\gamma(k) \leq F^{\alpha}_{J,T-K}(\lambda_0)$ where $F^{\alpha}_{J,T-K}(\lambda_0)$ denotes the critical point of the noncentral F with λ_0 as noncentrality parameter, then the hypothesis $\lambda \leq \lambda_0$ or, equivalently, $\partial R[\mathbf{b}_R(k)]/\partial k \leq 0$ would be accepted.

Thus we have proposed a stepwise procedure to approximate that value k which gives the minimum of the risk $R[\mathbf{b}_R(k)]$.

Using the above results and reasoning we can immediately propose a test for the hypothesis that the restricted least squares estimator $\mathbf{b}_R = \mathbf{b}_R(\infty)$ (see 19)) has at most the same risk as one of the estimators of the class $\{\mathbf{b}_R(k): 0 \leq k < \infty\}$.

Note that

$$\lambda_0 \to \tfrac{1}{2} \quad \text{as } k \to 0$$

and

$$\lambda_0 \to \{1 + (\lambda_1/\lambda_J)^{1/2}\}^{-1} = \lambda_0(\infty) \quad \text{as } k \to \infty. \tag{3.3.49}$$

Furthermore, $\partial \lambda_0/\partial k < 0$ so that λ_0 is a decreasing function of k. If we let $k \to \infty$ in (17), we get

$$\sigma^{-2}\mathbf{g}'(\boldsymbol{R}\boldsymbol{U}\boldsymbol{R}')^{-1}\mathbf{g} \leq 1 \tag{3.3.50}$$

as a necessary and sufficient condition for the restricted LSE $\mathbf{b}_R$ to dominate the OLSE $\mathbf{b}_0 = \mathbf{b}_R(0)$.

The left-hand term in (50) is just the noncentrality parameter of the statistic

$$\gamma(\infty) = \frac{(\mathbf{r} - \boldsymbol{R}\mathbf{b}_0)'(\boldsymbol{R}\boldsymbol{U}\boldsymbol{R}')^{-1}(\mathbf{r} - \boldsymbol{R}\mathbf{b})}{\mathbf{y}'(\boldsymbol{I} - \boldsymbol{X}\boldsymbol{U}\boldsymbol{X}')\mathbf{y}} \frac{T-K}{J} \tag{3.3.51}$$

and the null hypothesis to be tested is (see (48))

$$\lambda \leq \lambda_0(\infty) = \{1 + (\lambda_1/\lambda_J)^{1/2}\}^{-1}. \tag{3.3.52}$$

Summarizing, we can state that the test procedure should be used as an exploratory device supporting the process of model building by finding the risk-minimizing k. This approach will give an idea of the interval of k where the mixed estimator has the best performance.

3.4 TEST PROCEDURES

3.4.1 Testing the Compatibility of Prior and Sample Information

Assuming auxiliary information on $\boldsymbol{\beta}$ such as (3.1.2) leads to the mixed estimator $\hat{\boldsymbol{\beta}}(\sigma^2)$ (3.1.9) which has smaller variance than the GLSE $\mathbf{b}$ whatever the restrictions $\mathbf{r} = \boldsymbol{R}\boldsymbol{\beta} + \boldsymbol{\phi}$ mean. Thus the decrease of variance is ensured algebraically, regardless of the statistical background of the imposed prior model $\mathbf{r} = \boldsymbol{R}\boldsymbol{\beta} + \boldsymbol{\phi}$. Before accepting the mixed estimator, however, we wish to *check the possibility that prior and sample information may be in conflict with each other.* Theil (1963) developed a procedure to test the null-hypothesis

$$H_0\text{: prior and sample information are in agreement.} \tag{3.4.1}$$

Under this hypothesis we have two independent estimators of the $J \times 1$-vector $\boldsymbol{R\beta}$:

first, the prior estimator $\mathbf{r}$

and

second, the GLS estimator $\boldsymbol{R}\mathbf{b}$.

If both types of information are in agreement, we would expect their difference

$$\boldsymbol{\delta} = \mathbf{r} - \boldsymbol{R}\mathbf{b} = \boldsymbol{\phi} - \boldsymbol{RS}^{-1}\boldsymbol{X}'\boldsymbol{W}^{-1}\boldsymbol{\varepsilon} \tag{3.4.2}$$

to be close to zero.

The matrix of second moments of $\boldsymbol{\delta}$ is

$$E\boldsymbol{\delta\delta}' = \boldsymbol{V} + \sigma^2 \boldsymbol{RS}^{-1}\boldsymbol{R}'. \tag{3.4.3}$$

As this matrix is positive definite we may write (see Theorem A.2ii)

$$E\boldsymbol{\delta\delta}' = \boldsymbol{Z}^{1/2}\boldsymbol{Z}^{1/2}$$

and thus we conclude that

$$\boldsymbol{\delta}'\boldsymbol{Z}^{-1/2} \sim (\mathbf{0}, \boldsymbol{I}).$$

Let us now assume normality conditions, that is

$$\boldsymbol{\varepsilon} \sim N_T(\mathbf{0}, \sigma^2 \boldsymbol{W}), \quad \boldsymbol{\phi} \sim N_J(\mathbf{0}, \boldsymbol{V}). \tag{3.4.4}$$

Then $\boldsymbol{\delta}'\boldsymbol{Z}^{-1/2} \sim N_J(\mathbf{0}, \boldsymbol{I})$ and, moreover, the natural test statistic $\boldsymbol{\delta}'\boldsymbol{Z}^{-1}\boldsymbol{\delta}$ follows a central χ^2-distribution:

$$\begin{aligned} \gamma &= \boldsymbol{\delta}'\boldsymbol{Z}^{-1/2}\boldsymbol{Z}^{-1/2}\boldsymbol{\delta} \\ &= (\mathbf{r} - \boldsymbol{R}\mathbf{b})'[\boldsymbol{V} + \sigma^2 \boldsymbol{RS}^{-1}\boldsymbol{R}']^{-1}(\mathbf{r} - \boldsymbol{R}\mathbf{b}) \sim \chi^2_J. \end{aligned} \tag{3.4.5}$$

As σ^2 is unknown, for practical purposes we will have to work with

$$\hat{\gamma} = (\mathbf{r} - \boldsymbol{R}\mathbf{b})'[\boldsymbol{V} + s^2 \boldsymbol{RS}^{-1}\boldsymbol{R}']^{-1}(\mathbf{r} - \boldsymbol{R}\mathbf{b}) \tag{3.4.6}$$

which has asymptotically the same distribution. It is usual to call $\hat{\gamma}$ the *compatibility statistic*. If a critical point $\chi^2_{1-\alpha}$ of the χ^2-distribution is chosen according to

$$P(\chi^2 \geq \chi^2_{1-\alpha}) = \alpha$$

where α is the fixed error probability of the first kind, then the null hypothesis (1) will be accepted for $\hat{\gamma} < \chi^2_{1-\alpha}$. (For another type of compatibility see Chipman and Rao (1964)).

3.4.2 Testing Linear Restrictions—the Biased Restricted Estimator

In the previous subsection we proposed a procedure for testing the hypothesis 'prior and sample information are in agreement' where the 'degree of agreement' was measured by the statistic $\hat{\gamma}$. A somewhat different question is whether a specified set of linear restrictions is true. The truth or falsity of such (exact) restrictions will be checked by tests such as those in Section 2.3.5. If a hypothesis such as '$R\beta = \mathbf{r}$' has to be rejected, one possible conclusion might be to assume

$$\mathbf{r} = R\beta + \phi \quad \text{or} \quad \mathbf{r} = R\beta + \mathbf{g} + \phi,$$

as was done in (3.3.6). However, perhaps a more important question is whether a specific set of restrictions leads to estimators which dominate the unrestricted GLSE $\mathbf{b}$ where in general these restrictions may be misspecified.

'Philosophically speaking every model is misspecified. Thus the question is whether a model contains a structure which comes close to the true one.' (Menges, 1971)

In the following, the assumed misspecification of the restrictions is given the form

$$\mathbf{r} \neq R\beta \tag{3.4.7}$$

or, expressed by the previously used difference,

$$\mathbf{g} = \mathbf{r} - R\beta \neq \mathbf{0}. \tag{3.4.8}$$

Now we have two possibilities: first, to use the unrestricted GLSE $\mathbf{b}$; or secondly, to use the conditional restricted estimator $\tilde{\mathbf{b}}_R$ (2.3.71) which no longer is unbiased.

Let us use the abbreviation

$$D = S^{-1}R'(RS^{-1}R')^{-1}. \tag{3.4.9}$$

Then the restricted LSE $\tilde{\mathbf{b}}_R$ becomes

$$\tilde{\mathbf{b}}_R = \mathbf{b} + D(\mathbf{r} - R\mathbf{b})$$
$$E\tilde{\mathbf{b}}_R = \beta + D(\mathbf{r} - R\beta) = \beta + D\mathbf{g}$$

and, therefore, we get

$$\text{bias } \tilde{\mathbf{b}}_R = D\mathbf{g} \neq \mathbf{0}. \tag{3.4.10}$$

The dispersion matrix is not influenced by assumption (7):

$$V(\tilde{\mathbf{b}}_R) = \sigma^2 S^{-1} - \sigma^2 D(RS^{-1}R')D' \tag{3.4.11}$$

and, moreover, $V(\mathbf{b}) - V(\tilde{\mathbf{b}}_R) \geq \mathbf{0}$ holds regardless of whether or not $\mathbf{g} \neq \mathbf{0}$.

Comparing the unbiased GLSE $\mathbf{b}$ and the biased estimator $\tilde{\mathbf{b}}_R$ which has smaller variance, we have to weigh the advantages against the disadvantages of both estimators. As criteria of 'betterness' we will take three versions of the mean square error criterion.

Mean-Square-Error Criterion I

Definition 3.1 (Strong MSE Criterion) An estimator $\hat{\beta}_1$ of β is said to be MSE I-better than another estimator $\hat{\beta}_2$ if the difference

$$\Delta(\hat{\beta}_2, \hat{\beta}_1) = \text{MSE}(\hat{\beta}_2) - \text{MSE}(\hat{\beta}_1) \tag{3.4.12}$$

is nonnegative definite.

As MSE $(\mathbf{b}) = V(\mathbf{b}) = \sigma^2 S^{-1}$ and by $\mathbf{g} \neq \mathbf{0}$ (and, for example, (2.1.21)),

$$\begin{aligned}\text{MSE}(\tilde{\mathbf{b}}_R) &= V(\tilde{\mathbf{b}}_R) + (\text{bias } \tilde{\mathbf{b}}_R)(\text{bias } \tilde{\mathbf{b}}_R)' \\ &= \sigma^2 S^{-1} - \sigma^2 D(RS^{-1}R')D' + D\mathbf{g}\mathbf{g}'D'\end{aligned}$$

we may conclude that

$$\begin{aligned}\Delta(\mathbf{b}, \tilde{\mathbf{b}}_R) &= \sigma^2 D[(RS^{-1}R') - \sigma^{-2}\mathbf{g}\mathbf{g}']D' \\ &= \sigma^2 DC'^{1/2}[I - \sigma^{-2}C^{-1/2}\mathbf{g}\mathbf{g}'C^{-1/2}]C^{1/2}D' \end{aligned} \tag{3.4.13}$$

using the fact that the positive definite matrix $RS^{-1}R'$ has the representation (see Theorem A.2)

$$RS^{-1}R' = C^{1/2}C^{1/2}, \text{ say,}$$

where $C^{1/2}$ is regular, and $C^{-1/2}C^{-1/2} = (RS^{-1}R')^{-1}$.

Thus $\tilde{\mathbf{b}}_R$ is MSE I-better than $\mathbf{b}$ if and only if

$$I - \sigma^{-2}C^{-1/2}\mathbf{g}\mathbf{g}'C^{-1/2} \geq \mathbf{0}. \tag{3.4.14}$$

Using Theorem A.17, (14) holds if and only if

$$\lambda = \sigma^{-2}\mathbf{g}'(RS^{-1}R')^{-1}\mathbf{g} \leq 1 \tag{3.4.15}$$

(for an alternative proof see Toro-Vizcarrondo and Wallace, 1968).

Mean-Square-Error Criterion II

Definition 3.2 (First weak MSE Criterion). $\hat{\beta}_1$ is said to be MSE II-better than $\hat{\beta}_2$ if it has smaller scalar MSE (see (2.1.22)), i.e. if

$$\begin{aligned}&E(\hat{\beta}_2 - \beta)'(\hat{\beta}_2 - \beta) - E(\hat{\beta}_1 - \beta)'(\hat{\beta}_1 - \beta) \\ &\quad = \text{tr}\{\text{MSE}(\hat{\beta}_2) - \text{MSE}(\hat{\beta}_1)\} \geq 0.\end{aligned} \tag{3.3.16}$$

Clearly, if $\hat{\beta}_1$ is MSE I-better than $\hat{\beta}_2$, then this holds for the MSE II-criterion, too. Applying the MSE II-criterion on $\mathbf{b}$ and $\tilde{\mathbf{b}}_R$ we get as the condition to prefer $\tilde{\mathbf{b}}_R$:

$$\begin{aligned}&\text{tr}\{\text{MSE}(\mathbf{b}) - \text{MSE}(\tilde{\mathbf{b}}_R)\} = \text{tr}\,\Delta(\mathbf{b}, \tilde{\mathbf{b}}_R) \\ &= \sigma^2 \text{tr}\, D(RS^{-1}R')D' - \mathbf{g}'D'D\mathbf{g} \geq 0\end{aligned} \tag{3.4.17}$$

that is,

$$\begin{aligned}\mathbf{g}'D'D\mathbf{g} &\leq \sigma^2 \text{ tr } D(RS^{-1}R')D' \\ &= \text{tr}\,(V(\mathbf{b}) - V(\tilde{\mathbf{b}}_R)).\end{aligned} \tag{3.4.18}$$

Thus the biased estimator $\tilde{\mathbf{b}}_R$ dominates $\mathbf{b}$ according to the MSE II-criterion if its squared bias is less than the total decrease in variance of all K estimated β-components.

If we use the relation

$$D'SD = (RS^{-1}R')^{-1}$$

we have (see Wallace, 1972)

$$\mathbf{g}'D'SD\mathbf{g} = \mathbf{g}'(RS^{-1}R')^{-1}\mathbf{g} = \sigma^2\lambda \tag{3.4.19}$$

where λ was defined in (15). By Theorem A.13,

$$d_K \le \frac{\mathbf{g}'D'SD\mathbf{g}}{\mathbf{g}'D'D\mathbf{g}} \le d_1 \quad (\text{for } \mathbf{g}'D' \ne \mathbf{0}) \tag{3.4.20}$$

where $d_1 \ge d_2 \ge \ldots \ge d_K > 0$ are the eigenvalues of $S = X'W^{-1}X$. Then a sufficient condition for (17) to hold is

$$\lambda \le \lambda_0 \quad \text{where} \quad \lambda_0 = d_K \operatorname{tr} S^{-1}R(R'S^{-1}R)^{-1}R'S^{-1} \tag{3.4.21}$$

(see also Wallace, 1972, and Möller, 1976, p. 15). Yancey *et al.* (1973) used reparametrization of the model $\mathbf{y} = X\boldsymbol{\beta} + \boldsymbol{\varepsilon}$ and (8) to get tighter bounds for λ which depend on the unknown vector $\mathbf{g}$. The relationship of the scalar MSE risk of GLSE and the restricted LSE involve a region for λ where $\operatorname{tr}\Delta(\mathbf{b}, \tilde{\mathbf{b}}_R) \le 0$, a region for λ where $\operatorname{tr}\Delta(\mathbf{b}, \tilde{\mathbf{b}}_R) \ge 0$, and a region for λ where uncertainty exists about the comparison of tr MSE ($\mathbf{b}$) and tr MSE ($\tilde{\mathbf{b}}_R$). Without knowing the value of $\mathbf{g}$ there is no way to reduce the region of uncertainty further.

Mean-Square-Error Criterion III

Definition 3.3 (Second Weak MSE Criterion) $\hat{\boldsymbol{\beta}}_1$ is said to be MSE III-better than $\hat{\boldsymbol{\beta}}_2$ if

$$E(\hat{\boldsymbol{\beta}}_2 - \boldsymbol{\beta})'S(\hat{\boldsymbol{\beta}}_2 - \boldsymbol{\beta}) - E(\hat{\boldsymbol{\beta}}_1 - \boldsymbol{\beta})'S(\hat{\boldsymbol{\beta}}_1 - \boldsymbol{\beta}) \ge 0. \tag{3.4.22}$$

Comparing this with the general risk $R(\hat{\boldsymbol{\beta}}, A)$ (2.1.27) we see that, in (22), $A = S$. This special choice of A may be motivated by the canonical representation of the regression model (see (3.3.33) and Wallace, 1972).

Applying criterion (22) to our problem we see that $\tilde{\mathbf{b}}_R$ is MSE III-better than $\mathbf{b}$ if

$$\begin{aligned}\operatorname{tr} S\Delta(\mathbf{b}, \tilde{\mathbf{b}}_R) &= \sigma^2 \operatorname{tr} SS^{-1}R'(RS^{-1}R')^{-1}RS^{-1} \\ &\quad - \mathbf{g}'(RS^{-1}R')^{-1}RS^{-1}SS^{-1}R'(RS^{-1}R)^{-1}\mathbf{g} \\ &= \sigma^2 \operatorname{tr} I_J - \sigma^2\lambda \quad (\text{see } (19)) \\ &= \sigma^2(J - \lambda) \ge 0,\end{aligned}$$

that is, $\tilde{\mathbf{b}}_R$ is preferred if

$$\lambda \le J. \tag{3.4.23}$$

Summarizing the above results we see that the restricted LSE $\tilde{\mathbf{b}}_R$ is better in the sense of one of the three MSE-criteria if

$$\lambda = \sigma^{-2}(\mathbf{r} - R\boldsymbol{\beta})'(RS^{-1}R')^{-1}(\mathbf{r} - R\boldsymbol{\beta}) \tag{3.4.24}$$

is not greater than 1, λ_0 (21), or J, respectively. Now λ is just the noncentrality parameter of the test statistic

$$F = \frac{T-K}{Js^2}(\mathbf{r} - \boldsymbol{R}\mathbf{b})'(\boldsymbol{R}\boldsymbol{S}^{-1}\boldsymbol{R}')^{-1}(\mathbf{r} - \boldsymbol{R}\mathbf{b}). \qquad (3.4.25)$$

The test statistic F can be used to provide a uniformly most powerful test for the MSE-criteria. We test the null hypothesis

$$H_0\colon \lambda \leq 1 \text{ (or } \lambda \leq \lambda_0 \quad \text{or} \quad \lambda \leq J)$$

against the alternative

$$H_1\colon \lambda > 1 \text{ (or } \lambda > \lambda_0 \quad \text{or} \quad \lambda > J, \text{ respectively).}$$

The results connected with the test statistic F (25) are summarized in Figure 3.4.1. (*Hint* In the literature sometimes $\lambda/2$ with λ defined in (24) is taken as the noncentrality parameter of the F-statistic. Take this into account before using the corresponding tables.)

Criterion	Critical value of λ	Distribution of F under null hypothesis
Restrictions $\mathbf{r} = \boldsymbol{R}\boldsymbol{\beta}$ are true	$\lambda = 0$	central $F_{J,T-K}$
MSE $(\tilde{\mathbf{b}}_R) \leq$ MSE$(\mathbf{b})$ (MSE-criterion)	$\lambda \leq 1$	noncentral $F_{J,T-K}$ (1) (tabulated in Wallace and Toro-Vizcarrondo, 1969)
tr MSE$(\tilde{\mathbf{b}}_R) \leq$ tr MSE$(\mathbf{b})$ (first weak MSE-criterion)	$\lambda \leq \lambda_0$ (3.4.21)	noncentral $F_{J,T-K}(\lambda_0)$ (use approximation in Goodnight and Wallace, 1972)
tr$\boldsymbol{S}$MSE $(\tilde{\mathbf{b}}_R) \leq$ tr$\boldsymbol{S}$MSE$(\mathbf{b})$ (second weak MSE-criterion)	$\lambda \leq J$	*noncentral* $F_{J,T-K}(J)$ (tabulated in Goodnight and Wallace, 1972)

Figure 3.4.1 Alternative criteria and tests for linear restrictions

3.5 ALTERNATIVE ESTIMATORS

3.5.1 Mixing Exact and Stochastic Restrictions

The assumption of the positive definiteness of V, the covariance matrix of $\boldsymbol{\phi}$ (3.1.3), is fundamental for existence and optimality of the mixed estimator. In practice it may be happen that a subvector $\boldsymbol{\beta}_1$ of $\boldsymbol{\beta}' = (\boldsymbol{\beta}_1, \boldsymbol{\beta}_2)$ follows stochastic prior information of type (3.1.2), i.e.

$$\mathbf{r}_1 = \boldsymbol{R}_1\boldsymbol{\beta}_1 + \boldsymbol{\phi}, \quad \boldsymbol{\phi} \sim (\mathbf{0}, \boldsymbol{V}_1) \qquad (3.5.1)$$

where the remaining subvector $\boldsymbol{\beta}_2$ of $\boldsymbol{\beta}$ follows an exact prior restriction:

$$\mathbf{r}_2 = \boldsymbol{R}_2\boldsymbol{\beta}_2. \tag{3.5.2}$$

Combining (1) and (2) results in

$$\begin{pmatrix}\mathbf{r}_1\\ \mathbf{r}_2\end{pmatrix} = \begin{pmatrix}\boldsymbol{R}_1\\ \boldsymbol{R}_2\end{pmatrix}\boldsymbol{\beta} + \begin{pmatrix}\boldsymbol{\phi}\\ \mathbf{0}\end{pmatrix} \tag{3.5.3}$$

with the singular covariance matrix

$$E\begin{pmatrix}\boldsymbol{\phi}\\ \mathbf{0}\end{pmatrix}(\boldsymbol{\phi}', \quad \boldsymbol{0}') = \begin{pmatrix}\boldsymbol{V}_1 & \boldsymbol{0}\\ \boldsymbol{0} & \boldsymbol{0}\end{pmatrix} = \boldsymbol{V}. \tag{3.5.4}$$

Thus estimation under auxiliary information as

$$\mathbf{r} = \boldsymbol{R}\boldsymbol{\beta} + \boldsymbol{\phi}, \quad \boldsymbol{\phi} \sim (0, \boldsymbol{V}), \text{ rank } \boldsymbol{V} < J \tag{3.5.5}$$

becomes interesting. Then the following theorem holds (Toutenburg, 1968 and 1975a, p. 99).

Theorem 3.5

In the family $\{\hat{\boldsymbol{\beta}} = \boldsymbol{C}_1'\mathbf{y} + \boldsymbol{C}_2'\mathbf{r}\}$ of heterogeneous estimators of $\boldsymbol{\beta}$,

$$\mathbf{b}_R(\boldsymbol{V}) = \mathbf{b} + \boldsymbol{S}^{-1}\boldsymbol{R}'[\sigma^{-2}\boldsymbol{V} + \boldsymbol{R}\boldsymbol{S}^{-1}\boldsymbol{R}']^{-1}(\mathbf{r} - \boldsymbol{R}\mathbf{b}) \tag{3.5.6}$$

is best linear unbiased under restriction (5) and has dispersion matrix

$$\boldsymbol{V}[\mathbf{b}_R(\boldsymbol{V})] = \sigma^2\boldsymbol{S}^{-1} - \sigma^2\boldsymbol{S}^{-1}\boldsymbol{R}'[\sigma^{-2}\boldsymbol{V} + \boldsymbol{R}\boldsymbol{S}^{-1}\boldsymbol{R}']^{-1}\boldsymbol{R}\boldsymbol{S}^{-1}. \tag{3.5.7}$$

Proof The special heterogeneous set-up

$$\hat{\boldsymbol{\beta}} = \underset{K,T}{\boldsymbol{C}_1'}\mathbf{y} + \underset{K,J}{\boldsymbol{C}_2'}\mathbf{r} = (\boldsymbol{C}_1'\boldsymbol{X} + \boldsymbol{C}_2'\boldsymbol{R})\boldsymbol{\beta} + \boldsymbol{C}_1'\boldsymbol{\varepsilon} + \boldsymbol{C}_2'\boldsymbol{\phi} \tag{3.5.8}$$

yields the condition of unbiasedness

$$\boldsymbol{C}_1'\boldsymbol{X} + \boldsymbol{C}_2'\boldsymbol{R} - \boldsymbol{I}_K = \boldsymbol{0}. \tag{3.5.9}$$

Then the MSE-risk of $\hat{\boldsymbol{\beta}}$ (8) becomes

$$\boldsymbol{R}(\hat{\boldsymbol{\beta}}) = \text{tr } \boldsymbol{A}[\sigma^2\boldsymbol{C}_1'\boldsymbol{W}\boldsymbol{C}_1 + \boldsymbol{C}_2'\boldsymbol{V}\boldsymbol{C}_2]. \tag{3.5.10}$$

Minimizing $R(\hat{\boldsymbol{\beta}})$ under condition (9) requires the solution of

$$\min_{C_1, C_2, \lambda}\left\{\boldsymbol{R}(\hat{\boldsymbol{\beta}}) - 2\sum_{i=1}^{K}\lambda_i'(\boldsymbol{C}_1'\boldsymbol{X} + \boldsymbol{C}_2'\boldsymbol{R} - \boldsymbol{I})_{(i)}\right\} \tag{3.5.11}$$

$$= \min_{C_1, C_2, \lambda} \tilde{R}(\hat{\boldsymbol{\beta}}), \text{ say,}$$

where $\underset{K,K}{\lambda} = (\lambda_1, \ldots, \lambda_K)$ is a matrix of Lagrangian multipliers and $(\ )_{(i)}$ denotes the ith column of the matrix inside the brackets. This results in the normal equations (see Theorems A.50, A.53)

$$\frac{1}{2}\frac{\partial \tilde{R}(\hat{\boldsymbol{\beta}})}{\partial \boldsymbol{C}_1'} = \sigma^2 \boldsymbol{A}\boldsymbol{C}_1' \boldsymbol{W} - \boldsymbol{\lambda}\boldsymbol{X}' = \boldsymbol{0} \tag{3.5.12}$$

$$\frac{1}{2}\frac{\partial \tilde{R}(\hat{\boldsymbol{\beta}})}{\partial \boldsymbol{C}_2'} = \boldsymbol{A}\boldsymbol{C}_2' \boldsymbol{V} - \boldsymbol{\lambda}\boldsymbol{R}' = \boldsymbol{0} \tag{3.5.13}$$

$$\frac{1}{2}\frac{\partial \tilde{R}(\hat{\boldsymbol{\beta}})}{\partial \boldsymbol{\lambda}} = \boldsymbol{C}_1' \boldsymbol{X} + \boldsymbol{C}_2' \boldsymbol{R} - \boldsymbol{I} = \boldsymbol{0}. \tag{3.5.14}$$

$\hat{\boldsymbol{\beta}} = \mathbf{b}_R(\boldsymbol{V})$ is the unique solution of this system.

Comparing $\mathbf{b}_R(\boldsymbol{V})$ with other estimators gives the following:

(i) Connection to the unrestricted GLSE $\mathbf{b}$: the restricted estimator $\mathbf{b}_R(\boldsymbol{V})$ is more efficient than $\mathbf{b}$ in the sense of smaller variance:

$$\boldsymbol{V}(\mathbf{b}) - \boldsymbol{V}[\mathbf{b}_R(\boldsymbol{V})] = \sigma^2 \boldsymbol{S}^{-1}\boldsymbol{R}'[\sigma^{-2}\boldsymbol{V} + \boldsymbol{R}\boldsymbol{S}^{-1}\boldsymbol{R}']^{-1}\boldsymbol{R}\boldsymbol{S}^{-1} \geq \boldsymbol{0}. \tag{3.5.15}$$

(ii) Connection to the conditional restricted LSE $\tilde{\mathbf{b}}_R$: if the auxiliary stochastic restriction (1) tends to exact information, i.e. if $\boldsymbol{V} \to \boldsymbol{0}$, then

$$\mathbf{b}_R(\boldsymbol{V}) \xrightarrow[\boldsymbol{V}\to\boldsymbol{0}]{} \tilde{\mathbf{b}}_R = \mathbf{b} + \boldsymbol{S}^{-1}\boldsymbol{R}'(\boldsymbol{R}\boldsymbol{S}^{-1}\boldsymbol{R}')^{-1}(\mathbf{r} - \boldsymbol{R}\mathbf{b}) \tag{3.5.16}$$

(see (2.3.71)).

3.5.2 Auxiliary Estimates and Piecewise Regression

We now consider the estimation of a parameter vector $\boldsymbol{\beta}$ for part of which a prior estimator exists. This was described in Section 3.1.1 (example (ii)). That is, the subvector $\boldsymbol{\beta}_1$ of $\boldsymbol{\beta}' = (\boldsymbol{\beta}_1', \boldsymbol{\beta}_2')$ has a prior estimate

$$\mathbf{b}_1^* = \boldsymbol{\beta}_1 + \boldsymbol{\phi}, \quad \boldsymbol{\phi} \sim (\mathbf{0}, \boldsymbol{V}). \tag{3.5.17}$$

The piecewise procedure works as follows. Suppose that the matrix $\boldsymbol{X}$ was divided into submatrices $\boldsymbol{X}_1$ and $\boldsymbol{X}_2$, i.e. $\boldsymbol{X} = (\boldsymbol{X}_1, \boldsymbol{X}_2)$, where the dimensions of $\boldsymbol{X}_1$ and $\boldsymbol{X}_2$ correspond to those of $\boldsymbol{\beta}_1$ and $\boldsymbol{\beta}_2$. This gives

$$\mathbf{y} = \boldsymbol{X}_1\boldsymbol{\beta}_1 + \boldsymbol{X}_2\boldsymbol{\beta}_2 + \boldsymbol{\varepsilon}. \tag{3.5.18}$$

We use $\mathbf{y}-$ to denote the vector after fitting $\mathbf{y}$ by the prior estimate $\mathbf{b}_1^*$. In other words

$$\begin{aligned}\mathbf{y}- &= \mathbf{y} - \boldsymbol{X}_1\mathbf{b}_1^* = \boldsymbol{X}_1(\boldsymbol{\beta}_1 - \mathbf{b}_1^*) + \boldsymbol{X}_2\boldsymbol{\beta}_2 + \boldsymbol{\varepsilon} \\ &= \boldsymbol{X}_2\boldsymbol{\beta}_2 + (\boldsymbol{\varepsilon} - \boldsymbol{X}_1\boldsymbol{\phi})\end{aligned} \tag{3.5.19}$$

where the new disturbance vector has mean zero and dispersion $\tilde{\boldsymbol{W}}$:

$$\tilde{\boldsymbol{\varepsilon}} = (\boldsymbol{\varepsilon} - \boldsymbol{X}_1\boldsymbol{\phi}) \sim (\mathbf{0}, \sigma^2\boldsymbol{W} + \boldsymbol{X}_1\boldsymbol{V}\boldsymbol{X}_1') = (\mathbf{0}, \tilde{\boldsymbol{W}}), \text{ say.} \tag{3.5.20}$$

Hence the GLSE of $\boldsymbol{\beta}_2$ in model (19) is

$$\mathbf{b}_2^* = (\boldsymbol{X}_2'\tilde{\boldsymbol{W}}^{-1}\boldsymbol{X}_2)^{-1}\boldsymbol{X}_2'\tilde{\boldsymbol{W}}^{-1}\mathbf{y}-. \tag{3.5.21}$$

This has dispersion matrix

$$V(\mathbf{b}_2^*) = (X_2' \tilde{W}^{-1} X_2)^{-1}.$$

Thus the piecewise estimator of $\boldsymbol{\beta}$ is defined as

$$\mathbf{b}^* = \begin{pmatrix} \mathbf{b}_1^* \\ \mathbf{b}_2^* \end{pmatrix} \tag{3.5.22}$$

and has dispersion matrix

$$V(\mathbf{b}^*) = \begin{pmatrix} V & X_1' V X_2 V(\mathbf{b}_2^*) \\ V(\mathbf{b}_2^*) X_2' V X_1 & V(\mathbf{b}_2^*) \end{pmatrix}. \tag{3.5.23}$$

Exact knowledge of $\boldsymbol{\beta}_1$

In this case $\boldsymbol{\phi} = \mathbf{0}$ (and $V = \boldsymbol{0}$). Then we have $\tilde{\boldsymbol{\varepsilon}} = \boldsymbol{\varepsilon}$, $\tilde{W} = \sigma^2 W$ and $\mathbf{y} - = \mathbf{y} - X_1 \boldsymbol{\beta}_1$. Denoting $X_i' W^{-1} X_j = S_{ij} (i, j = 1, 2)$ we get the piecewise estimator

$$\mathbf{b}^* = \begin{pmatrix} \boldsymbol{\beta}_1 \\ S_{22}^{-1} X_2' W^{-1} (\mathbf{y} - X_1 \boldsymbol{\beta}_1) \end{pmatrix} \tag{3.5.24}$$

with dispersion matrix

$$V(\mathbf{b}^*) = \begin{pmatrix} \boldsymbol{0} & \boldsymbol{0} \\ \boldsymbol{0} & \sigma^2 S_{22}^{-1} \end{pmatrix}. \tag{3.5.25}$$

Comparing $\mathbf{b}^*$ and the unrestricted GLSE $\mathbf{b}$ gives (see Bibby and Toutenburg, 1978, p. 105)

$$V(\mathbf{b}) - V(\mathbf{b}^*) \geq \boldsymbol{0}. \tag{3.5.26}$$

That is, the piecewise estimator $\mathbf{b}^*$ is more efficient than the GLSE if a subvector of $\boldsymbol{\beta}$ is known exactly. This result is not surprising but it yields the basis for the comparison of $\mathbf{b}^*$ with $\mathbf{b}$ and $\mathbf{b}_R(V)$ (6), respectively, when $\boldsymbol{\beta}_1$ is not known exactly.

If we define

$$\Delta(V) = V(\mathbf{b}) - V(\mathbf{b}^*) \tag{3.5.27}$$

then $\Delta(\boldsymbol{0}) \geq \boldsymbol{0}$ is given in (26). Then we may conclude that whenever $\mathbf{a}'\Delta(\boldsymbol{0})\mathbf{a} > 0$, the inequality $\mathbf{a}'\Delta(V)\mathbf{a} > 0$ is satisfied throughout a certain 'neighbourhood' of $V = \boldsymbol{0}$. (Here $\mathbf{a} \neq \mathbf{0}$ is any fixed vector.)

That is, the subvector $\boldsymbol{\beta}_1$ must be known sufficiently well, but not necessary exactly, in order to make $\mathbf{b}^*$ better than $\mathbf{b}$.

If we write the prior estimate (17) equivalently as $\mathbf{r} = R\boldsymbol{\beta} + \boldsymbol{\phi}$ with $\mathbf{r} = \mathbf{b}_1^*$ and $R = (I, \boldsymbol{0})$, then in the case of exact knowledge ($\boldsymbol{\phi} = \mathbf{0}$) we have the estimator

	Model	Estimator	Gains in efficiency
(i)	$\mathbf{y} = \boldsymbol{X\beta} + \varepsilon,\ \varepsilon \sim (\mathbf{0}, \sigma^2 \boldsymbol{W})$ $\mathbf{r} = \boldsymbol{R\beta} + \boldsymbol{\phi}, \boldsymbol{\phi} \sim (\mathbf{0}, \boldsymbol{V})$ rank $\boldsymbol{V} = J$	Mixed estimator (unbiased, not practicable) $\hat{\boldsymbol{\beta}}(\sigma^2) = (\sigma^{-2}\boldsymbol{S} + \boldsymbol{R}'\boldsymbol{V}^{-1}\boldsymbol{R})^{-1} \times (\sigma^{-2}\boldsymbol{X}'\boldsymbol{W}^{-1}\mathbf{y} + \boldsymbol{R}'\boldsymbol{V}^{-1}\mathbf{r})$ $\boldsymbol{V}[\hat{\boldsymbol{\beta}}(\sigma^2)] = (\sigma^{-2}\boldsymbol{S} + \boldsymbol{R}'\boldsymbol{V}^{-1}\boldsymbol{R})^{-1}$	$\boldsymbol{V}(\mathbf{b}) - \boldsymbol{V}[\hat{\boldsymbol{\beta}}(\sigma^2)] \geq \boldsymbol{0}$
(ii)	as model (i)	Modified mixed estimator (asymptotically unbiased) $\hat{\boldsymbol{\beta}}(s^2) = (s^{-2}\boldsymbol{S} + \boldsymbol{R}'\boldsymbol{V}^{-1}\boldsymbol{R}^{-1} \times (s^{-2}\boldsymbol{X}'\boldsymbol{W}^{-1}\mathbf{y} + \boldsymbol{R}'\boldsymbol{V}^{-1}\mathbf{r})$	$\boldsymbol{V}(\mathbf{b}) - \boldsymbol{V}[\hat{\boldsymbol{\beta}}(s^2)] \geq \boldsymbol{0}$ if the sample size is large enough asymptotic dispersion: $\boldsymbol{V}[\hat{\boldsymbol{\beta}}(\sigma^2)]$
(iii)	as model (i) and $\sigma_1^2 < \sigma^2 < \sigma_2^2$	Modified mixed estimator (unbiased) $\hat{\boldsymbol{\beta}}(c) = (c\boldsymbol{S} + \boldsymbol{R}'\boldsymbol{V}^{-1}\boldsymbol{R})^{-1}(c\boldsymbol{X}'\boldsymbol{W}^{-1}\mathbf{y} + \boldsymbol{R}'\boldsymbol{V}^{-1}\mathbf{r})$ $\hat{\boldsymbol{\beta}}(c) \xrightarrow[c \to 0]{} \tilde{\mathbf{b}}_R,\ \hat{\boldsymbol{\beta}}(c) \xrightarrow[c \to \infty]{} \mathbf{b}$ $\boldsymbol{V}[\hat{\boldsymbol{\beta}}(c)] = \boldsymbol{M}_c^{-1}(c^2\sigma^2\boldsymbol{S} + \boldsymbol{R}'\boldsymbol{V}^{-1}\boldsymbol{R})\boldsymbol{M}_c^{-1}$	$\boldsymbol{V}(\mathbf{b}) - \boldsymbol{V}[\hat{\boldsymbol{\beta}}(c)] \geq \boldsymbol{0}$ under conditions of Theorems 3.2 and 3.3.
(iv)	as model (i)	'Mixed' ordinary least squares $\mathbf{b}_R^* = (\boldsymbol{X}'\boldsymbol{X} + \boldsymbol{R}'\boldsymbol{R})^{-1}(\boldsymbol{X}'\mathbf{y} + \boldsymbol{R}'\mathbf{r})$ $\boldsymbol{V}(\mathbf{b}_R^*) = (\boldsymbol{X}'\boldsymbol{X} + \boldsymbol{R}'\boldsymbol{R})^{-1}(\sigma^2\boldsymbol{X}'\boldsymbol{W}\boldsymbol{X} + \boldsymbol{R}'\boldsymbol{V}\boldsymbol{R})(\boldsymbol{X}'\boldsymbol{X} + \boldsymbol{R}'\boldsymbol{R})^{-1}$	$\boldsymbol{V}[\hat{\boldsymbol{\beta}}(c)] - \boldsymbol{V}(\mathbf{b}_R^*) \geq \boldsymbol{0}$ if var $(\varepsilon) \approx$ var $(\boldsymbol{\phi})$ and both are small enough

Figure 3.5.1 Connection of mixed estimators in the case of full rank of prior information

	Model	Estimator	Gains in Efficiency
(i)	$\mathbf{y} = \boldsymbol{X}\beta + \varepsilon,\ \varepsilon \sim (\mathbf{0}, \sigma^2\boldsymbol{W})$	General restricted least squares (unbiased)	$\boldsymbol{V}[\mathbf{b}_R(\boldsymbol{V})] \leq \boldsymbol{V}(\mathbf{b})$
	$\mathbf{r} = \boldsymbol{R}\beta + \phi,\ \phi \sim (\mathbf{0}, \boldsymbol{V})$	$\mathbf{b}_R(\boldsymbol{V}) = \mathbf{b} + \boldsymbol{S}^{-1}\boldsymbol{R}'(\sigma^{-2}\boldsymbol{V} + \boldsymbol{R}\boldsymbol{S}^{-1}\boldsymbol{R}')^{-1}(\boldsymbol{r} - \boldsymbol{R}\mathbf{b})$	
	rank $\boldsymbol{V} < J$	$\boldsymbol{V}[\mathbf{b}_R(\boldsymbol{V})] = \sigma^2\boldsymbol{S}^{-1} - \sigma^2\boldsymbol{S}^{-1}\boldsymbol{R}'(\sigma^{-2}\boldsymbol{V} + \boldsymbol{R}\boldsymbol{S}^{-1}\boldsymbol{R}')^{-1}\boldsymbol{R}\boldsymbol{S}^{-1}$	
		$\mathbf{b}_R(\boldsymbol{V}) \xrightarrow[\boldsymbol{V} \to \boldsymbol{0}]{} \tilde{\mathbf{b}}_R$ (2.3.71)	
(ii)	as in (i), but	Piecewise estimator (unbiased)	$\boldsymbol{V}(\mathbf{b}^*) \leq \boldsymbol{V}(\mathbf{b})$ and
	$\mathbf{r} = \mathbf{b}^*_1,\ \boldsymbol{R} = (\boldsymbol{I}, \boldsymbol{0})$	$\mathbf{b}^* = \begin{pmatrix} \mathbf{b}^*_1 \\ \mathbf{b}^*_2 \end{pmatrix}$ (3.5.22)	$\boldsymbol{V}(\mathbf{b}^*) \leq \boldsymbol{V}(\mathbf{b}_R(\boldsymbol{V}))$ if
	(auxiliary estimate)	$\boldsymbol{V}(\mathbf{b}^*)$ (3.5.23)	$E\phi\phi' = \boldsymbol{V} = \boldsymbol{0}$ or $\boldsymbol{V} \approx \boldsymbol{0}$

Figure 3.5.2 Estimators in the case of mixing exact and stochastic restrictions

$\mathbf{b}_R(\boldsymbol{0}) = \mathbf{\grave{b}}_R$ in its corresponding special form:

$$\mathbf{b}_R(\boldsymbol{0}) = \mathbf{b} + \boldsymbol{S}^{-1}\begin{pmatrix}\boldsymbol{I}\\ \boldsymbol{0}\end{pmatrix}\left[(\boldsymbol{I}, \boldsymbol{0})\boldsymbol{S}^{-1}\begin{pmatrix}\boldsymbol{I}\\ \boldsymbol{0}\end{pmatrix}\right]^{-1}\begin{pmatrix}\boldsymbol{I}\\ \boldsymbol{0}\end{pmatrix}\begin{pmatrix}\boldsymbol{\beta}_1 - \mathbf{b}_1\\ \boldsymbol{\beta}_2 - \mathbf{b}_2\end{pmatrix} \tag{3.5.28}$$

$$= \begin{pmatrix}\boldsymbol{\beta}_1\\ \mathbf{b}_2 - \boldsymbol{V}'_{12}\boldsymbol{V}_{11}^{-1}(\mathbf{b}_1 - \boldsymbol{\beta}_1)\end{pmatrix}$$

where $\mathbf{b}_1$ and $\mathbf{b}_2$ are the components of the GLSE $\mathbf{b}$ which correspond to $\boldsymbol{\beta}_1$ and $\boldsymbol{\beta}_2$ (see (2.2.46)). The dispersion matrix (7) becomes

$$\boldsymbol{V}[\mathbf{b}_R(\boldsymbol{0})] = \begin{pmatrix}\boldsymbol{0} & \boldsymbol{0}\\ \boldsymbol{0} & \boldsymbol{V}_{22} - \boldsymbol{V}'_{12}\boldsymbol{V}_{11}^{-1}\boldsymbol{V}_{12}\end{pmatrix}. \tag{3.5.29}$$

As derived in Section 2.2.5, the dispersion matrix of $\mathbf{b} = \begin{pmatrix}\mathbf{b}_1\\ \mathbf{b}_2\end{pmatrix}$ is

$$\boldsymbol{V}(\mathbf{b}) = \sigma^2\boldsymbol{S}^{-1} = \begin{pmatrix}\boldsymbol{V}_{11} & \boldsymbol{V}_{12}\\ \boldsymbol{V}'_{12} & \boldsymbol{V}_{22}\end{pmatrix}$$

$$= \begin{pmatrix}\sigma^2\boldsymbol{S}_{11}^{-1} + \boldsymbol{V}_{12}\boldsymbol{V}_{22}^{-1}\boldsymbol{V}'_{12} & -\boldsymbol{S}_{11}^{-1}\boldsymbol{S}_{12}\boldsymbol{V}_{22}\\ -\boldsymbol{V}_{22}\boldsymbol{S}_{21}\boldsymbol{S}_{11}^{-1} & \sigma^2(\boldsymbol{S}_{22} - \boldsymbol{S}_{21}\boldsymbol{S}_{11}^{-1}\boldsymbol{S}_{12})^{-1}\end{pmatrix}. \tag{3.5.30}$$

Hence we deduce from (25) and (29) that in the case $\boldsymbol{\phi} = \mathbf{0}$ (i.e. $\boldsymbol{V} = \boldsymbol{0}$)

$$\boldsymbol{\Delta}(\boldsymbol{0}) = \boldsymbol{V}[\mathbf{b}_R(\boldsymbol{0})] - \boldsymbol{V}(\mathbf{b}^*)$$

$$= \begin{pmatrix}\boldsymbol{0} & \boldsymbol{0}\\ \boldsymbol{0} & \boldsymbol{V}_{22} - \boldsymbol{V}'_{12}\boldsymbol{V}_{11}^{-1}\boldsymbol{V}_{12} - \sigma^2\boldsymbol{S}_{22}^{-1}\end{pmatrix}.$$

This matrix may be shown to be nonnegative definite (see Bibby and Toutenburg, 1978, p. 106), i.e. we have $\boldsymbol{\Delta}(\boldsymbol{0}) \geq \boldsymbol{0}$. Thus, $\boldsymbol{a}'\boldsymbol{\Delta}(\boldsymbol{V})\mathbf{a} > 0$ is satisfied in a 'neighbourhood' of $\boldsymbol{V} = \boldsymbol{0}$ whenever $\boldsymbol{a}'\boldsymbol{\Delta}(\boldsymbol{0})\mathbf{a} > 0$ ($\mathbf{a} \neq \mathbf{0}$ any fixed vector).

Note The connections between the various estimators of this chapter are summarized in Figures 3.5.1 and 3.5.2.

IV Minimax Estimation

4.1 THE IDEA AND DERIVATION OF THE ESTIMATOR

4.1.1 Introduction

Minimax estimation in linear models has recently received attention in statistical literature. If one has prior information on the unknown parameter vector $\boldsymbol{\beta}$ such that $\boldsymbol{\beta}$ may be assumed to lie in a *concentration ellipsoid*, the resulting unbiased minimax-linear estimator (MILE) has the form of a ridge-type estimator (see Section 2.3.3) and has a smaller quadratic risk than the unbiased GLSE. The MILE was first derived by Kuks and Olman (1972) for the case where the loss matrix has rank one. Läuter (1975) generalized this result for the case of an arbitrary nonnegative (or positive) definite loss matrix where the result shows the dependence of MILE on the loss matrix. Hoffmann (1977) characterized the admissibility of a linear estimator with respect to the case where a constrained parameter set is given by an ellipsoid. Liski (1979) gives an interpretation of MILE as a special restricted least squares estimator.

A fairly extensive discussion of the problems of minimax-estimation can be found in Rao (1976), Bunke (1975), Bibby and Toutenburg (1978, Chapter 10), Toutenburg and Roedel (1978, Chapter 4) and Toutenburg (1975c, 1976, 1977a, 1980a, b).

4.1.2 Inequality Restrictions—The Concentration Ellipsoid

In many practical problems one has prior information on the components of the parameter vector $\boldsymbol{\beta}$ of the regression model $\mathbf{y} = \boldsymbol{X}\boldsymbol{\beta} + \boldsymbol{\varepsilon}$. If that prior information allows a linearized formulation, methods of least squares under linear restrictions are available (see Chapters II, III, and IV). On the other hand, nonlinear restrictions on $\boldsymbol{\beta}$ may be of interest. We shall concentrate on prior information on $\boldsymbol{\beta}$ which has the following form.

Let $\boldsymbol{\beta}$ be prior constrained to lie in the convex set

$$B = \{\boldsymbol{\beta}\colon \boldsymbol{\beta}'\boldsymbol{T}\boldsymbol{\beta} \leq k\} \tag{4.1.1}$$

where $k \geq 0$ is a given constant term and $\boldsymbol{T}$ is a known $K \times K$-matrix, which in general shall be assumed to be positive definite. In other words, we assume that some known constraint exists on the length of $\boldsymbol{\beta}$, and this is expressed by the concentration ellipsoid in (1). Such prior information is almost always true, especially if k is chosen 'large enough'. The larger k, the less binding is the constraint in (1). If $k \to \infty$, $B \to E^K$ and the constraint vanishes. Geometrically, the constrained parameter set (1) is an ellipsoid centred at the origin. More generally, the following translated parameter set seems to be more realistic:

$$B = \{\boldsymbol{\beta}: (\boldsymbol{\beta} - \boldsymbol{\beta}_0)' \boldsymbol{T}(\boldsymbol{\beta} - \boldsymbol{\beta}_0) \leq k\} \tag{4.1.2}$$

where $\boldsymbol{\beta}_0$ is the known centre of the ellipsoid.

Let, for instance, a special type of prior information, namely the *component-by-component restriction*, on $\boldsymbol{\beta}$ be given:

$$a_i \leq \beta_i \leq b_i \quad i = 1, \ldots, K \tag{4.1.3}$$

which is of high relevance in many models of economics. The interval limits a_i and b_i are known. These inequalities build a cuboid in E^K which may be written, equivalently, as

$$\frac{|\beta_i - (a_i + b_i)/2|}{\frac{1}{2}(b_i - a_i)} \leq 1 \quad i = 1, \ldots, K. \tag{4.1.4}$$

Now, we shall construct an ellipsoid $(\boldsymbol{\beta} - \boldsymbol{\beta}_0)' \boldsymbol{T}(\boldsymbol{\beta} - \boldsymbol{\beta}_0) = 1$ which contains the cuboid (4), and which fulfils the following conditions:

(i) The ellipsoid has the same centre $\boldsymbol{\beta}_0 = \frac{1}{2}(a_1 + b_1, \ldots, a_K + b_K)$ as the cuboid.
(ii) The axes of the ellipsoid are parallel to the coordinate axes or, in other words, the matrix $\boldsymbol{T}$ of the ellipsoid is diagonal, i.e. $\boldsymbol{T} = \text{diag}(t_1, \ldots, t_K)$.
(iii) The surface of the ellipsoid contains all corner points of the cuboid.
(iv) The ellipsoid shall have minimal volume $V = \pi \prod_{i=1}^{K} t_i^{-1}$.

Condition (iii) means that

$$\sum_{i=1}^{K} \left(\frac{a_i - b_i}{2}\right)^2 t_i = 1 \tag{4.1.5}$$

We take (5) into consideration as a linear restriction by using a Lagrangian multiplier λ which gives the optimization problem

$$\min_{\{t_i\}} \left\{ \prod_{i=1}^{K} t_i^{-1} - \lambda \left[\sum_{i=1}^{K} \left(\frac{a_i - b_i}{2}\right)^2 t_i - 1 \right] \right\} = \min_{\{t_i\}} \tilde{V}. \tag{4.1.6}$$

The first derivatives of $\tilde{V}$ with respect to $t_j (j = 1, \ldots, K)$ and λ are

$$\frac{\partial \tilde{V}}{\partial t_j} = -t_j^{-2} \prod_{i \neq j} t_i - \lambda \left(\frac{a_j - b_j}{2}\right)^2 = 0$$

$$\frac{\partial \tilde{V}}{\partial \lambda} = \sum_{i=1}^{K} \left(\frac{a_i - b_i}{2}\right)^2 t_i - 1 = 0.$$

Therefore we have

$$\lambda = -t_j^{-2} \prod_{i \neq j} t_i \left(\frac{2}{a_j - b_j} \right)^2 \quad \text{for all } j = 1, \ldots, K$$

from which follows

$$t_i \left(\frac{a_i - b_i}{2} \right)^2 = t_j \left(\frac{a_j - b_j}{2} \right)^2$$

and, as a conclusion,

$$\sum_{i=1}^{K} \left(\frac{a_i - b_i}{2} \right)^2 t_i = K t_j \left(\frac{a_j - b_j}{2} \right)^2 = 1 \quad (j \text{ fixed, all } j = 1, \ldots, K).$$

This gives

$$t_j = \frac{4}{K} (a_j - b_j)^{-2} \quad (j = 1, \ldots, K).$$

So the required ellipsoid $(\boldsymbol{\beta} - \boldsymbol{\beta}_0)' \boldsymbol{T} (\boldsymbol{\beta} - \boldsymbol{\beta}_0) = 1$ has

$$\left\{ \begin{array}{l} \boldsymbol{\beta}_0' = \frac{1}{2}(a_1 + b_1, \ldots, a_K + b_K), \\ \boldsymbol{T} = \text{diag}\,(4/K)\,((b_1 - a_1)^{-2}, \ldots, (b_K - a_K)^{-2}). \end{array} \right\} \tag{4.1.7}$$

The original prior information '$\boldsymbol{\beta}$ is contained in the cuboid (3)' now is weakened to the restriction $\boldsymbol{\beta} \in B = \{\boldsymbol{\beta}: (\boldsymbol{\beta} - \boldsymbol{\beta}_0)' \boldsymbol{T} (\boldsymbol{\beta} - \boldsymbol{\beta}_0) \leq 1\}$ where $\boldsymbol{\beta}_0$ and $\boldsymbol{T}$ are determined by (7) (see Toutenburg and Roeder, 1978).

4.1.3 The Minimax Principle

We turn now to consider the method of minimax-linear estimation (MILE) in the familiar generalized linear regression model

$$\mathbf{y} = \boldsymbol{X\beta} + \boldsymbol{\varepsilon}, \quad \boldsymbol{\varepsilon} \sim (\mathbf{0}, \sigma^2 \boldsymbol{W}) \tag{4.1.8}$$

where $\boldsymbol{W}$ is assumed to be known. The quadratic risk function (see (2.1.27))

$$R(\hat{\boldsymbol{\beta}}, \boldsymbol{A}) = \text{tr}\, \boldsymbol{A} E(\hat{\boldsymbol{\beta}} - \boldsymbol{\beta})(\hat{\boldsymbol{\beta}} - \boldsymbol{\beta})' \tag{4.1.9}$$

is unbounded in E^K but now it is bounded on B for fixed k. Hence the application of the minimax principle to the risk considerations is possible if $\boldsymbol{\beta} \in B$ may be assumed to hold (here B is either the set (1) or the set (2)). Let a class $\mathscr{C}$ of estimators be given; then the *criterion of minimax-linear estimation* founded by Kuks and Olman (see Kuks and Olman 1971, 1972; Kuks, 1972) leads to the following definition.

Definition 4.1 An estimator $\mathbf{b}^* \in \mathscr{C}$ is said to be a minimax-linear estimator (MILE) of $\boldsymbol{\beta}$ if

$$\min_{\hat{\boldsymbol{\beta}}} \sup_{\boldsymbol{\beta} \in B} R(\hat{\boldsymbol{\beta}}, \boldsymbol{A}) = \sup_{\boldsymbol{\beta} \in B} R(\mathbf{b}^*, \boldsymbol{A}) \tag{4.1.10}$$

holds.

So the quality of an estimator to be MILE depends on the class $\mathscr{C}$ of possible estimators. We shall concentrate on the class of homogeneous linear estimators $\mathscr{C} = \{\hat{\boldsymbol{\beta}} = \boldsymbol{C}'\mathbf{y}\}$. Then we deduce

$$\hat{\boldsymbol{\beta}} - \boldsymbol{\beta} = (\boldsymbol{C}'\boldsymbol{X} - \boldsymbol{I})\boldsymbol{\beta} + \boldsymbol{C}'\boldsymbol{\varepsilon}$$

and

$$R(\hat{\boldsymbol{\beta}}, \boldsymbol{A}) = \sigma^2 \operatorname{tr} \boldsymbol{A}\boldsymbol{C}'\boldsymbol{W}\boldsymbol{C} + \boldsymbol{\beta}'\boldsymbol{T}^{1/2}\tilde{\boldsymbol{A}}\boldsymbol{T}^{1/2}\boldsymbol{\beta} \tag{4.1.11}$$

where

$$\tilde{\boldsymbol{A}} = \boldsymbol{T}^{-1/2}(\boldsymbol{C}'\boldsymbol{X} - \boldsymbol{I})'\boldsymbol{A}(\boldsymbol{C}'\boldsymbol{X} - \boldsymbol{I})\boldsymbol{T}^{-1/2} \tag{4.1.12}$$

is used as an abbreviation ($\tilde{\boldsymbol{A}}$ is a $K \times K$-matrix).

4.1.4 The Presentation of MILE for a General Loss Matrix

$\hat{\boldsymbol{\beta}} = \boldsymbol{C}'\mathbf{y}$ and $R(\hat{\boldsymbol{\beta}}, \boldsymbol{A})$ from (11) give for the restriction $\boldsymbol{\beta}'\boldsymbol{T}\boldsymbol{\beta} \leq k$ (1) the relation

$$\sup_{\boldsymbol{\beta}'\boldsymbol{T}\boldsymbol{\beta} \leq k} = R(\hat{\boldsymbol{\beta}}, \boldsymbol{A}) = \sigma^2 \operatorname{tr} \boldsymbol{A}\boldsymbol{C}'\boldsymbol{W}\boldsymbol{C} + k\,\lambda_{\max}(\tilde{\boldsymbol{A}}) \tag{4.1.13}$$

where $\lambda_{\max}(\tilde{\boldsymbol{A}})$ is the largest eigenvalue of $\tilde{\boldsymbol{A}}$. This is proved by the extremal property of the largest eigenvalue, namely

$$\max_{\boldsymbol{\beta}} \frac{\boldsymbol{\beta}'\boldsymbol{T}^{1/2}\tilde{\boldsymbol{A}}\boldsymbol{T}^{1/2}\boldsymbol{\beta}}{\boldsymbol{\beta}'\boldsymbol{T}\boldsymbol{\beta}} = \lambda_{\max}(\tilde{\boldsymbol{A}}) \tag{4.1.14}$$

(see Theorem A.13; for an alternative proof of (13) the reader is referred to Toutenburg, 1976).

The MILE is obtained by finding the $\boldsymbol{C}$ which minimizes (13). However, since $\lambda_{\max}(\boldsymbol{A})$ depends on $\boldsymbol{C}$, this problem is complicated for rank $\boldsymbol{A} > 1$. Kuks and Olman (1972) have developed an iterative minimization procedure to indicate a point of maximum of the risk function. Läuter (1975) gave an explicit presentation of a MILE in the case of regular $\boldsymbol{A}$. The solution is based on the relations between minimax decision functions and Bayesian decision functions for the most unfavourable prior distributions (see also Humak, 1977, p. 137).

Let

$$\tilde{\boldsymbol{S}} = \sigma^{-2}\boldsymbol{T}^{-1/2}\boldsymbol{S}\boldsymbol{T}^{-1/2} \tag{4.1.15}$$

(where $\boldsymbol{S} = \boldsymbol{X}'\boldsymbol{W}^{-1}\boldsymbol{X}$) and

$$\boldsymbol{F} = \tilde{\boldsymbol{S}}^{-1}\boldsymbol{T}^{-1/2}\boldsymbol{A}\boldsymbol{T}^{-1/2}\tilde{\boldsymbol{S}}^{-1} \tag{4.1.16}$$

and let $\boldsymbol{A}$ be at least nonnegative definite. Then the following theorem holds.

Theorem 4.1 (Läuter, 1975)

Let there exist a $K \times K$ matrix $\boldsymbol{V} \geq \boldsymbol{0}$ and a scalar $v > 0$ such that

(i) $\dfrac{1}{\sqrt{v}}(\boldsymbol{F} + \boldsymbol{V})^{1/2} \geq \tilde{\boldsymbol{S}}^{-1}$,

(ii) $\sqrt{v}(F+V)^{-1/2}V = \tilde{S}V$

(iii) $\frac{1}{\sqrt{v}} \operatorname{tr}[(F+V)^{1/2}] = 1 + \operatorname{tr}\tilde{S}^{-1}$.

Using V and v taken from (i) to (iii), we define

$$D_v = \frac{1}{\sqrt{v}}(F+V)^{1/2} - \tilde{S}^{-1}. \tag{4.1.17}$$

Then $\mathbf{b}^* = C^{*\prime}\mathbf{y}$ is a MILE of $\boldsymbol{\beta}$ in the model $\mathbf{y} = X\boldsymbol{\beta} + \boldsymbol{\varepsilon}$, $\boldsymbol{\varepsilon} \sim (\mathbf{0}, \sigma^2 W)$ under the restriction $\boldsymbol{\beta}' T \boldsymbol{\beta} \leq 1$, where

$$C^{*\prime} = T^{-1/2} D_v T^{-1/2} X'(\sigma^2 W + X T^{-1/2} D_v T^{-1/2} X')^{-1}. \tag{4.1.18}$$

For the complicated proof of this theorem the reader is referred to the original paper of Läuter (1975).

The calculation of the matrix C^* (18) is difficult in the general case. Moreover, we see from D_v that C^* in general depends on the matrix A of the risk function.

In the following special case Läuter (1975) gave an explicit formula for a MILE. Let $T = k^{-1} I$ (i.e. $\boldsymbol{\beta}'\boldsymbol{\beta} \leq k$) and $A = I$ (i.e. the risk is of the scalar MSE-type, see (2.1.22)). Then a MILE of $\boldsymbol{\beta}$ is given by

$$\mathbf{b}^* = \frac{k}{k + \operatorname{tr} S^{-1}} \mathbf{b} \tag{4.1.19}$$

where $\mathbf{b} = S^{-1} X' W^{-1} \mathbf{y}$ is the GLSE.

4.2 THE MILE FOR A LOSS MATRIX OF RANK ONE

4.2.1 The Presentation of MILE

We shall confine ourselves to the assumption rank $A = 1$.

Without loss of generality we may write $A = \mathbf{a}\mathbf{a}'$ where $\mathbf{a}$ is a nonzero $K \times 1$-vector. Then it follows from (4.1.13) and (4.1.14) that

$$\sup_{\boldsymbol{\beta}' T \boldsymbol{\beta} \leq k} R(\hat{\boldsymbol{\beta}}, \mathbf{a}\mathbf{a}') = \sigma^2 \mathbf{a}' C' W C \mathbf{a} + k \lambda_{\max}(\tilde{A}) \tag{4.2.1}$$

where now $\tilde{A}$ (4.1.12) has the simplified form

$$\tilde{A} = [T^{-1/2}(C'X - I)'\mathbf{a}][T^{-1/2}(C'X - I)'\mathbf{a}]' = \tilde{\mathbf{a}}\tilde{\mathbf{a}}' \tag{4.2.2}$$

with rank $\tilde{A} = 1$, $\lambda_{\max}(\tilde{A}) = \tilde{\mathbf{a}}'\tilde{\mathbf{a}}$. Then we have

$$\sup_{\boldsymbol{\beta}' T \boldsymbol{\beta} \leq k} R(\hat{\boldsymbol{\beta}}, \mathbf{a}\mathbf{a}') = \sigma^2 \mathbf{a}' C' W C \mathbf{a} + k \mathbf{a}'(C'X - I) T^{-1} (C'X - I)' \mathbf{a}. \tag{4.2.3}$$

The minimization of (3) with respect to C gives (see Theorem A.52)

$$\frac{1}{2} \frac{\mathrm{d}}{\mathrm{d}C} \left[\sup_{\boldsymbol{\beta}' T \boldsymbol{\beta} \leq k} R(\hat{\boldsymbol{\beta}}, \mathbf{a}\mathbf{a}') \right] = (\sigma^2 W + k X T^{-1} X') C \mathbf{a}\mathbf{a}' - k X T^{-1} \mathbf{a}\mathbf{a}'.$$

This is zero when

$$C^* = k(\sigma^2 W + kXT^{-1}X')^{-1}XT^{-1}. \tag{4.2.4}$$

Transposing C^* (4) and multiplying from the left by the matrix $(\sigma^2 T + kS)$ with $S = X'W^{-1}X$ gives

$$\begin{aligned}(\sigma^2 T + kS)C^{*\prime} &= k\sigma^2 X'W^{-1}W(\sigma^2 W + kXT^{-1}X')^{-1} \\ &\quad + kX'W^{-1}(kXT^{-1}X')(\sigma^2 W + kXT^{-1}X')^{-1} \\ &= kX'W^{-1}\end{aligned}$$

from which we get

$$C^{*\prime} = (k^{-1}\sigma^2 T + S)^{-1}X'W^{-1} \tag{4.2.5}$$

or, with the useful abbreviation

$$D = (k^{-1}\sigma^2 T + S) \tag{4.2.6}$$

we have

$$C^{*\prime} = D^{-1}X'W^{-1}.$$

This yields a biased estimator whose properties are described in the following theorem.

Theorem 4.2

In the model $\mathbf{y} = X\boldsymbol{\beta} + \boldsymbol{\varepsilon}$, $\boldsymbol{\varepsilon} \sim (\mathbf{0}, \sigma^2 W)$, with constraint $\boldsymbol{\beta}' T\boldsymbol{\beta} \le k$, the minimax linear estimator with $A = \mathbf{aa}'$ as loss matrix in (4.1.9) is

$$\begin{aligned}\mathbf{b}^* = C^{*\prime}\mathbf{y} &= (k^{-1}\sigma^2 T + S)^{-1}X'W^{-1}\mathbf{y} \\ &= D^{-1}X'W^{-1}\mathbf{y}, \text{ say.}\end{aligned} \tag{4.2.7}$$

This estimator has

$$\text{bias}\, \mathbf{b}^* = -k^{-1}\sigma^2 D^{-1}T\boldsymbol{\beta}, \tag{4.2.8}$$

dispersion matrix

$$V(\mathbf{b}^*) = \sigma^2 D^{-1}SD^{-1} \tag{4.2.9}$$

and minimax risk

$$\sup_{\boldsymbol{\beta}' T\boldsymbol{\beta} \le k} R(\mathbf{b}^*, \mathbf{aa}') = \sigma^2 \mathbf{a}' D^{-1}\mathbf{a}. \tag{4.2.10}$$

If the ellipsoid is not centred at the origin but at $\boldsymbol{\beta}_0 \ne \mathbf{0}$, i.e. if the prior restriction (4.1.2) is given, then the following theorem holds.

Theorem 4.3

In the model and for the loss matrix as in Theorem 4.2, with constraint $(\boldsymbol{\beta} - \boldsymbol{\beta}_0)' T(\boldsymbol{\beta} - \boldsymbol{\beta}_0) \le k$ the MILE of $\boldsymbol{\beta}$ is

$$\mathbf{b}^*(\boldsymbol{\beta}_0) = \boldsymbol{\beta}_0 + D^{-1}X'W^{-1}(\mathbf{y} - X\boldsymbol{\beta}_0). \tag{4.2.11}$$

This estimator has

$$\text{bias } \mathbf{b}^*(\boldsymbol{\beta}_0) = -k^{-1}\sigma^2 \boldsymbol{D}^{-1}\boldsymbol{T}(\boldsymbol{\beta}-\boldsymbol{\beta}_0) \tag{4.2.12}$$

$$V[\mathbf{b}^*(\boldsymbol{\beta}_0)] = V(\mathbf{b}^*) \qquad \text{(as given in (9))} \tag{4.2.13}$$

and

$$\sup_{(\boldsymbol{\beta}-\boldsymbol{\beta}_0)'\boldsymbol{T}(\boldsymbol{\beta}-\boldsymbol{\beta}_0)\leq k} R\,[\mathbf{b}^*(\boldsymbol{\beta}_0), \mathbf{aa}'] = \sigma^2\mathbf{a}'\boldsymbol{D}^{-1}\mathbf{a}. \tag{4.2.14}$$

The proof of this theorem is similar to that of Theorem 4.2, where the transformation $\boldsymbol{\beta}-\boldsymbol{\beta}_0 = \tilde{\boldsymbol{\beta}}$ may be used. It follows that the change of the centre of the concentration ellipsoid has influence on the estimator and its bias, whereas the dispersion and the minimax risk of the estimator are not influenced by $\boldsymbol{\beta}_0$.

For convenience we evaluate the *unconstrained* R-risk of the MILE $\mathbf{b}^*$:

$$\begin{aligned} R(\mathbf{b}^*, \mathbf{aa}') &= \operatorname{tr} \mathbf{aa}'\{V(\mathbf{b}^*) + (\text{bias } \mathbf{b}^*)(\text{bias } \mathbf{b}^*)'\} \\ &= \sigma^2\mathbf{a}'\boldsymbol{D}^{-1}(\boldsymbol{S}+k^{-2}\sigma^2\,\boldsymbol{T}\boldsymbol{\beta}\boldsymbol{\beta}'\boldsymbol{T})\boldsymbol{D}^{-1}\mathbf{a}. \end{aligned} \tag{4.2.15}$$

4.2.2 Connection of MILE with GLSE

From the unbiasedness condition $\boldsymbol{C}'\boldsymbol{X}-\boldsymbol{I}=\boldsymbol{0}$ and (4.1.12) it follows that

$$\sup_{\boldsymbol{\beta}'\boldsymbol{T}\boldsymbol{\beta}\leq k} R(\boldsymbol{C}'\mathbf{y}, \boldsymbol{A}) = R(\boldsymbol{C}'\mathbf{y}, \boldsymbol{A}) \quad \text{if } \boldsymbol{C}'\boldsymbol{X}-\boldsymbol{I}=\boldsymbol{0}. \tag{4.2.16}$$

Hence we have the following result.

Theorem 4.4

In the case of unbiased estimation the MILE and the R-optimal estimator coincide.

In other words, *the GLSE* **b** *may be interpreted as a minimax estimator*. This result may be seen from another point of view, as follows.

If there is no restriction, or in other words, if $B = E^K$ or equivalently $k = \infty$ holds in (1), we have $\lim_{k\to\infty} \boldsymbol{D} = \boldsymbol{S}$ and therefore

$$\lim_{k\to\infty} \mathbf{b}^* = \mathbf{b} = \boldsymbol{S}^{-1}\boldsymbol{X}'\boldsymbol{W}^{-1}\mathbf{y}. \tag{4.2.17}$$

Thus, the GLSE may be understood as an unrestricted MILE.

Now, let us investigate the gain in efficiency in estimating $\boldsymbol{\beta}$ by the MILE $\mathbf{b}^*$ which uses the prior information $\boldsymbol{\beta}\in B$. The unrestricted estimator is the GLSE **b** which has, by Theorem 4.4:

$$\sup_{\boldsymbol{\beta}'\boldsymbol{T}\boldsymbol{\beta}\leq k} R(\mathbf{b}, \mathbf{aa}') = R(\mathbf{b}, \mathbf{aa}') = \sigma^2\mathbf{a}'\boldsymbol{S}^{-1}\mathbf{a}. \tag{4.2.18}$$

So the MILE $\mathbf{b}^*$ (7) is better than $\mathbf{b}$ with respect to the minimax risk:

$$R(\mathbf{b}, \mathbf{aa}') - \sup_{\beta' T \beta \leq k} R(\mathbf{b}^*, \mathbf{aa}') = \sigma^2 \mathbf{a}' \{S^{-1} - (k^{-1}\sigma^2 T + S)^{-1}\} \mathbf{a} \geq 0 \quad (4.2.19)$$

(because the matrix in brackets { } is nonnegative definite).

If we now compare the minimax estimator and the GLSE $\mathbf{b}$ with respect to the unrestricted R-risk we get (using (15)):

$$\begin{aligned} R(\mathbf{b}) - R(\mathbf{b}^*) &= \sigma^2 \mathbf{a}'[S^{-1} - D^{-1}(S + k^{-2}\sigma^2 T\beta\beta' T)D^{-1}]\mathbf{a} \quad (4.2.20) \\ &= \sigma^2 \mathbf{a}' D^{-1}[DS^{-1}D - S - k^{-2}\sigma^2 T\beta\beta' T]D^{-1}\mathbf{a} \\ &= \sigma^2 \mathbf{a}' D^{-1} B D^{-1} \mathbf{a}, \text{ say.} \end{aligned}$$

The MILE $\mathbf{b}^*$ is therefore R-better than $\mathbf{b}$ if and only if the matrix $B \geq 0$. We write it as

$$\begin{aligned} B &= k^{-2}\sigma^4 T[\{S^{-1} + 2k\sigma^{-2}T^{-1}\} - \sigma^{-2}\beta\beta']T \\ &= k^{-2}\sigma^4 T C^{1/2}[I - \sigma^{-2}C^{-1/2}\beta\beta' C^{-1/2}]C^{1/2}T \end{aligned}$$

where $C = S^{-1} + 2k\sigma^{-2}T^{-1}$.

Now $B \geq 0$ if and only if the matrix in brackets $[\ ] \geq 0$. Applying Theorem A.17 we have the result that the MILE $\mathbf{b}^*$ is better than the GLSE $\mathbf{b}$ with respect to the unrestricted risk $R(\hat{\beta}, \mathbf{aa}')$ if and only if

$$\sigma^{-2}\beta'(S^{-1} + 2k\sigma^{-2}T^{-1})^{-1}\beta \leq 1 \quad (4.2.21)$$

which may also be written as

$$\sigma^{-2}\beta' T^{1/2}(T^{1/2}S^{-1}T^{1/2} + 2k\sigma^{-2}I)^{-1}T^{1/2}\beta \leq 1. \quad (4.2.22)$$

Note As a sufficient condition for $B \geq 0$ we have

$$2T^{-1} - k^{-1}\beta\beta' \geq 0$$

which gives for $T = I$, by Theorem A.17, that

$$k^{-1} \leq \frac{2}{\beta'\beta}$$

is sufficient to make $\mathbf{b}^*$ R-better than $\mathbf{b}$. As may be seen from (2.3.36), relation (22) is similar to the improvement region of the ridge estimator.

4.2.3 Comparison with the R-optimal Biased Estimator

First, we look for the homogeneous biased R-optimal estimator $\hat{\beta}_2$ (2.4.6) which may be written in the following equivalent form (see (2.4.8)):

$$\hat{\beta}_2 = (\sigma^2 + \beta' S \beta)^{-1}\beta\beta' X' W^{-1}\mathbf{y}$$

and which has risk (using (2.4.12))

$$R(\hat{\beta}_2, \mathbf{aa}') = \sigma^2(\sigma^2 + \beta' S\beta)^{-1}\mathbf{a}'\beta\beta'\mathbf{a}. \quad (4.2.23)$$

Comparing the two biased estimators $\hat{\beta}_2$ and $\mathbf{b}^*$ (7) with respect to the R-risk leads to the result

$$R(\mathbf{b}^*) - R(\hat{\beta}_2) \geq 0 \tag{4.2.24}$$

as $\hat{\beta}_2$ is R-optimal in the class of homogeneous biased linear estimators (see Section 2.4.2, and Bibby and Toutenburg, 1978, p. 152, for the explicit difference (24)).

As a point of more interest, we now shall compare the two estimators with respect to the minimax risk.

Let us define within this chapter the difference of minimax risks of two estimators $\tilde{\beta}_1, \tilde{\beta}_2$ by

$$\Delta^k(\tilde{\beta}_1, \tilde{\beta}_2) = \sup_{\beta' T\beta \leq k} R(\tilde{\beta}_1, \mathbf{aa}') - \sup_{\beta' T\beta \leq k} R(\tilde{\beta}_2, \mathbf{aa}'). \tag{4.2.25}$$

If $\Delta^k(\tilde{\beta}_1, \tilde{\beta}_2) \geq 0$ then $\tilde{\beta}_2$ is better than $\tilde{\beta}_1$ with respect to the minimax risk. Using this abbreviation, we have from Theorem 4.2 that

$$\Delta^k(\hat{\beta}_2, \mathbf{b}^*) \geq 0$$

must hold. A more detailed investigation leads to the following result. Using (23) we may evaluate

$$\sup_{\beta' T\beta \leq k} \beta'\mathbf{aa}'\beta = k\mathbf{a}' T^{-1}\mathbf{a},$$

$$\sup_{\beta' T\beta \leq k} \beta' S\beta = k\lambda_{\max}(T^{-1/2} S T^{-1/2}),$$

and

$$\sup_{\beta' T\beta \leq k} R(\hat{\beta}_2, \mathbf{aa}') \geq \sigma^2 \sup_{\beta' T\beta \leq k} \beta'\mathbf{aa}'\beta \left[\sup_{\beta' T\beta \leq k} (\sigma^2 + \beta' S\beta)\right]^{-1}$$

$$= \frac{\sigma^2 k\mathbf{a}' T^{-1}\mathbf{a}}{\sigma^2 + k\lambda_{\max}(T^{-1/2} S T^{-1/2})} = k^*\mathbf{a}' T^{-1}\mathbf{a} \tag{4.2.26}$$

with $k^{*-1} = k^{-1} + \sigma^{-2}\lambda_{\max}(T^{-1/2} S T^{-1/2})$.

Therefore $k^*\mathbf{a}' T^{-1}\mathbf{a}$ is a lower bound for the minimax risk $\sup_{\beta' T\beta \leq k} R(\hat{\beta}_2, \mathbf{aa}')$. To ensure that it is also an upper bound for the minimax risk of $\mathbf{b}^*$ the following relation must hold (see (6) and (10)):

$$k^*\mathbf{a}' T^{-1}\mathbf{a} \geq \mathbf{a}'(k^{-1} T + \sigma^{-2} S)^{-1}\mathbf{a}.$$

This holds, whenever

$$(k^{-1} - k^{*-1}) T + \sigma^{-2} S = \sigma^{-2} S - T\lambda_{\max}(T^{-1/2} S T^{-1/2}) \tag{4.2.27}$$

is a nonnegative definite matrix. This is true, for example, when $T = S$ and $\sigma^2 < 1$.

4.2.4 MILE and Ridge Estimator

Now, we shall investigate the connection between MILE and the well-known class of biased estimators, namely the ridge estimators (see Section 2.3.3).

The ridge estimator in the generalized regression model is defined as

$$\mathbf{b}(c) = [c\boldsymbol{I} + \boldsymbol{S}]^{-1}\boldsymbol{X}'\boldsymbol{W}^{-1}\mathbf{y} = \boldsymbol{D}_I^{-1}\boldsymbol{X}'\boldsymbol{W}^{-1}\mathbf{y}, \text{ say} \qquad (4.2.28)$$

where $c \geq 0$. This estimator is biased:

$$\text{bias } \mathbf{b}(c) = -c\boldsymbol{D}_I^{-1}\boldsymbol{\beta}$$

and has

$$V(\mathbf{b}(c)) = \sigma^2 \boldsymbol{D}_I^{-1}\boldsymbol{S}\boldsymbol{D}_I^{-1}$$

and

$$R[\mathbf{b}(c), \mathbf{aa}'] = \sigma^2 \mathbf{a}'\boldsymbol{D}_I^{-1}\boldsymbol{S}\boldsymbol{D}_I^{-1}\mathbf{a} + [c\mathbf{a}'\boldsymbol{D}_I^{-1}\boldsymbol{\beta}]^2.$$

The minimax risk of $\mathbf{b}(c)$ is

$$\sup_{\boldsymbol{\beta}'\boldsymbol{T}\boldsymbol{\beta} \leq k} R[\mathbf{b}(c), \mathbf{aa}'] = \mathbf{a}'\boldsymbol{D}_I^{-1}(\sigma^2\boldsymbol{S} + c^2 k\boldsymbol{T}^{-1})\boldsymbol{D}_I^{-1}\mathbf{a}. \qquad (4.2.29)$$

Comparison of ridge and minimax estimators gives, by Theorem 4.2,

$$\Delta^k(\mathbf{b}(c), \mathbf{b}^*) = \mathbf{a}'\boldsymbol{D}_I^{-1}(\sigma^2\boldsymbol{S} + c^2 k\boldsymbol{T}^{-1} - \boldsymbol{D}_I\boldsymbol{D}^{-1}\boldsymbol{D}_I)\boldsymbol{D}_I^{-1}\mathbf{a} \geq 0.$$

As may be seen from the formula (28), the ridge estimator may be interpreted as a minimax estimator with respect to the restriction $\boldsymbol{\beta}'\boldsymbol{\beta} \leq c^{-1}\sigma^2$. That is, *minimax linear estimation is a more general method than ridge regression.*

4.3 MINIMAX ESTIMATION UNDER INCORRECT PRIOR INFORMATION

4.3.1 Incorrectness Caused by Rotation and/or Changing the Length of the Axes

The aim of minimax estimation, as well as the other relevant restrictive estimators (see Chapters II and III), is to overcome the restrained usefulness of the GLSE by using prior information as in (4.1.1) or (4.1.2). The goodness of MILE depends on the assumption that the prior information is correct. But it is clear that a misidentification of prior restrictions may happen. The influence of incorrect chosen prior regions therefore has to be investigated.

The situation may be as follows. The regression model $\mathbf{y} = \boldsymbol{X}\boldsymbol{\beta} + \boldsymbol{\varepsilon}$ is such that $\boldsymbol{\beta}$ is contained in the ellipsoid

$$\boldsymbol{\beta}'\boldsymbol{T}\boldsymbol{\beta} \leq k \quad \text{(correct prior information)}. \qquad (4.3.1)$$

Then, by Theorem 4.2, $\mathbf{b}^*$ (4.2.7) is the correct MILE and $\mathbf{b}^*$ is better than the unrestricted GLSE $\mathbf{b}$ with respect to the minimax risk (see (4.2.7) and (4.2.28)):

$$\Delta^k(\mathbf{b}, \mathbf{b}^*) = \sigma^2\mathbf{a}'\{\boldsymbol{S}^{-1} - \boldsymbol{D}^{-1}\}\mathbf{a} \geq 0. \qquad (4.3.2)$$

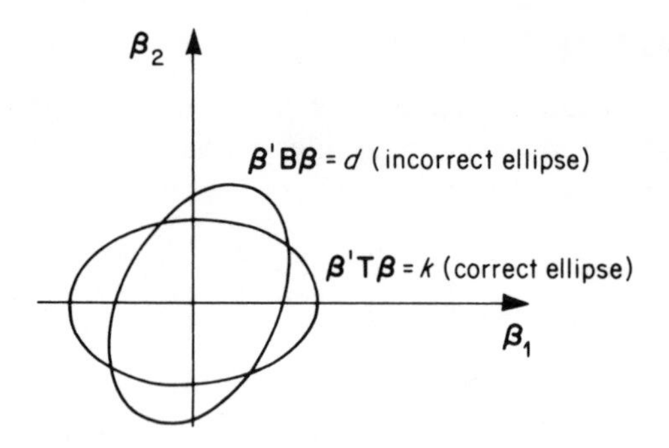

Figure 4.3.1 Incorrectly specified prior region, case 1. Rotation and changed length of the axes of the correct ellipse

Now, in practice, the model builder may choose an incorrect ellipsoid

$$\boldsymbol{\beta}' \boldsymbol{B} \boldsymbol{\beta} \leq d \quad \text{(incorrect prior information)} \tag{4.3.3}$$

A typical situation is given in Figure 4.3.1. What happens is that the estimator

$$\begin{aligned} \mathbf{b}_B &= (d^{-1}\sigma^2 \boldsymbol{B} + \boldsymbol{S})^{-1} \boldsymbol{X}' \boldsymbol{W}^{-1} \mathbf{y} \\ &= \boldsymbol{D}_B^{-1} \boldsymbol{X}' \boldsymbol{W}^{-1} \mathbf{y}, \text{ say,} \end{aligned} \tag{4.3.4}$$

has been used. This estimator is based on the incorrect ellipsoid (3) and has

$$\text{bias } \mathbf{b}_B = -d^{-1}\sigma^2 \boldsymbol{D}_B^{-1} \boldsymbol{B}\boldsymbol{\beta}, \tag{4.3.5}$$

$$V(\mathbf{b}_B) = \sigma^2 \boldsymbol{D}_B^{-1} \boldsymbol{S} \boldsymbol{D}_B^{-1}. \tag{4.3.6}$$

As the ellipsoid (1) was assumed to be the correct one, the risk of $\mathbf{b}_B$ has to be computed with respect to this correct prior information:

$$\sup_{\boldsymbol{\beta}' T \boldsymbol{\beta} \leq k} R(\mathbf{b}_B, \mathbf{a}\mathbf{a}') = \sigma^2 \mathbf{a}' \boldsymbol{D}_B^{-1} [\boldsymbol{S} + \sigma^2 k d^{-2} \boldsymbol{B} \boldsymbol{T}^{-1} \boldsymbol{B}] \boldsymbol{D}_B^{-1} \mathbf{a}. \tag{4.3.7}$$

By the optimality property of the MILE $\mathbf{b}^*$ we have immediately

$$\Delta^k(\mathbf{b}_B, \mathbf{b}^*) = \sigma^2 \mathbf{a}' \boldsymbol{D}_B^{-1} [\boldsymbol{S} + \sigma^2 k d^{-2} \boldsymbol{B} \boldsymbol{T}^{-1} \boldsymbol{B} - \boldsymbol{D}_B \boldsymbol{D}^{-1} \boldsymbol{D}_B] \boldsymbol{D}_B^{-1} \mathbf{a} \geq 0. \tag{4.3.8}$$

Let us give this loss for two special cases of incorrect prior information.

(i) Let $k = d = 1$ and $\boldsymbol{B} = c\boldsymbol{T}$ with $c > 0$. In other words, the incorrectness is caused by changing the length of each axis of the ellipsoid by the same factor c. Then (8) becomes

$$\Delta^1(\mathbf{b}_{cT}, \mathbf{b}^*) = \sigma^4 (c-1)^2 \mathbf{a}' \boldsymbol{D}_{cT}^{-1} \boldsymbol{T} [\boldsymbol{T}^{-1} - \sigma^2 \boldsymbol{D}^{-1}] \boldsymbol{T} \boldsymbol{D}_{cT}^{-1} \mathbf{a} \geq 0. \tag{4.3.9}$$

The matrix in brackets [ ] is nonnegative definite (see Theorem A.12).

$$\Delta^1(\mathbf{b}_{cT}, \mathbf{b}^*) = 0 \text{ holds if } c = 1.$$

(ii) Let again $k = d = 1$ and $\boldsymbol{B} = \boldsymbol{CTC}$ with $\boldsymbol{C} = \text{diag}(\sqrt{c_1}, \ldots, \sqrt{c_K})$ with

$c_i > 0$. This means that incorrectness is caused by changing the length of each axis by a corresponding factor c_i. Then (8) becomes

$$\Delta^1(\mathbf{b}_{CTC}, \mathbf{b}^*) = \sigma^4 \mathbf{a}' D_{CTC}^{-1}(CTC - T)[T^{-1} - \sigma^2 D^{-1}](CTC - T) D_{CTC}^{-1} \mathbf{a} \geq 0. \tag{4.3.10}$$

So we have to conclude that incorrect specified prior information yields an estimator $\mathbf{b}_B$ which is less efficient than $\mathbf{b}^*$. The question arises whether this estimator $\mathbf{b}_B$ is better than the GLSE $\mathbf{b}$. This is just the problem of robustness of the minimax estimation method. We have to investigate whether

$$\Delta^k(\mathbf{b}, \mathbf{b}_B) = \sigma^2 \mathbf{a}' D_B^{-1}[D_B S^{-1} D_B - S - \sigma^2 k d^{-2} B T^{-1} B] D_B^{-1} \mathbf{a} \tag{4.3.11}$$

is nonnegative.

$\Delta^k(\mathbf{b}, \mathbf{b}_B) \geq 0$ holds if the matrix $[\ ] \geq \mathbf{0}$. This gives

$$\begin{aligned} [\ ] &= (d^{-1}\sigma^2 B + S) S^{-1} (d^{-1}\sigma^2 B + S) - S - \sigma^2 k d^{-2} B T^{-1} B \\ &= \sigma^2 d^{-1} B \{\sigma^2 d^{-1} S^{-1} - k d^{-1} T^{-1} + 2B^{-1}\} B. \end{aligned} \tag{4.3.12}$$

So $[\ ] \geq \mathbf{0}$ if the matrix $\{\ \} \geq \mathbf{0}$. We now investigate this for some special cases.

(i) If $\sigma^2 S^{-1} - k T^{-1} \geq \mathbf{0}$ then (12) holds for all B. Let, for instance, $T = S$, then $\sigma^2 \geq k$ is a sufficient condition for $\Delta^k(\mathbf{b}, \mathbf{b}_B) \geq 0$.

(ii) The matrix $\{\ \}$ of (12) may be written as

$$(\sigma^2 d^{-1} S^{-1} + B^{-1}) + (B^{-1} - k d^{-1} T^{-1}).$$

The first matrix is $\geq \mathbf{0}$. $[\ ] \geq \mathbf{0}$ is therefore fulfilled if the matrix difference $d B^{-1} - k T^{-1} \geq \mathbf{0}$ or, equivalently, if

$$k^{-1} T - d^{-1} B \geq \mathbf{0}. \tag{4.3.13}$$

To give a geometrical interpretation of the sufficient condition (13), we assume without loss of generality that both prior regions $\boldsymbol{\beta}' T \boldsymbol{\beta} \leq k$ (correct ellipsoid) and $\boldsymbol{\beta}' B \boldsymbol{\beta} \leq d$ (incorrect ellipsoid) are spheres. Thus we have $T = B = I$ and $k = r_1^2$, $d = r_2^2$ with r_1 and r_2 the radii of the spheres. Then (13) implies $r_1 \leq r_2$, or, in other words, the incorrect prior region contains the correct prior region. This gives the following theorem.

Theorem 4.5

Let the correct prior region $\boldsymbol{\beta}' T \boldsymbol{\beta} \leq k$ be fully contained in the incorrect prior region $\boldsymbol{\beta}' B \boldsymbol{\beta} \leq d$. Then the estimator $\mathbf{b}_B$ (4) is better than the GLSE $\mathbf{b}$ in the sense that $\Delta^k(\mathbf{b}, \mathbf{b}_B) \geq 0$.

(iii) Let the incorrectness be caused by a wrong 'radius' of the prior region, only. That is, each axis of the ellipsoid is transformed by the same factor. This leads to the assumption $B = T$. Now, $\Delta^k(\mathbf{b}, \mathbf{b}_B) \geq 0$ iff the matrix (12) is nonnegative definite. Using $B = T$, this is true iff

$$\sigma^2 d^{-1} T^{1/2} \{\sigma^2 d^{-1} T^{1/2} S^{-1} T^{1/2} + (2 - k d^{-1}) I\} T^{1/2} \geq \mathbf{0} \tag{4.3.14}$$

or, iff the matrix in brackets $\{\ \} \geq \boldsymbol{0}$, or, equivalently, if its eigenvalues are nonnegative. Now, if $0 \leq \lambda_1 \leq \ldots \leq \lambda_K$ denote the eigenvalues of $\boldsymbol{T}^{1/2}\boldsymbol{S}^{-1}\boldsymbol{T}^{1/2}$, the eigenvalues of the matrix $\{\ \}$ in (14) are evaluated to be

$$\sigma^2 d^{-1}\lambda_i + (2 - kd^{-1}) \quad (i = 1, \ldots, K)$$

which have to be nonnegative. This gives

$$d \geq \frac{k - \sigma^2 \lambda_i}{2} \quad (i = 1, \ldots, K)$$

so that

$$d \geq \frac{k - \sigma^2 \lambda_{\min}(\boldsymbol{T}^{1/2}\boldsymbol{S}^{-1}\boldsymbol{T}^{1/2})}{2} \tag{4.3.15}$$

is necessary and sufficient to ensure $\Delta^k(\mathbf{b}, \mathbf{b}_{\boldsymbol{B}}) \geq 0$ in the case $\boldsymbol{B} = \boldsymbol{T}$.

Note In accordance with Theorem 4.5, the condition $d \geq k$ is sufficient to fulfil (15).

4.3.2 Incorrectness Caused by Translation

The situation is more complicated—and also more realistic—if we have also a translation of the centre of the ellipsoid (see Figure 4.3.2). In other words, the correct prior information is $\boldsymbol{\beta}'\boldsymbol{T}\boldsymbol{\beta} \leq k$ as before, but the actually used prior information may be described by the ellipsoid

$$(\boldsymbol{\beta} - \boldsymbol{\beta}_0)'\boldsymbol{B}(\boldsymbol{\beta} - \boldsymbol{\beta}_0) \leq d \text{ (incorrect prior information).} \tag{4.3.16}$$

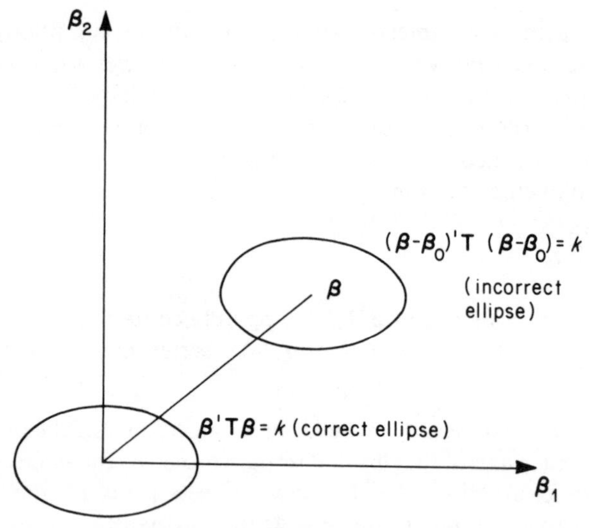

Figure 4.3.2 Incorrectly specified prior region, case 2. Centre translated

This means that we have used the estimator (see (4.2.11))

$$\mathbf{b}_B(\boldsymbol{\beta}_0) = \boldsymbol{\beta}_0 + \boldsymbol{D}_B^{-1} \boldsymbol{X}' \boldsymbol{W}^{-1}(\mathbf{y} - \boldsymbol{X}\boldsymbol{\beta}_0) \tag{4.3.17}$$

which has

$$\text{bias } \mathbf{b}_B(\boldsymbol{\beta}_0) = -\sigma^2 d^{-1} \boldsymbol{D}_B^{-1} \boldsymbol{B}(\boldsymbol{\beta} - \boldsymbol{\beta}_0) \tag{4.3.18}$$

and dispersion matrix

$$V[\mathbf{b}_B(\boldsymbol{\beta}_0)] = \sigma^2 \boldsymbol{D}_B^{-1} \boldsymbol{S} \boldsymbol{D}_B^{-1}. \tag{4.3.19}$$

Now we calculate the minimax risk of this estimator where the supremum has to be taken with respect to the correct prior region $\boldsymbol{\beta}' \boldsymbol{T} \boldsymbol{\beta} \leq k$. This gives

$$\begin{aligned}
&\sup_{\boldsymbol{\beta}' \boldsymbol{T}\boldsymbol{\beta} \leq k} [\mathbf{a}' \text{ bias } \mathbf{b}_B(\boldsymbol{\beta}_0)]^2 \\
&\quad = \sup_{\boldsymbol{\beta}' \boldsymbol{T}\boldsymbol{\beta} \leq k} [\mathbf{a}' \sigma^2 d^{-1} \boldsymbol{D}_B^{-1} \boldsymbol{B} \boldsymbol{T}^{-1/2} \boldsymbol{T}^{1/2} (\boldsymbol{\beta} - \boldsymbol{\beta}_0)]^2 \\
&\quad = \sup_{\boldsymbol{\beta}' \boldsymbol{T}\boldsymbol{\beta} \leq k} [\tilde{\mathbf{a}}' (\tilde{\boldsymbol{\beta}} - \tilde{\boldsymbol{\beta}}_0)]^2
\end{aligned}$$

where

$$\begin{aligned}
\tilde{\mathbf{a}}' &= \mathbf{a}' \sigma^2 d^{-1} \boldsymbol{D}_B^{-1} \boldsymbol{B} \boldsymbol{T}^{-1/2} \\
\tilde{\boldsymbol{\beta}} &= \boldsymbol{T}^{1/2} \boldsymbol{\beta}, \quad \tilde{\boldsymbol{\beta}}_0 = \boldsymbol{T}^{1/2} \boldsymbol{\beta}_0.
\end{aligned}$$

Now we have

$$\begin{aligned}
[\tilde{\mathbf{a}}' (\tilde{\boldsymbol{\beta}} - \tilde{\boldsymbol{\beta}}_0)]^2 &= (\tilde{\mathbf{a}}' \tilde{\boldsymbol{\beta}} - \tilde{\mathbf{a}}' \tilde{\boldsymbol{\beta}}_0)^2 \\
&= (\tilde{\mathbf{a}}' \tilde{\boldsymbol{\beta}})^2 + (\tilde{\mathbf{a}}' \tilde{\boldsymbol{\beta}}_0)^2 - 2(\tilde{\mathbf{a}}' \tilde{\boldsymbol{\beta}})(\tilde{\mathbf{a}}' \tilde{\boldsymbol{\beta}}_0) \\
&\leq (\tilde{\mathbf{a}}' \tilde{\boldsymbol{\beta}})^2 + (\tilde{\mathbf{a}}' \tilde{\boldsymbol{\beta}}_0)^2 + 2|\tilde{\mathbf{a}}' \tilde{\boldsymbol{\beta}}|\,|\tilde{\mathbf{a}}' \tilde{\boldsymbol{\beta}}_0| \\
&\leq (\tilde{\boldsymbol{\beta}}' \tilde{\boldsymbol{\beta}})(\tilde{\mathbf{a}}' \tilde{\mathbf{a}}) + (\tilde{\mathbf{a}}' \tilde{\boldsymbol{\beta}}_0)^2 + 2\sqrt{\tilde{\boldsymbol{\beta}}' \tilde{\boldsymbol{\beta}} \tilde{\mathbf{a}}' \tilde{\mathbf{a}}}\,|\tilde{\mathbf{a}}' \tilde{\boldsymbol{\beta}}_0| \\
&\leq k(\tilde{\mathbf{a}}' \tilde{\mathbf{a}}) + (\tilde{\mathbf{a}}' \tilde{\boldsymbol{\beta}}_0)^2 + 2\sqrt{k \tilde{\mathbf{a}}' \tilde{\mathbf{a}}}\,|\tilde{\mathbf{a}}' \tilde{\boldsymbol{\beta}}_0| \\
&= (\sqrt{k \tilde{\mathbf{a}}' \tilde{\mathbf{a}}} + |\tilde{\mathbf{a}}' \tilde{\boldsymbol{\beta}}_0|)^2
\end{aligned} \tag{4.3.20}$$

and so (20) is just

$$\sup_{\boldsymbol{\beta}' \boldsymbol{T}\boldsymbol{\beta} \leq k} [\mathbf{a}' \text{ bias } \mathbf{b}_B(\boldsymbol{\beta}_0)]^2.$$

Geometrical Interpretation

Let the correct prior information be of type $\boldsymbol{\beta}' \boldsymbol{\beta} \leq k^{-1}$ (i.e. $\boldsymbol{T} = k^2 \boldsymbol{I}$). Then we have

$$\sup_{\boldsymbol{\beta}' \boldsymbol{\beta} \leq k^{-1}} (\boldsymbol{\beta} - \boldsymbol{\beta}_0)' (\boldsymbol{\beta} - \boldsymbol{\beta}_0) = [(\boldsymbol{\beta}_0' \boldsymbol{\beta}_0)^{1/2} + k^{-1/2}]^2.$$

Figure 4.3.3 shows this situation.

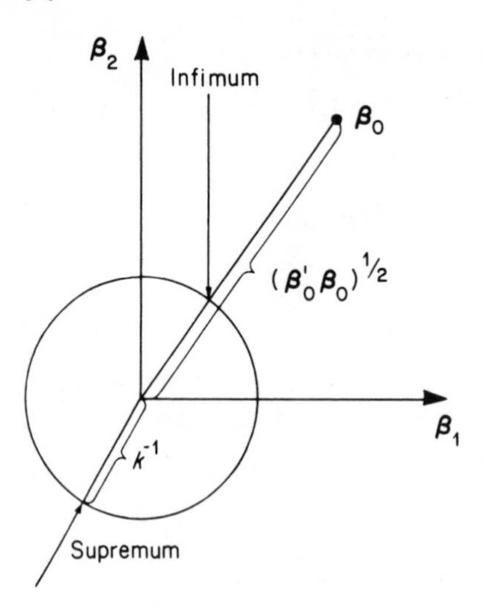

Figure 4.3.3. Geometrical interpretation of $\sup_{\beta'\beta \le k^{-1}} (\beta-\beta_0)'(\beta-\beta_0)$

Using (19) and (20) the minimax risk of $\mathbf{b}_B(\beta_0)$ follows as

$$\begin{aligned}
&\sup_{\beta' T\beta \le k} R[\mathbf{b}_B(\beta_0), \mathbf{aa}'] \\
&= \sigma^2 \mathbf{a}' D_B^{-1}[S + k\sigma^2 d^{-2} BT^{-1}B + \sigma^2 d^{-2} B\beta_0\beta_0' B] D_B^{-1}\mathbf{a} \\
&\quad + 2(k\sigma^4 d^{-2}\mathbf{a}' D_B^{-1} BT^{-1} BD_B^{-1}\mathbf{a})^{1/2} |\sigma^2 d^{-1}\mathbf{a}' D_B^{-1} B\beta_0|. \qquad (4.3.21)
\end{aligned}$$

For the last term we calculate as an upper bound

$$\begin{aligned}
|\mathbf{a}' D_B^{-1} BT^{-1/2} T^{1/2}\beta_0| &\le |\mathbf{a}' D_B^{-1} BT^{-1/2}| \, |T^{1/2}\beta_0| \\
&\le (\mathbf{a}' D_B^{-1} BT^{-1} BD_B^{-1}\mathbf{a})^{1/2} (\beta_0' T\beta_0)^{1/2}
\end{aligned}$$

and so we have

$$\begin{aligned}
\sup_{\beta' T\beta \le k} R[\mathbf{b}_B(\beta_0), \mathbf{aa}'] &\le \sigma^2 \mathbf{a}' D_B^{-1}[S + \sigma^2 d^{-2} B\beta_0\beta_0' B + \\
&\quad + \sqrt{k}\sigma^2 d^{-2} BT^{-1}B(\sqrt{k} + 2(\beta_0' T\beta_0)^{1/2})] D_B^{-1}\mathbf{a}. \qquad (4.3.22)
\end{aligned}$$

As in Section 4.3.1, the question arises of whether the used incorrect prior region (16) gives an estimator which is better compared with the GLSE $\mathbf{b}$. In other words, we have to prove if $\Delta^k[\mathbf{b}, \mathbf{b}_B(\beta_0)] \ge 0$. By (4.2.18) and (22) we have

$$\begin{aligned}
\sigma^{-2}\Delta^k[\mathbf{b}, \mathbf{b}_B(\beta_0)] &\ge \mathbf{a}' D_B^{-1}[D_B S^{-1} D_B - S - \sigma^2 d^{-2} B\beta_0\beta_0' B \\
&\quad - \sqrt{k}\sigma^2 d^{-2} BT^{-1}B(\sqrt{k} + 2(\beta_0' T\beta_0)^{1/2})] D_B^{-1}\mathbf{a} \\
&= \mathbf{a}' D_B^{-1} U D_B^{-1}\mathbf{a}, \text{ say.} \qquad (4.3.23)
\end{aligned}$$

If on the right-hand side of (23) the matrix U is nonnegative definite, then this is sufficient for $\Delta^k[\mathbf{b}, \mathbf{b}_B(\beta_0)] \ge 0$.

With $D_B = (d^{-1}\sigma^2 B + S)$ we get

$$U = \sigma^2 d^{-1} B\{\sigma^2 d^{-1} S^{-1} + 2B^{-1} - d^{-1}\beta_0\beta_0' - d^{-1}(k + 2\sqrt{k}(\beta_0' T\beta_0)^{1/2})T^{-1}\}B. \tag{4.3.24}$$

Note If $\beta_0 = 0$, (24) is just the matrix (12).

In order to give explicitly a condition for $U \geq 0$, we shall confine ourselves to the case $B = T$, i.e. where incorrectness is caused by translation of the ellipsoid's centre as well as by symmetrical distortion (i.e. changing the length of the axes by the same factor $\sqrt{d/k}$). This case seems to be of practical relevance. Using $B = T$, U (24) becomes

$$\begin{aligned} U &= \sigma^2 d^{-2} T^{1/2}\{\sigma^2 T^{1/2} S^{-1} T^{1/2} - T^{1/2}\beta_0\beta_0' T^{1/2} \\ &\quad + (2d - k - 2\sqrt{k}(\beta_0' T\beta_0)^{1/2})I\} T^{1/2} \\ &= \sigma^2 d^{-2} T^{1/2}\tilde{U} T^{1/2}, \text{ say.} \end{aligned}$$

Now, $\tilde{U} \geq 0$ implies

$$\lambda_i(\tilde{U}) = \sigma^2 \lambda_i(T^{1/2} S^{-1} T^{1/2} - \sigma^{-2} T^{1/2}\beta_0\beta_0' T^{1/2}) + (2d - k - 2\sqrt{k\beta_0' T\beta_0}) \geq 0. \tag{4.3.25}$$

Here $\lambda_i(\tilde{U})$ and $\lambda_i(T^{1/2}\tilde{S}T^{1/2})$ with $\tilde{S} = S^{-1} - \sigma^{-2}\beta_0\beta_0'$ are the eigenvalues of $\tilde{U}$ and $T^{1/2}\tilde{S}T^{1/2}$, respectively.

Therefore (25) is true iff

$$d \geq \tfrac{1}{2}(k + 2\sqrt{k\beta_0' T\beta_0} - \sigma^2 \lambda_{\min}(T^{1/2}\tilde{S}T^{1/2})). \tag{4.3.26}$$

Note (i) If $\beta_0 = 0$, the relations (26) and (15) coincide.
(ii) In contrast to the results in Section 4.3.1, the condition $d \geq k$ is not sufficient for the robustness of minimax estimation.

Theorem 4.6

If the prior information $(\beta - \beta_0)' T(\beta - \beta_0) \leq d$ (incorrect region) instead of $\beta' T\beta \leq k$ (correct region) is used to evaluate a minimax estimator $\mathbf{b}_T(\beta_0)$ (see (17) with $B = T$), then this estimator has a smaller minimax risk compared with the GLSE when the 'radius' d of the incorrect region fulfils (26).

4.3.3 Substitution of σ^2

Unfortunately the minimax estimators $\mathbf{b}^*$ (4.2.7) as well as $\mathbf{b}^*(\beta_0)$ (4.2.11) contain the unknown parameter σ and therefore they are not practicable.

A way of avoiding the unknown σ^2 in the MILE is to use a restriction on the 'coefficient of variation' β/σ such as

$$\beta' T\beta \leq k\sigma^2. \tag{4.3.27}$$

We see that with $\boldsymbol{T}^* = \boldsymbol{T}\sigma^{-2}$ this can also be written $\boldsymbol{\beta}'\boldsymbol{T}^*\boldsymbol{\beta} \leq k$ as in (4.1.1). This leads immediately to the following result.

Theorem 4.7

In the generalized regression model $\mathbf{y} = \boldsymbol{X\beta}+\boldsymbol{\varepsilon}$, $\boldsymbol{\varepsilon} \sim (\mathbf{0}, \sigma^2\boldsymbol{W})$, the MILE under the constraint (27) is

$$(k^{-1}\boldsymbol{T}+\boldsymbol{S})^{-1}\boldsymbol{X}'\boldsymbol{W}^{-1}\mathbf{y}. \tag{4.3.28}$$

An alternative way of overcoming the unknown σ^2 is to use sample or prior information to substitute an estimator for σ. One could propose substitution by s, the well-known estimator of σ given in Section 2.2.1 as

$$s^2 = (\mathbf{y}-\boldsymbol{X}\mathbf{b})'\boldsymbol{W}^{-1}(\mathbf{y}-\boldsymbol{XB})(T-K)^{-1}.$$

But s is stochastic, so that an explicit calculation of the risk function of the resulting MILE would be very difficult (Theil, 1963, has given approximations to such a risk function in the case of the mixed estimator).

Therefore we confine ourselves to the substitution of σ^2 by a nonstochastic value $c > 0$ (Toutenburg, 1975c). This gives the estimator

$$\begin{aligned}\mathbf{b}_c^* &= (k^{-1}c\boldsymbol{T}+\boldsymbol{S})^{-1}\boldsymbol{X}'\boldsymbol{W}^{-1}\mathbf{y} \qquad (4.3.29)\\ &= \boldsymbol{D}_c^{-1}\boldsymbol{X}'\boldsymbol{W}^{-1}\mathbf{y}, \text{ say,}\end{aligned}$$

which has

$$\text{bias } \mathbf{b}_c^* = -k^{-1}c\boldsymbol{D}_c^{-1}\boldsymbol{T\beta}, \tag{4.3.30}$$

$$V(\mathbf{b}_c^*) = \sigma^2\boldsymbol{D}_c^{-1}\boldsymbol{S}\boldsymbol{D}_c^{-1}, \tag{4.3.31}$$

and

$$\sup_{\boldsymbol{\beta}'\boldsymbol{T\beta}\leq k} R(\mathbf{b}_c^*, \mathbf{aa}') = \sigma^2\mathbf{a}'\boldsymbol{D}_c^{-1}(\boldsymbol{S}+k^{-1}c^2\sigma^{-2}\boldsymbol{T})\boldsymbol{D}_c^{-1}\mathbf{a}. \tag{4.3.32}$$

The relation

$$\Delta^k(\mathbf{b}_c^*, \mathbf{b}^*) \geq 0.$$

follows immediately from Theorem 4.2. Equality holds iff $c = \sigma^2$.

We first seek values of c such that $\mathbf{b}_c^*$ is better than the GLSE $\mathbf{b}$ in the sense that $\Delta^k(\mathbf{b}, \mathbf{b}_c^*) \geq 0$ holds. As the substitution of σ^2 by c may be interpreted as an incorrect chosen prior ellipsoid, namely

$$\boldsymbol{\beta}'\boldsymbol{B\beta} \leq d \quad \text{with } \boldsymbol{B} = \boldsymbol{T}, d = k\sigma^2c^{-1}, \tag{4.3.33}$$

we may use the results of the above sections to investigate the difference $\Delta^k(\mathbf{b}, \mathbf{b}_c^*)$.

Replacing d in (15) by $k\sigma^2c^{-1}$ gives

$$\Delta^k(\mathbf{b}, \mathbf{b}_c^*) \geq 0 \quad \text{iff} \quad c \leq \frac{2\sigma^2}{1-\sigma^2k^{-1}\lambda_{\min}(\boldsymbol{T}^{1/2}\boldsymbol{S}^{-1}\boldsymbol{T}^{1/2})} \tag{4.3.34}$$

(Kuks and Olman, 1972, gave a similar condition). As the eigenvalues of the matrix $\boldsymbol{T}^{1/2}\boldsymbol{S}^{-1}\boldsymbol{T}^{1/2}$ are nonnegative, the condition

$$c \leq 2\sigma^2 \tag{4.3.35}$$

is sufficient to fulfil (34).

To realize (34) or (35), which are just conditions for the robustness of the minimax estimator, we must have prior information on σ^2. These ideas have been developed by Toutenburg (1976) (see also Bibby and Toutenburg, 1978, p. 154).

We now assume that prior knowledge gives an interval such that

$$0 < \sigma_1^2 < \sigma^2 < \sigma_2^2 < \infty \tag{4.3.36}$$

where σ_1^2, σ_2^2 are known. Now calculate (see (32))

$$\frac{\partial}{\partial c}\left\{\sup_{\beta' T\beta \leq k} R(\mathbf{b}_c^*, \mathbf{aa}')\right\} = -2\sigma^2 c^{-1}(1 - c\sigma^{-2})(\tilde{\boldsymbol{T}} - \tilde{\boldsymbol{T}}\boldsymbol{D}_c^{-1}\tilde{\boldsymbol{T}})$$

where $(\tilde{\boldsymbol{T}} - \tilde{\boldsymbol{T}}\boldsymbol{D}_c^{-1}\tilde{\boldsymbol{T}}) \geq \boldsymbol{0}$, $\tilde{\boldsymbol{T}} = k^{-1}c\boldsymbol{T}$. Hence we deduce that $\sup_{\beta' T\beta \leq k} R(\mathbf{b}_c^*, \mathbf{aa}')$ is minimal when $c = \sigma^2$ (in accordance with Theorem 4.2), and, moreover,

$$\sup_{\beta' T\beta \leq k} R(\mathbf{b}_c^*, \mathbf{aa}') \text{ is monotonically increasing for } c > \sigma^2,$$

$$\sup_{\beta' T\beta \leq k} R(\mathbf{b}_c^*, \mathbf{aa}') \text{ is monotonically decreasing for } c < \sigma^2. \tag{4.3.37}$$

The three conditions (35), (36), and (37) are combined optimally when $c = 2\sigma_1^2$. This gives the following theorem.

Theorem 4.8

Given a lower bound such that $\sigma_1^2 < \sigma^2$, then

$$\mathbf{b}_{2\sigma_1^2}^* = (2\sigma_1^2 k^{-1}\boldsymbol{T} + \boldsymbol{S})^{-1}\boldsymbol{X}'\boldsymbol{W}^{-1}\mathbf{y} \tag{4.3.38}$$

is better than the GLSE $\mathbf{b}$ with respect to the minimax risk, and, moreover, it is optimal in the sense that

$$\max_{c \leq 2\sigma_1^2} \Delta^k(\mathbf{b}, \mathbf{b}_c^*) = \Delta^k(\mathbf{b}, \mathbf{b}_{2\sigma_1^2}^*) \geq 0.$$

Using the prior information $\sigma_1^2 < \sigma^2$ therefore gives a practicable estimator.

The determination of σ_1^2 may be simple if the disturbances vector is normally distributed, i.e. if $\boldsymbol{\varepsilon} \sim N(\boldsymbol{0}, \sigma^2\boldsymbol{W})$. In that case we have

$$s^2(T-K) \sim \sigma^2\chi_{T-K}^2$$

(where $s^2 = (\mathbf{y} - \boldsymbol{X}\mathbf{b})'\boldsymbol{W}^{-1}(\mathbf{y} - \boldsymbol{X}\mathbf{b})(T-K)^{-1}$). Let χ_α^2 denote the α-quantile of the χ_{T-K}^2-distribution. Then we have

$$P\left(\frac{s^2(T-K)}{\chi_\alpha^2} < \sigma^2\right) = \alpha. \tag{4.3.39}$$

So we may choose α and determine

$$\sigma_1^2 \leq \frac{s^2(T-K)}{\chi_\alpha^2} \tag{4.3.40}$$

as a natural lower bound of σ^2.

4.3.4 A Numerical Example

We confine ourselves to a classical regression model with $K = 3$ regressors

$$\mathbf{y} = \boldsymbol{X}_1\boldsymbol{\beta}_1 + \boldsymbol{X}_2\boldsymbol{\beta}_2 + \boldsymbol{X}_3\boldsymbol{\beta}_3 + \boldsymbol{\varepsilon},$$
$$E\boldsymbol{\varepsilon} = \mathbf{0},\ E\boldsymbol{\varepsilon}\boldsymbol{\varepsilon}' = \sigma^2\boldsymbol{I}.$$

We assume that we have the following $T = 6$-dimensional sample (see Goldberger, 1964, p. 160):

$$\mathbf{y} = \begin{bmatrix} 0 \\ 2 \\ 1 \\ 2 \\ -1 \\ 1 \end{bmatrix} \qquad \boldsymbol{X} = \begin{bmatrix} 1 & -1 & 0 \\ 1 & 0 & 1 \\ 1 & 1 & 0 \\ 1 & 2 & 1 \\ 1 & 0 & -1 \\ 1 & 0 & 0 \end{bmatrix}$$

and we calculate (Toutenburg, 1970d)

$$\boldsymbol{X}'\boldsymbol{X} = \begin{bmatrix} 6 & 2 & 1 \\ 2 & 6 & 2 \\ 1 & 2 & 3 \end{bmatrix}, \quad \boldsymbol{X}'\mathbf{y} = \begin{bmatrix} 5 \\ 5 \\ 5 \end{bmatrix}, \quad |\boldsymbol{X}'\boldsymbol{X}| = 74$$

$$(\boldsymbol{X}'\boldsymbol{X})^{-1} = \frac{1}{74}\begin{bmatrix} 14 & -4 & -2 \\ -4 & 17 & -10 \\ -2 & -10 & 32 \end{bmatrix}$$

$$\mathbf{b}_0 = \frac{1}{74}\begin{bmatrix} 40 \\ 15 \\ 100 \end{bmatrix} = \begin{bmatrix} 0.541 \\ 0.203 \\ 1.352 \end{bmatrix}$$

$$\hat{\mathbf{y}} = \boldsymbol{X}\mathbf{b}_0 = \frac{1}{74}\begin{bmatrix} 25 \\ 140 \\ 55 \\ 170 \\ -60 \\ 40 \end{bmatrix}, \quad \hat{\boldsymbol{\varepsilon}} = \mathbf{y} - \boldsymbol{X}\mathbf{b}_0 = \frac{1}{74}\begin{bmatrix} -25 \\ 8 \\ 19 \\ -22 \\ -14 \\ 34 \end{bmatrix}$$

$$s^2 = \hat{\boldsymbol{\varepsilon}}'\hat{\boldsymbol{\varepsilon}}(T-K)^{-1} = \frac{2886}{74^2 \times 3} = \frac{13}{74}.$$

If $\boldsymbol{\varepsilon}$ is normally distributed, s^2 is $\sigma^2(T-K)^{-1}\chi_{T-K}^2$ distributed and a lower bounded confidence interval will be of the form (see (39))

$$P\left(\frac{s^2(T-K)}{\chi^2_\alpha} < \sigma^2\right) = \alpha.$$

$\alpha = 0.05$ gives the lower bound as 0.0674. Based on this we may choose $\sigma_1^2 = 0.05$.

Let it be prior known that the following relations hold:

$$\beta_1^2 \leq 1, \quad \beta_2^2 \leq 1, \quad \beta_3^2 \leq 2.$$

This gives $\boldsymbol{\beta}' \boldsymbol{T} \boldsymbol{\beta} \leq 1$ with

$$\boldsymbol{T} = \begin{bmatrix} 0.33 & 0 & 0 \\ 0 & 0.33 & 0 \\ 0 & 0 & 0.17 \end{bmatrix}$$

$$\mathbf{b}^*_{2\sigma_1^2} = \mathbf{b}^*_{0.1} = (0.1\,\boldsymbol{T} + \boldsymbol{X}'\boldsymbol{X})^{-1}\boldsymbol{X}'\mathbf{y}, \quad |0.1\,\boldsymbol{T} + \boldsymbol{X}'\boldsymbol{X}| = 79.58,$$

$$D_{0.1}^{-1} = (0.1\,\boldsymbol{T} + \boldsymbol{X}'\boldsymbol{X})^{-1} = (79.58)^{-1} \begin{bmatrix} 14.21 & -4.04 & -2.03 \\ -4.04 & 17.21 & -10.06 \\ -2.03 & -10.06 & 32.36 \end{bmatrix}$$

$$\mathbf{b}^*_{0.1} = \begin{bmatrix} 0.511 \\ 0.195 \\ 1.274 \end{bmatrix}.$$

So we can calculate the two risk functions; for example, in the case $\mathbf{a}' = (1, 1, 1)$ we get

$$R(\mathbf{b}_0, \mathbf{aa}') = \frac{31}{74}\sigma^2 = 0.459\sigma^2$$

and, for $c = 2\sigma_1^2$,

$$\sup_{\boldsymbol{\beta}'\boldsymbol{T}\boldsymbol{\beta} \leq 1} R(\mathbf{b}^*_{2\sigma_1^2}, \mathbf{aa}') = 0.374\sigma^2 + 0.060\sigma_1^4.$$

Thus the required difference of the risks becomes

$$\begin{aligned} \Delta^1(\mathbf{b}_0, \mathbf{b}^*_{2\sigma_1^2}) &= \sigma^2(0.085 - 0.060\sigma_1^4/\sigma^2) \\ &\geq \sigma^2(0.085 - 0.060\sigma_1^2) \end{aligned}$$

as $\sigma_1^2 < \sigma^2$ was assumed. Inserting the chosen value $\sigma_1^2 = 0.05$ gives

$$\Delta^1(\mathbf{b}_0, \mathbf{b}^*_{0.1}) \geq 0.082\sigma^2.$$

In other words, the relative decrease in minimax risk is about 18%.

4.4 ROBUSTNESS OF THE MINIMAX ESTIMATOR—A SIMULATION APPROACH

4.4.1 Introduction

In the three preceding sections we have investigated some aspects of robustness of the MILE. Robustness of the minimax estimator was understood as the behaviour of some incorrect estimators $\tilde{\boldsymbol{\beta}}$ compared with the GLSE $\mathbf{b}$. The minimax estimator $\mathbf{b}^*$ was defined to be robust with respect to changes of the prior ellipsoid if $\Delta^k(\mathbf{b}, \tilde{\boldsymbol{\beta}}) \geq 0$ where $\tilde{\boldsymbol{\beta}}$ was one of the estimators $\mathbf{b}_{\boldsymbol{B}}^*$ (4.3.4), $\mathbf{b}_{\boldsymbol{B}}^*(\boldsymbol{\beta}_0)$ (4.3.17), and $\mathbf{b}_c^*$(4.3.29), respectively.

Now we want to investigate another aspect of robustness of the MILE $\mathbf{b}^*$. This will be done by setting up a simulation experiment, as described in the following.

As was pointed out in Section 4.3, the optimality of the MILE $\mathbf{b}^*$ (4.2.7) as well as its superiority over the GLSE $\mathbf{b}$ depends on the correctness of the used prior information. In practice the choice of the prior ellipsoid or of a cuboid as in (4.1.3) has to be based on theoretical investigations, natural restrictions, or on prior samples. In particular, the choice of the length of prior intervals $a_i \leq \beta_i \leq b_i$ as well as the choice of the centre of the cuboid (4.1.4) is often an arbitrary action. But the ellipsoid (4.1.7) which is based on the cuboid (4.1.4) influences the MILE $\mathbf{b}^*$ as well as its moments and its risk.

For the prior ellipsoid, we may evaluate the following.

(i) Case of Low Prior Information

Let the prior information

$$a_i \leq \beta_i \leq b_i \quad (i = 1, \ldots, K) \tag{4.4.1}$$

be given; this builds the cuboid (4.1.4). The greater the intervals (1) the less binding is the prior restriction of the components of $\boldsymbol{\beta}$. If the lengths $(b_i - a_i)$ $(i = 1, \ldots, K)$ tend to infinity we see from (4.1.7) that for the matrix of the corresponding ellipsoid $\boldsymbol{T} = \operatorname{diag}(t_1, \ldots, t_K)$ it follows that $t_i \to 0$ $(i = 1, \ldots, K)$. This gives in accordance with (4.2.17)

$$\lim_{\substack{t_i \to 0 \\ (\forall i)}} \mathbf{b}^* = \boldsymbol{S}^{-1}\boldsymbol{X}'\boldsymbol{W}^{-1}\mathbf{y} = \mathbf{b}. \tag{4.4.2}$$

(ii) Case of Strong Prior Information

The smaller the prior intervals (1) the greater is their influence on the resulting minimax estimator. If the lengths of the intervals tend to zero we have $t_i \to \infty$ and therefore

$$\lim_{\substack{t_i \to \infty \\ (\forall i)}} \mathbf{b}^* = \boldsymbol{\beta}_0 = \boldsymbol{\beta}. \tag{4.4.3}$$

In other words, the trivial estimator $\hat{\boldsymbol{\beta}}_1 = \boldsymbol{\beta}$ implies full knowledge of the parameters (and therefore it is without any practical usefulness). On the other hand, the GLSE $\mathbf{b}$ is based on the sample information only. Thus the minimax estimator $\mathbf{b}^*$ is arranged between these two limit cases of prior information.

As far as the arrangement $\boldsymbol{\beta}_0' = 1/2\,(a_1 + b_1, \ldots, a_K + b_K)$ is concerned (this is the centre of the cuboid (1) as well as the centre of the corresponding ellipsoid (4.1.7)), we know from Section 4.2 that $\boldsymbol{\beta}_0$ influences the numerical value of the estimator $\mathbf{b}^*(\boldsymbol{\beta}_0)$ as well as bias $\mathbf{b}^*(\boldsymbol{\beta}_0)$, but that $\boldsymbol{\beta}_0$ is without any influence on the dispersion and the minimax risk of $\mathbf{b}^*(\boldsymbol{\beta}_0)$. This is true if the prior ellipsoid is specified correctly. Then we have (see (4.2.14))

$$\begin{aligned}\Delta_k(\mathbf{b}, \mathbf{b}^*(\boldsymbol{\beta}_0)) &= R(\mathbf{b}, \mathbf{aa}') - \sup_{(\boldsymbol{\beta}-\boldsymbol{\beta}_0)' T(\boldsymbol{\beta}-\boldsymbol{\beta}_0) \leq k} R(\mathbf{b}^*(\boldsymbol{\beta}_0), \mathbf{aa}') \\ &= \sigma^2 \mathbf{a}'(\boldsymbol{S}^{-1} - \boldsymbol{D}^{-1})\mathbf{a} \geq 0\end{aligned} \tag{4.4.4}$$

which is independent of the (correctly specified) centre $\boldsymbol{\beta}_0$.

The centre $\boldsymbol{\beta}_0$ has to be assumed nonstochastic. If, as a theoretical experiment, we replaced $\boldsymbol{\beta}_0$ in the formula for $\mathbf{b}^*(\boldsymbol{\beta}_0)$ (4.2.11) by a stochastic vector, in particular by the GLSE $\mathbf{b}$, we would get

$$\begin{aligned}\lim_{\boldsymbol{\beta}_0 \to \mathbf{b}} \mathbf{b}^*(\boldsymbol{\beta}_0) &= \mathbf{b} + \boldsymbol{D}^{-1}\boldsymbol{X}'\boldsymbol{W}^{-1}(\boldsymbol{I} - \boldsymbol{X}\boldsymbol{S}^{-1}\boldsymbol{X}'\boldsymbol{W}^{-1})\mathbf{y} \\ &= \mathbf{b}\end{aligned} \tag{4.4.5}$$

as the matrix product $\boldsymbol{X}'\boldsymbol{W}^{-1}(\boldsymbol{I} - \boldsymbol{X}\boldsymbol{S}^{-1}\boldsymbol{X}'\boldsymbol{W}^{-1}) = \boldsymbol{0}$. This result may be interpreted in such a way that a centre $\boldsymbol{\beta}_0$ of the ellipsoid $(\boldsymbol{\beta}-\boldsymbol{\beta}_0)'\boldsymbol{T}(\boldsymbol{\beta}-\boldsymbol{\beta}_0) \leq k$ which is nearly the GLSE $\mathbf{b}$ shall lead to a high numerical correlation of both estimators $\mathbf{b}^*(\boldsymbol{\beta}_0)$ and $\mathbf{b}$. In other words, the smaller the difference of the vectors $\boldsymbol{\beta}_0$ and $\mathbf{b}$, the less the expected gain in efficiency of the minimax estimator compared with the unrestricted GLSE $\mathbf{b}$.

4.4.2 The Simulation Model

The following is a theoretical regression model which is part of the econometric model of the GDR (Toutenburg, 1977c; Schiele, 1980):

$$\mathbf{y} = (\mathbf{i}, \mathbf{x})\boldsymbol{\beta} + \boldsymbol{\varepsilon}, \quad \boldsymbol{\varepsilon} \sim (\mathbf{0}, \sigma^2 \boldsymbol{I}) \tag{4.4.6}$$

where for the regressor $\mathbf{x}$ two samples are used which are typical for economic time series:

$$\mathbf{x}^{(1)} = (1.0, 1.5, 2.0, 2.2, 2.5, 2.6, 2.8, 3.1, 3.2) \tag{4.4.7}$$

$$\mathbf{x}^{(2)} = (1.8, 1.6, 1.5, 1.7, 1.9, 2.1, 2.2, 2.3, 2.5) \tag{4.4.8}$$

(i.e. the sample size is $T = 9$).

For the disturbance vector $\boldsymbol{\varepsilon}$ there are produced five alternative time-series $\boldsymbol{\varepsilon}^{(1)}, \ldots, \boldsymbol{\varepsilon}^{(5)}$ (which are taken from tables in Kendall and Babington-Smith, 1951). If the true parameters are assumed as $\beta_1 = 1, \beta_2 = 2$ the possible combinations $(\mathbf{x}^{(i)}, \boldsymbol{\varepsilon}^{(j)}), i = 1, 2, j = 1, \ldots, 5$ may be used to produce the corresponding ten vectors $\mathbf{y}^{(i,j)}$ and to calculate the OLSE $\mathbf{b}_0 = (\boldsymbol{X}'\boldsymbol{X})^{-1}\boldsymbol{X}'\mathbf{y}$ (see Figure 4.4.1) as well as lower bounds $\sigma_1^2 < \sigma^2$ (see (4.3.40)). Based on these assumptions and data the simulation of prior intervals

$$\begin{aligned} a_1 &\leq \beta_1 \leq b_1 \\ a_2 &\leq \beta_2 \leq b_2 \end{aligned} \tag{4.4.9}$$

	$\mathbf{x}^{(1)}, \varepsilon^{(1)}$	$\mathbf{x}^{(1)}, \varepsilon^{(2)}$	$\mathbf{x}^{(1)}, \varepsilon^{(3)}$	$\mathbf{x}^{(1)}, \varepsilon^{(4)}$	$\mathbf{x}^{(1)}, \varepsilon^{(5)}$	$\mathbf{x}^{(2)}, \varepsilon^{(1)}$	$\mathbf{x}^{(2)}, \varepsilon^{(2)}$	$\mathbf{x}^{(2)}, \varepsilon^{(3)}$	$\mathbf{x}^{(2)}, \varepsilon^{(4)}$	$\mathbf{x}^{(2)}, \varepsilon^{(5)}$
$\hat{\beta}_1$	1.3805	0.8492	0.9960	0.6811	0.8607	1.3856	1.0072	0.9731	0.6891	0.6488
$\hat{\beta}_2$	1.8292	1.9772	2.0007	2.1457	2.1224	1.7946	1.8921	2.0125	2.1689	2.2537
$s(\hat{\beta}_2)$	0.1568	0.1858	0.0195	0.1469	0.1089	0.3558	0.3975	0.0416	0.3308	0.2353
$\sigma(\hat{\beta}_2)$	0.1708	0.0228	0.0007	0.1457	0.1224	0.2054	0.1079	0.0125	0.1689	0.2537
$R^{(2)}$	0.9511	0.9418	0.9993	0.9683	0.9819	0.7842	0.7640	0.9970	0.8599	0.9291

Figure 4.4.1 Ordinary least squares of $\boldsymbol{\beta}$ in the simulation model.
$s(\hat{\boldsymbol{\beta}}_2)$: estimated standard deviation of $\hat{\boldsymbol{\beta}}_2$
$\sigma(\hat{\boldsymbol{\beta}}_2)$: true standard deviation of $\hat{\boldsymbol{\beta}}_2$
R^2: coefficient of determination (see (2.1.13))

is done where the 5 combinations of interval lengths

$$(200, 8), (200, 4), (400, 4), (400, 8), (100, 4) \tag{4.4.10}$$

and 15 different positions of the centre $\boldsymbol{\beta}_0$ of the rectangle are used. In this way we get 750 values of the minimax estimator

$$\mathbf{b}^*_{2\sigma_1^2} = \boldsymbol{\beta}_0 + (2\sigma_1^2 k^{-1}\boldsymbol{T} + \boldsymbol{X}'\boldsymbol{X})^{-1}\boldsymbol{X}'(\mathbf{y} - \boldsymbol{X}\boldsymbol{\beta}_0) \tag{4.4.11}$$

($\boldsymbol{W} = \boldsymbol{I}$ and $\boldsymbol{\beta}_0 \neq \mathbf{0}$ in (4.3.38)) which are given in Figures 4.4.2(i)–4.4.6(i).

If the interval lengths of the rectangle are fixed but its centre is moved then the minimax estimator based on the ellipsoid (4.1.7) changes without its risk being changed. Thus the minimax risk is not appropriate for measuring the dependence of the numerical values of $\mathbf{b}^*_{2\sigma_1^2}$ on the centre of the prior region. To do this we calculate the deviation of $\mathbf{b}^*_{2\sigma_1^2}$ from the true parameter vector $\boldsymbol{\beta}$. This is done in Figures 4.4.2(ii)–4.4.6(ii); Figures 4.4.2(iii)–4.4.6(iii) show the percentage deviations $\boldsymbol{\beta}_{(i)}^{-1}(\mathbf{b}^*_{2\sigma_1^2} - \boldsymbol{\beta})_{(i)} \times 100\%$ $(i = 1, 2)$.

If we investigate the percentage decrease of the variances of the minimax estimators' components compared with the corresponding variances of the OLSE $\mathbf{b}_0$ we get

$$\Delta V_i = \frac{V(\mathbf{b}_0)_i - V(\mathbf{b}_2^*\sigma_1^2)_i}{V(\mathbf{b}_0)_i} \cdot 100\% \quad (i = 1, 2) \tag{4.4.12}$$

where $V(\mathbf{b}_0)_i$ is the ith component of

$$\sigma^{-2} V(\mathbf{b}_0) = (\boldsymbol{X}'\boldsymbol{X})^{-1}$$

and $V(\mathbf{b}^*_{2\sigma_1^2})_i$ is the ith component of

$$\sigma^{-2} V(\mathbf{b}^*_{2\sigma_1^2}) = \boldsymbol{D}^{-1}_{2\sigma_1^2}\boldsymbol{X}'\boldsymbol{X}\boldsymbol{D}^{-1}_{2\sigma_1^2} \quad \text{(see (4.3.31))}.$$

The results are given in Figure 4.4.7.

As an analogue we investigate the percentage decrease of the minimax risk. By (4.2.18) we have

$$\sigma^{-2} R(\mathbf{b}_0, \mathbf{aa}') = \mathbf{a}'(\boldsymbol{X}'\boldsymbol{X})^{-1}\mathbf{a}. \tag{4.4.13}$$

Number of experiment	$\boldsymbol{\beta}_0$	$\mathbf{x}^{(1)}, \varepsilon^{(1)}$	$\mathbf{x}^{(1)}, \varepsilon^{(2)}$	$\mathbf{x}^{(1)}, \varepsilon^{(3)}$	$\mathbf{x}^{(1)}, \varepsilon^{(4)}$	$\mathbf{x}^{(1)}, \varepsilon^{(5)}$	$\mathbf{x}^{(2)}, \varepsilon^{(1)}$	$\mathbf{x}^{(2)}, \varepsilon^{(2)}$	$\mathbf{x}^{(2)}, \varepsilon^{(3)}$	$\mathbf{x}^{(2)}, \varepsilon^{(4)}$	$\mathbf{x}^{(2)}, \varepsilon^{(5)}$
OLSE $\mathbf{b}_0$		1.3805 1.8292	0.8492 1.9772	0.9960 2.0007	0.6811 2.1457	0.8607 2.1224	1.3856 1.7946	1.0072 1.8921	0.9731 2.0125	0.6891 2.1689	0.6488 2.2537
1	1 2	1.3799 1.8295	0.8491 1.9772	0.9960 2.0007	0.6815 2.1455	0.8609 2.1223	1.3826 1.7961	1.0052 1.8931	0.9731 2.0125	0.6914 2.1678	0.6502 2.2530
2	2 2	1.3799 1.8295	0.8491 1.9772	0.9960 2.0007	0.6815 2.1455	0.8609 2.1223	1.3826 1.7961	1.0052 1.8930	0.9731 2.0125	0.6914 2.1677	0.6502 2.2530
3	0 2	1.3799 1.8295	0.8490 1.9772	0.9960 2.0007	0.6815 2.1455	0.8609 2.1223	1.3826 1.7961	1.0052 1.8931	0.9731 2.0125	0.6913 2.1678	0.6501 2.2530
4	3 2	1.3799 1.8295	0.8491 1.9772	0.9960 2.0007	0.6815 2.1455	0.8609 2.1223	1.3827 1.7960	1.0054 1.8930	0.9731 2.0125	0.6914 2.1677	0.6502 2.2530
5	−1 2	1.3799 1.8295	0.8490 1.9772	0.9960 2.0007	0.6815 2.1455	0.8609 2.1223	1.3825 1.7962	1.0051 1.8931	0.9731 2.0125	0.6913 2.1678	0.6501 2.2531
6	4 2	1.3799 1.8295	0.8491 1.9772	0.9960 2.0007	0.6816 2.1455	0.8609 2.1223	1.3828 1.7960	1.0054 1.8930	0.9731 2.0125	0.6915 2.1677	0.6502 2.2530
7	−2 2	1.3799 1.8295	0.8490 1.9772	0.9960 2.0007	0.6814 2.1455	0.8609 2.1223	1.3825 1.7962	1.0050 1.8931	0.9731 2.0125	0.6912 2.1678	0.6501 2.2531
8	1 1	1.3833 1.8280	0.8538 1.9752	0.9960 2.0007	0.6846 2.1442	0.8626 2.1216	1.3971 1.7887	1.0236 1.8836	0.9734 2.0124	0.7045 2.1610	0.6554 2.2504
9	1 3	1.3765 1.8310	0.8443 1.9792	0.9961 2.0007	0.6784 2.1468	0.8592 2.1231	1.3681 1.8035	0.9868 1.9025	0.9729 2.0126	0.6782 2.1745	0.6449 2.2557
10	1 0.5	1.3850 1.8272	0.8562 1.9741	0.9960 2.0006	0.6861 2.1435	0.8634 2.1212	1.4043 1.7850	1.0328 1.8790	0.9735 2.0124	0.7111 2.1577	0.6581 2.2490
11	1 3.5	1.3748 1.8317	0.8419 1.9802	0.9960 2.0008	0.6769 2.1475	0.8584 2.1234	1.3609 1.8072	0.9777 1.9072	0.9728 2.0127	0.6716 2.1779	0.6422 2.2571
12	2 1	1.3833 1.8280	0.8538 1.9751	0.9960 2.0007	0.6846 2.1442	0.8626 2.1216	1.3971 1.7887	1.0237 1.8836	0.9733 2.0124	0.7046 2.1610	0.6554 2.2503
13	0 1	1.3833 1.8280	0.8538 1.9752	0.9960 2.0006	0.6845 2.1442	0.8626 2.1216	1.3970 1.7887	1.0235 1.8837	0.9734 2.0124	0.7045 2.1611	0.6554 2.2504
14	2 3	1.3765 1.8310	0.8443 1.9792	0.9960 2.0007	0.6785 2.1468	0.8592 2.1231	1.3682 1.8035	0.9869 1.9024	0.9729 2.0126	0.6782 2.1745	0.6449 2.2557
15	0 3	1.3765 1.8310	0.8443 1.9792	0.9960 2.0007	0.6784 2.1468	0.8592 2.1231	1.3681 1.8035	0.9868 1.9025	0.9729 2.0126	0.6782 2.1745	0.6448 2.2558

Figure 4.4.2(i) Minimax estimator $\mathbf{b}^*_{2\sigma_1^2}$ if the prior intervals have lengths (200, 8)

Number of experiment	β_0	$\mathbf{x}^{(1)}, \varepsilon^{(1)}$	$\mathbf{x}^{(1)}, \varepsilon^{(2)}$	$\mathbf{x}^{(1)}, \varepsilon^{(3)}$	$\mathbf{x}^{(1)}, \varepsilon^{(4)}$	$\mathbf{x}^{(1)}, \varepsilon^{(5)}$	$\mathbf{x}^{(2)}, \varepsilon^{(1)}$	$\mathbf{x}^{(2)}, \varepsilon^{(2)}$	$\mathbf{x}^{(2)}, \varepsilon^{(3)}$	$\mathbf{x}^{(2)}, \varepsilon^{(4)}$	$\mathbf{x}^{(2)}, \varepsilon^{(5)}$
OLSE $\mathbf{b}_0$		1.3805 1.8292	0.8492 1.9772	0.9960 2.0007	0.6811 2.1457	0.8607 2.1224	1.3856 1.7946	1.0072 1.8921	0.9731 2.0125	0.6891 2.1689	0.6488 2.2537
1	1 2	1.3782 1.8302	0.8487 1.9773	0.9960 2.0007	0.6828 2.1450	0.8615 2.1221	1.3740 1.8005	0.9995 1.8960	0.9731 2.0125	0.6979 2.1645	0.6541 2.2510
2	2 2	1.3782 1.8302	0.8487 1.9773	0.9960 2.0007	0.6828 2.1450	0.8615 2.1221	1.3740 1.8005	0.9996 1.8960	0.9731 2.0125	0.6979 2.1644	0.6542 2.2510
3	0 2	1.3782 1.8302	0.8487 1.9773	0.9960 2.0007	0.6828 2.1449	0.8615 2.1221	1.3739 1.8006	0.9994 1.8960	0.9731 2.0125	0.6978 2.1645	0.6541 2.2510
4	3 2	1.3782 1.8302	0.8488 1.9773	0.9960 2.0007	0.6829 2.1450	0.8615 2.1221	1.3741 1.8005	0.9996 1.8959	0.9731 2.0125	0.6979 2.1644	0.6542 2.2510
5	−1 2	1.3782 1.8303	0.8487 1.9774	0.9960 2.0007	0.6828 2.1449	0.8615 2.1221	1.3739 1.8006	0.9994 1.8960	0.9731 2.0125	0.6978 2.1645	0.6541 2.2510
6	4 2	1.3782 1.8302	0.8488 1.9773	0.9960 2.0007	0.6829 2.1450	0.8616 2.1221	1.3741 1.8005	0.9997 1.8959	0.9732 2.0125	0.6980 2.1644	0.6542 2.2510
7	−2 2	1.3781 1.8303	0.8487 1.9774	0.9960 2.0007	0.6828 2.1449	0.8615 2.1221	1.3738 1.8006	0.9993 1.8961	0.9731 2.0125	0.6978 2.1645	0.6541 2.2510
8	1 1	1.3917 1.8244	0.8677 1.9692	0.9962 2.0006	0.6951 2.1397	0.8683 2.1191	1.4306 1.7716	1.0710 1.8594	0.9740 2.0121	0.7495 2.1381	0.6751 2.2403
9	1 3	1.3646 1.8360	0.8298 1.9855	0.9958 2.0008	0.6706 2.1502	0.8547 2.1250	1.3173 1.8295	0.9280 1.9326	0.9723 2.0130	0.6462 2.1908	0.6332 2.2617
10	1 0.5	1.3985 1.8215	0.8771 1.9651	0.9963 2.0006	0.7012 2.1370	0.8717 2.1177	1.4589 1.7571	1.1067 1.8411	0.9744 2.0118	0.7753 2.1249	0.6856 2.2349
11	1 3.5	1.3578 1.8389	0.8203 1.9896	0.9957 2.0008	0.6645 2.1529	0.8513 2.1265	1.2890 1.8440	0.8923 1.9508	0.9719 2.0132	0.6204 2.2040	0.6227 2.2671
12	2 1	1.3918 1.8244	0.8677 1.9692	0.9962 2.0006	0.6951 2.1397	0.8683 2.1191	1.4306 1.7716	1.0711 1.8594	0.9740 2.0121	0.7495 2.1380	0.6751 2.2403
13	0 1	1.3917 1.8244	0.8677 1.9692	0.9962 2.0006	0.6950 2.1397	0.8683 2.1191	1.4305 1.7716	1.0709 1.8595	0.9740 2.0121	0.7494 2.1381	0.6751 2.2403
14	2 3	1.3646 1.8360	0.8298 1.9855	0.9958 2.0008	0.6706 2.1502	0.8547 2.1250	1.3174 1.8295	0.9281 1.9325	0.9723 2.0130	0.6463 2.1908	0.6332 2.2617
15	0 3	1.3646 1.8360	0.8298 1.9855	0.9958 2.0008	0.6706 2.1502	0.8547 2.1250	1.3173 1.8295	0.9279 1.9326	0.9723 2.0130	0.6462 2.1909	0.6331 2.2617

Figure 4.4.3(i) Minimax estimator $\mathbf{b}^*_{2\sigma_1^2}$ if the prior intervals have lengths (200, 4)

Number of experiment	β_0	$\mathbf{x}^{(1)}, \varepsilon^{(1)}$	$\mathbf{x}^{(1)}, \varepsilon^{(2)}$	$\mathbf{x}^{(1)}, \varepsilon^{(3)}$	$\mathbf{x}^{(1)}, \varepsilon^{(4)}$	$\mathbf{x}^{(1)}, \varepsilon^{(5)}$	$\mathbf{x}^{(2)}, \varepsilon^{(1)}$	$\mathbf{x}^{(2)}, \varepsilon^{(2)}$	$\mathbf{x}^{(2)}, \varepsilon^{(3)}$	$\mathbf{x}^{(2)}, \varepsilon^{(4)}$	$\mathbf{x}^{(2)}, \varepsilon^{(5)}$
OLSE $\mathbf{b}_0$		1.3805 1.8292	0.8492 1.9772	0.9960 2.0007	0.6811 2.1457	0.8607 2.1224	1.3856 1.7946	1.0072 1.8921	0.9731 2.0125	0.6891 2.1689	0.6488 2.2537
1	1 2	1.3782 1.8302	0.8487 1.9773	0.9960 2.0007	0.6828 2.1450	0.8615 2.1221	1.3740 1.8005	0.9995 1.8960	0.9731 2.0125	0.6978 2.1644	0.6541 2.2510
2	2 2	1.3782 1.8302	0.8487 1.9773	0.9960 2.0007	0.6828 2.1450	0.8615 2.1221	1.3740 1.8005	0.9995 1.8960	0.9731 2.0125	0.6978 2.1644	0.6541 2.2510
3	0 2	1.3782 1.8302	0.8487 1.9773	0.9960 2.0007	0.6828 2.1450	0.8615 2.1221	1.3740 1.8005	0.9995 1.8960	0.9731 2.0125	0.6978 2.1645	0.6541 2.2510
4	3 2	1.3782 1.8302	0.8487 1.9773	0.9960 2.0007	0.6828 2.1450	0.8615 2.1221	1.3740 1.8005	0.9995 1.8960	0.9731 2.0125	0.6979 2.1644	0.6542 2.2510
5	−1 2	1.3782 1.8302	0.8487 1.9774	0.9960 2.0007	0.6828 2.1450	0.8615 2.1220	1.3740 1.8005	0.9995 1.8960	0.9731 2.0125	0.6978 2.1645	0.6541 2.2510
6	4 2	1.3782 1.8302	0.8487 1.9773	0.9960 2.0007	0.6828 2.1450	0.8615 2.1221	1.3740 1.8005	0.9995 1.8960	0.9732 2.0125	0.6979 2.1644	0.6542 2.2510
7	−2 2	1.3781 1.8303	0.8487 1.9773	0.9960 2.0007	0.6828 2.1449	0.8615 2.1220	1.3740 1.8005	0.9994 1.8960	0.9731 2.0125	0.6978 2.1645	0.6541 2.2510
8	1 1	1.3917 1.8244	0.8677 1.9692	0.9962 2.0006	0.6951 2.1397	0.8683 2.1191	1.4306 1.7716	1.0710 1.8594	0.9740 2.0121	0.7495 2.1381	0.6751 2.2403
9	1 3	1.3646 1.8360	0.8298 1.9855	0.9958 2.0008	0.6706 2.1502	0.8547 2.1250	1.3173 1.8295	0.9280 1.9326	0.9723 2.0130	0.6462 2.1908	0.6332 2.2618
10	1 0.5	1.3985 1.8215	0.8772 1.9651	0.9963 2.0006	0.7012 2.1370	0.8717 2.1176	1.4589 1.7571	1.1067 1.8412	0.9744 2.0118	0.7753 2.1249	0.6856 2.2349
11	1 3.5	1.3579 1.8390	0.8203 1.9896	0.9957 2.0008	0.6645 2.1529	0.8513 2.1265	1.2890 1.8439	0.8922 1.9508	0.9719 2.0132	0.6204 2.2040	0.6227 2.2671
12	2 1	1.3917 1.8244	0.8677 1.9692	0.9962 2.0006	0.6951 2.1397	0.8683 2.1191	1.4306 1.7716	1.0710 1.8594	0.9740 2.0121	0.7495 2.1381	0.6751 2.2403
13	0 1	1.3917 1.8244	0.8677 1.9692	0.9962 2.0006	0.6951 2.1397	0.8683 2.1191	1.4306 1.7716	1.0710 1.8594	0.9740 2.0120	0.7494 2.1381	0.6751 2.2403
14	2 3	1.3646 1.8360	0.8298 1.9855	0.9958 2.0008	0.6706 2.1502	0.8547 2.1250	1.3174 1.8295	0.9280 1.9325	0.9723 2.0130	0.6462 2.1908	0.6332 2.2618
15	0 3	1.3646 1.8360	0.8298 1.9855	0.9958 2.0008	0.6706 2.1502	0.8547 2.1250	1.3173 1.8295	0.9280 1.9326	0.9723 2.0130	0.6462 2.1908	0.6332 2.2617

Figure 4.4.4(i) Minimax estimator $\mathbf{b}^*_{2\sigma_1^2}$ if the prior intervals have lengths (400, 4)

Number of experiment	β_o	$\mathbf{x}^{(1)}, \varepsilon^{(1)}$	$\mathbf{x}^{(1)}, \varepsilon^{(2)}$	$\mathbf{x}^{(1)}, \varepsilon^{(3)}$	$\mathbf{x}^{(1)}, \varepsilon^{(4)}$	$\mathbf{x}^{(1)}, \varepsilon^{(5)}$	$\mathbf{x}^{(2)}, \varepsilon^{(1)}$	$\mathbf{x}^{(2)}, \varepsilon^{(2)}$	$\mathbf{x}^{(2)}, \varepsilon^{(3)}$	$\mathbf{x}^{(2)}, \varepsilon^{(4)}$	$\mathbf{x}^{(2)}, \varepsilon^{(5)}$
GLSE $\mathbf{b}_o$		1.3805 1.8292	0.8492 1.9772	0.9960 2.0007	0.6811 2.1457	0.8607 2.1224	1.3856 1.7946	1.0072 1.8921	0.9731 2.0125	0.6891 2.1689	0.6488 2.2537
1	1 2	1.3781 1.8302	0.8487 1.9773	0.9960 2.0007	0.6828 2.1449	0.8615 2.1221	1.3739 1.8006	0.9995 1.8960	0.9731 2.0125	0.6979 2.1644	0.6542 2.2510
2	2 2	1.3782 1.8302	0.8488 1.9773	0.9960 2.0007	0.6829 2.1449	0.8616 2.1220	1.3740 1.8005	0.9997 1.8959	0.9732 2.0125	0.6981 2.1644	0.6542 2.2510
3	0 2	1.3781 1.8302	0.8487 1.9774	0.9960 2.0007	0.6828 2.1450	0.8615 2.1221	1.3737 1.8006	0.9993 1.8961	0.9731 2.0125	0.6977 2.1645	0.6541 2.2511
8	1 1	1.3917 1.8244	0.8676 1.9692	0.9962 2.0006	0.6950 2.1397	0.8683 2.1191	1.4305 1.7716	1.0710 1.8594	0.9740 2.0121	0.7495 2.1380	0.6751 2.2403

Figure 4.4.5(i) Minimax estimator $\mathbf{b}^*_{2\sigma_1^2}$ if the prior intervals have lengths (100, 4)

Number of experiment	β_o	$\mathbf{x}^{(1)}, \varepsilon^{(1)}$	$\mathbf{x}^{(1)}, \varepsilon^{(2)}$	$\mathbf{x}^{(1)}, \varepsilon^{(3)}$	$\mathbf{x}^{(1)}, \varepsilon^{(4)}$	$\mathbf{x}^{(1)}, \varepsilon^{(5)}$	$\mathbf{x}^{(2)}, \varepsilon^{(1)}$	$\mathbf{x}^{(2)}, \varepsilon^{(2)}$	$\mathbf{x}^{(2)}, \varepsilon^{(3)}$	$\mathbf{x}^{(2)}, \varepsilon^{(4)}$	$\mathbf{x}^{(2)}, \varepsilon^{(5)}$
GLSE $\mathbf{b}_o$		1.3805 1.8292	0.8492 1.9772	0.9960 2.0007	0.6811 2.1457	0.8607 2.1224	1.3856 1.7946	1.0072 1.8921	0.9731 2.0125	0.6891 2.1689	0.6488 2.2537
1	1 2	1.3799 1.8295	0.8490 1.9772	0.9960 2.0007	0.6815 2.1453	0.8609 2.1223	1.3826 1.7961	1.0052 1.8931	0.9731 2.0125	0.6913 2.1678	0.6502 2.2530
2	2 2	1.3799 1.8295	0.8490 1.9772	0.9960 2.0007	0.6815 2.1455	0.8609 2.1223	1.3826 1.7961	1.0053 1.8931	0.9731 2.0125	0.6914 2.1678	0.6501 2.2530
3	0 2	1.3799 1.8295	0.8490 1.9772	0.9960 2.0007	0.6815 2.1455	0.8609 2.1223	1.3826 1.7961	1.0052 1.8931	0.9731 2.0125	0.6913 2.1678	0.6502 2.2530
8	1 1	1.3833 1.8280	0.8538 1.9752	0.9960 2.0006	0.6846 2.1442	0.8626 2.1216	1.3971 1.7887	1.0236 1.8836	0.9734 2.0124	0.7045 2.1610	0.6554 1.2503

Figure 4.4.6(i) Minimax estimator $\mathbf{b}^*_{2\sigma_1^2}$ if the prior intervals have lengths (400, 8)

Number of experiment	$\beta_0-\beta$	$\mathbf{x}^{(1)}, \varepsilon^{(1)}$	$\mathbf{x}^{(1)}, \varepsilon^{(2)}$	$\mathbf{x}^{(1)}, \varepsilon^{(3)}$	$\mathbf{x}^{(1)}, \varepsilon^{(4)}$	$\mathbf{x}^{(1)}, \varepsilon^{(5)}$	$\mathbf{x}^{(2)}, \varepsilon^{(1)}$	$\mathbf{x}^{(2)}, \varepsilon^{(2)}$	$\mathbf{x}^{(2)}, \varepsilon^{(3)}$	$\mathbf{x}^{(2)}, \varepsilon^{(4)}$	$\mathbf{x}^{(2)}, \varepsilon^{(5)}$
$\mathbf{b}_0-\beta$		0.38 −0.17	−0.15 −0.02	−0.004 0.0007	−0.32 0.15	−0.14 0.12	0.39 −0.21	0.007 −0.11	−0.03 0.01	−0.31 0.17	−0.35 0.25
1	0 0	0.38 −0.17	−0.15 −0.02	−0.004 0.0007	−0.32 0.14	−0.14 0.12	0.38 −0.20	0.005 −0.11	−0.03 0.01	−0.31 0.17	−0.35 0.25
2	1 0	0.38 −0.17	−0.15 −0.02	−0.004 0.0007	−0.32 0.14	−0.14 0.12	0.38 −0.20	0.005 −0.11	−0.03 0.01	−0.31 0.17	−0.35 0.25
3	−1 0	0.38 −0.17	−0.15 −0.02	−0.004 0.0007	−0.32 0.14	−0.14 0.12	0.38 −0.20	0.005 −0.11	−0.03 0.01	−0.31 0.17	−0.35 0.25
4	2 0	0.38 −0.17	−0.15 −0.02	−0.004 0.0007	−0.32 0.14	−0.14 0.12	0.38 −0.20	0.005 −0.11	−0.03 0.01	−0.31 0.17	−0.35 0.25
5	−2 0	0.38 −0.17	−0.15 −0.02	−0.004 0.0007	−0.32 0.14	−0.14 0.12	0.38 −0.20	0.005 −0.11	−0.03 0.01	−0.31 0.17	−0.35 0.25
6	3 0	0.38 −0.17	−0.15 −0.02	−0.004 0.0007	−0.32 0.14	−0.14 0.12	0.38 −0.20	0.005 −0.11	−0.03 0.01	−0.31 0.17	−0.35 0.25
7	−3 0	0.38 −0.17	−0.15 −0.02	−0.0004 0.0007	−0.32 0.14	−0.14 0.12	0.38 −0.20	0.005 −0.11	−0.03 0.01	−0.31 0.17	−0.35 0.25
8	0 −1	0.38 −0.17	−0.15 −0.02	−0.004 0.0007	−0.32 0.14	−0.14 0.12	0.40 −0.21	0.02 −0.12	−0.03 0.01	−0.30 0.16	−0.34 0.25
9	0 1	0.38 −0.17	−0.16 −0.02	−0.004 0.0007	−0.32 0.15	−0.14 0.12	0.37 −0.20	−0.013 −0.098	−0.03 0.01	−0.32 0.17	−0.36 0.26
10	0 −1.5	0.38 −0.17	−0.14 −0.03	−0.004 0.0006	−0.32 0.14	−0.14 0.12	0.40 −0.22	0.03 −0.12	−0.03 0.01	−0.29 0.16	−0.34 0.25
11	0 1.5	0.37 −0.17	−0.16 −0.02	−0.004 0.0008	−0.32 0.15	−0.14 0.12	0.36 −0.19	−0.02 −0.093	−0.03 0.01	−0.33 0.18	−0.36 0.26
12	1 −1	0.38 −0.17	−0.15 −0.12	−0.004 0.0007	−0.32 0.14	−0.14 0.12	0.40 −0.21	0.02 −0.12	−0.03 0.01	−0.30 0.16	−0.34 0.25
13	−1 −1	0.38 −0.17	−0.15 −0.02	−0.004 0.0006	−0.32 0.14	−0.14 0.12	0.40 −0.21	0.02 −0.12	−0.03 0.01	−0.30 0.16	−0.34 0.25
14	1 1	0.38 −0.17	−0.16 −0.02	−0.004 0.0007	−0.32 0.15	−0.14 0.12	0.37 −0.20	−0.013 −0.097	−0.03 0.01	−0.32 0.17	−0.36 0.26
15	−1 1	0.38 −0.17	−0.16 −0.02	−0.004 0.0007	−0.32 0.15	−0.14 0.12	0.37 −0.20	−0.013 −0.098	−0.03 0.01	−0.32 0.17	−0.36 0.26

Figure 4.4.2(ii) Deviation $(\mathbf{b}^*_{2\sigma_1^2}-\boldsymbol{\beta})$ if the prior intervals have lengths (200, 8).

Number of experiment	$\beta_0-\beta$	$\mathbf{x}^{(1)}, \varepsilon^{(1)}$	$\mathbf{x}^{(1)}, \varepsilon^{(2)}$	$\mathbf{x}^{(1)}, \varepsilon^{(3)}$	$\mathbf{x}^{(1)}, \varepsilon^{(4)}$	$\mathbf{x}^{(1)}, \varepsilon^{(5)}$	$\mathbf{x}^{(2)}, \varepsilon^{(1)}$	$\mathbf{x}^{(2)}, \varepsilon^{(2)}$	$\mathbf{x}^{(2)}, \varepsilon^{(3)}$	$\mathbf{x}^{(2)}, \varepsilon^{(4)}$	$\mathbf{x}^{(2)}, \varepsilon^{(5)}$
$\mathbf{b}_0-\beta$		0.38 −0.17	−0.15 −0.02	−0.004 0.0007	−0.32 0.15	−0.14 0.12	0.39 −0.21	0.007 −0.11	−0.03 0.01	−0.31 0.17	−0.35 0.25
1	0 0	0.38 −0.17	−0.15 −0.02	−0.004 0.0007	−0.32 0.15	−0.14 0.12	0.37 −0.20	−0.0005 −0.10	−0.03 0.01	−0.30 0.16	−0.35 0.25
2	1 0	0.38 −0.17	−0.15 −0.02	−0.004 0.0007	−0.32 0.15	−0.14 0.12	0.37 −0.20	−0.0004 −0.10	−0.03 0.01	−0.30 0.16	−0.35 0.25
3	−1 0	0.38 −0.17	−0.15 −0.02	−0.004 0.0007	−0.32 0.14	−0.14 0.12	0.37 −0.20	−0.0006 −0.10	−0.03 0.01	−0.30 0.16	−0.35 0.25
4	2 0	0.38 −0.17	−0.15 −0.02	−0.004 0.0007	−0.32 0.15	−0.14 0.12	0.37 −0.20	−0.0004 −0.10	−0.03 0.01	−0.30 0.16	−0.35 0.25
5	−2 0	0.38 −0.17	−0.15 −0.02	−0.004 0.0007	−0.32 0.14	−0.14 0.12	0.37 −0.20	−0.0006 −0.10	−0.03 0.01	−0.30 0.16	−0.35 0.25
6	3 0	0.38 −0.17	−0.15 −0.02	−0.004 0.0007	−0.32 0.15	−0.14 0.12	0.37 −0.20	−0.0003 −0.10	−0.03 0.01	−0.30 0.16	−0.35 0.25
7	−3 0	0.38 −0.17	−0.15 −0.02	−0.004 0.0007	−0.32 0.14	−0.14 0.12	0.37 −0.20	−0.0007 −0.10	−0.03 0.01	−0.30 0.16	−0.35 0.25
8	0 −1	0.39 −0.18	−0.13 −0.03	−0.004 0.0006	−0.30 0.14	−0.13 0.12	0.43 −0.23	0.07 −0.14	−0.03 0.01	−0.25 0.14	−0.32 0.24
9	0 1	0.36 −0.16	−0.17 −0.01	−0.004 0.0008	−0.33 0.15	−0.15 0.13	0.32 −0.17	−0.07 −0.07	−0.03 0.01	−0.35 0.19	−0.37 0.26
10	0 −1.5	0.40 −0.18	−0.12 −0.03	−0.004 0.0006	−0.30 0.14	−0.13 0.12	0.46 −0.24	0.11 −0.16	−0.03 0.01	−0.22 0.12	−0.31 0.23
11	0 1.5	0.36 −0.16	−0.18 −0.01	−0.004 0.0008	−0.34 0.15	−0.15 0.13	0.29 −0.16	−0.11 −0.05	−0.03 0.01	−0.38 0.20	−0.38 0.27
12	1 −1	0.39 −0.18	−0.13 −0.03	−0.004 0.0006	−0.30 0.14	−0.13 0.12	0.43 −0.23	0.07 −0.14	−0.03 0.01	−0.25 0.14	−0.32 0.24
13	−1 −1	0.39 −0.18	−0.13 −0.03	−0.004 0.0006	−0.30 0.14	−0.13 0.12	0.43 −0.23	0.07 −0.14	−0.03 0.01	−0.25 0.14	−0.32 0.24
14	1 1	0.36 −0.16	−0.17 −0.01	−0.004 0.0008	−0.33 0.15	−0.15 0.13	0.32 −0.17	−0.07 −0.07	−0.03 0.01	−0.35 0.19	−0.37 0.26
15	−1 1	0.36 −0.16	−0.17 −0.01	−0.004 0.0008	−0.33 0.15	−0.15 0.13	0.32 −0.17	−0.07 −0.07	−0.03 0.01	−0.35 0.19	−0.37 0.26

Figure 4.4.3(ii) Deviation $(\mathbf{b}^*_{2\sigma_1^2}-\boldsymbol{\beta})$ if the prior intervals have lengths (200, 4)

Number of experiment	$\beta_0-\beta$	$\mathbf{x}^{(1)}, \varepsilon^{(1)}$	$\mathbf{x}^{(1)}, \varepsilon^{(2)}$	$\mathbf{x}^{(1)}, \varepsilon^{(3)}$	$\mathbf{x}^{(1)}, \varepsilon^{(4)}$	$\mathbf{x}^{(1)}, \varepsilon^{(5)}$	$\mathbf{x}^{(2)}, \varepsilon^{(1)}$	$\mathbf{x}^{(2)}, \varepsilon^{(2)}$	$\mathbf{x}^{(2)}, \varepsilon^{(3)}$	$\mathbf{x}^{(2)}, \varepsilon^{(4)}$	$\mathbf{x}^{(2)}, \varepsilon^{(5)}$
$\mathbf{b}_0-\boldsymbol{\beta}$		0.38 −0.17	−0.15 −0.02	−0.004 0.0007	−0.32 0.15	−0.14 0.12	0.39 −0.21	0.007 −0.11	−0.03 0.01	−0.31 0.17	−0.35 0.25
1	0 0	0.38 −0.17	−0.15 −0.02	−0.004 0.0007	−0.32 0.14	−0.14 0.12	0.37 0.20	−0.0005 −0.10	−0.03 0.01	−0.30 0.16	−0.35 0.25
2	1 0	0.38 −0.17	−0.15 −0.02	−0.004 0.0007	−0.32 0.14	−0.14 0.12	0.37 −0.20	−0.0005 −0.10	−0.03 0.01	−0.30 0.16	−0.35 0.25
3	−1 0	0.38 −0.17	−0.15 −0.02	−0.004 0.0007	−0.32 0.14	−0.14 0.12	0.37 −0.20	−0.0005 −0.10	−0.03 0.01	−0.30 0.16	−0.35 0.25
4	2 0	0.38 −0.17	−0.15 −0.02	−0.004 0.0007	−0.32 0.14	−0.14 0.12	0.37 −0.20	−0.0005 −0.10	−0.03 0.01	−0.30 0.16	−0.35 0.25
5	−2 0	0.38 −0.17	−0.15 −0.02	−0.004 0.0007	−0.32 0.14	−0.14 0.12	0.37 −0.20	−0.0005 −0.10	−0.03 0.01	−0.30 0.16	−0.35 0.25
6	3 0	0.38 −0.17	−0.15 −0.02	−0.004 0.0007	−0.32 0.14	−0.14 0.12	0.37 −0.20	−0.0005 −0.10	−0.03 0.01	−0.30 0.16	−0.35 0.25
7	−3 0	0.38 −0.17	−0.15 −0.02	−0.004 0.0007	−0.32 0.14	−0.14 0.12	0.37 −0.20	−0.0006 −0.10	−0.03 0.01	−0.30 0.16	−0.35 0.25
8	0 −1	0.39 −0.18	−0.13 −0.03	−0.004 0.0006	−0.30 0.14	−0.13 0.12	0.43 −0.23	0.07 −0.14	−0.03 0.01	−0.25 0.14	−0.32 0.24
9	0 1	0.36 −0.16	−0.17 −0.01	−0.004 0.0008	−0.33 0.15	−0.14 0.12	0.32 −0.17	−0.07 −0.07	−0.03 0.01	−0.35 0.19	−0.37 0.26
10	0 −1.5	0.40 −0.18	−0.12 −0.03	−0.004 0.0006	−0.30 0.14	−0.13 0.12	0.46 −0.24	0.11 −0.16	−0.03 0.01	−0.22 0.12	−0.31 0.23
11	0 1.5	0.36 −0.16	−0.18 −0.01	−0.004 0.0008	−0.34 0.15	−0.15 0.13	0.29 −0.16	−0.11 −0.05	−0.03 0.01	−0.38 0.20	−0.38 0.27
12	1 −1	0.39 −0.18	−0.13 −0.03	−0.004 0.0006	−0.30 0.14	−0.13 0.12	0.43 −0.23	0.07 −0.14	−0.03 0.01	−0.25 0.14	−0.32 0.24
13	−1 −1	0.39 −0.18	−0.13 −0.03	−0.004 0.0006	−0.30 0.14	−0.13 0.12	0.43 −0.23	0.07 −0.14	−0.03 0.01	−0.25 0.14	−0.32 0.24
14	1 1	0.36 −0.16	−0.17 −0.01	−0.004 0.0008	−0.33 0.15	−0.14 0.12	0.32 −0.17	−0.07 −0.07	−0.03 0.01	−0.35 0.19	−0.37 0.26
15	−1 1	0.36 −0.16	−0.17 −0.01	−0.004 0.0008	−0.33 0.15	−0.14 0.12	0.32 −0.17	−0.07 −0.07	−0.03 0.01	−0.35 0.19	−0.37 0.26

Figure 4.4.4(ii) Deviation $(\mathbf{b}^*_{2\sigma_1^2}-\boldsymbol{\beta})$ if the prior intervals have lengths (400, 4)

Number of experiment	$\beta_0-\beta$	$\mathbf{x}^{(1)}, \varepsilon^{(1)}$	$\mathbf{x}^{(1)}, \varepsilon^{(2)}$	$\mathbf{x}^{(1)}, \varepsilon^{(3)}$	$\mathbf{x}^{(1)}, \varepsilon^{(4)}$	$\mathbf{x}^{(1)}, \varepsilon^{(5)}$	$\mathbf{x}^{(2)}, \varepsilon^{(1)}$	$\mathbf{x}^{(2)}, \varepsilon^{(2)}$	$\mathbf{x}^{(2)}, \varepsilon^{(3)}$	$\mathbf{x}^{(2)}, \varepsilon^{(4)}$	$\mathbf{x}^{(2)}, \varepsilon^{(5)}$
$\mathbf{b}_0-\boldsymbol{\beta}$		0.38 −0.17	−0.15 −0.02	−0.004 0.0007	−0.32 0.15	−0.14 0.12	0.39 −0.21	0.007 −0.11	−0.03 0.01	−0.31 0.17	−0.35 0.25
1	0 0	0.38 −0.17	−0.15 −0.02	−0.004 0.0007	−0.32 0.14	−0.14 0.12	0.37 −0.20	−0.0005 −0.10	−0.03 0.01	−0.30 0.16	−0.35 0.25
2	1 0	0.38 −0.17	−0.15 −0.02	−0.004 0.0007	−0.32 0.14	−0.14 0.12	0.37 −0.20	−0.0003 −0.10	−0.03 0.01	−0.30 0.16	−0.35 0.25
3	−1 0	0.38 −0.17	−0.15 −0.02	−0.004 0.0007	−0.32 0.14	−0.14 0.12	0.37 −0.20	−0.0007 −0.10	−0.03 0.01	−0.30 0.16	−0.35 0.25
8	0 −1	0.39 −0.18	−0.13 −0.03	−0.004 0.0006	−0.30 0.14	−0.13 0.12	0.43 −0.23	0.07 −0.14	−0.03 0.01	−0.25 0.14	−0.32 0.24

Figure 4.4.5(ii) Deviation $(\mathbf{b}^*_{2\sigma_1^2}-\boldsymbol{\beta})$ if the prior intervals have lengths (100, 4)

Number of experiment	$\beta_0-\beta$	$\mathbf{x}^{(1)}, \varepsilon^{(1)}$	$\mathbf{x}^{(1)}, \varepsilon^{(2)}$	$\mathbf{x}^{(1)}, \varepsilon^{(3)}$	$\mathbf{x}^{(1)}, \varepsilon^{(4)}$	$\mathbf{x}^{(1)}, \varepsilon^{(5)}$	$\mathbf{x}^{(2)}, \varepsilon^{(1)}$	$\mathbf{x}^{(2)}, \varepsilon^{(2)}$	$\mathbf{x}^{(2)}, \varepsilon^{(3)}$	$\mathbf{x}^{(2)}, \varepsilon^{(4)}$	$\mathbf{x}^{(2)}, \varepsilon^{(5)}$
$\mathbf{b}_0-\boldsymbol{\beta}$		0.38 −0.17	−0.15 −0.02	−0.004 0.0007	−0.32 0.15	−0.14 0.12	0.39 −0.21	0.007 −0.11	−0.03 0.01	−0.31 0.17	−0.35 0.25
1	0 0	0.38 −0.17	−0.15 −0.02	−0.004 0.0007	−0.32 0.14	−0.14 0.12	0.38 −0.20	0.005 −0.11	−0.03 0.01	−0.31 0.17	−0.35 0.25
2	1 0	0.38 −0.17	−0.15 −0.02	−0.004 0.0007	−0.32 0.14	−0.14 0.12	0.38 −0.20	0.005 −0.11	−0.03 0.01	−0.31 0.17	−0.35 0.25
3	−1 0	0.38 −0.17	−0.15 −0.02	−0.004 0.0007	−0.32 0.14	−0.14 0.12	0.38 −0.20	0.005 −0.11	−0.03 0.01	−0.31 0.17	−0.35 0.25
8	0 −1	0.38 −0.17	−0.15 −0.02	−0.004 0.0006	−0.32 0.14	−0.14 0.12	0.40 −0.21	0.02 −0.12	−0.03 0.01	−0.30 0.16	−0.34 0.25

Figure 4.4.6(ii) Deviation $(\mathbf{b}^*_{2\sigma_1^2}-\boldsymbol{\beta})$ if the prior intervals have lengths (400, 8)

Number of experiment	$(\beta_0-\beta)$ (in %)	$\mathbf{x}^{(1)}, \varepsilon^{(1)}$	$\mathbf{x}^{(1)}, \varepsilon^{(2)}$	$\mathbf{x}^{(1)}, \varepsilon^{(3)}$	$\mathbf{x}^{(1)}, \varepsilon^{(4)}$	$\mathbf{x}^{(1)}, \varepsilon^{(5)}$	$\mathbf{x}^{(2)}, \varepsilon^{(1)}$	$\mathbf{x}^{(2)}, \varepsilon^{(2)}$	$\mathbf{x}^{(2)}, \varepsilon^{(3)}$	$\mathbf{x}^{(2)}, \varepsilon^{(4)}$	$\mathbf{x}^{(2)}, \varepsilon^{(5)}$
$(\mathbf{b}_0-\beta)$ (in %)		38.05 −8.54	−15.08 −1.14	−0.40 0.04	−31.89 7.29	−13.93 6.12	38.56 −10.27	0.72 −5.40	−2.69 0.63	−31.09 8.45	−35.12 12.69
1	0 0	37.99 −8.53	−15.09 −1.14	−0.40 0.04	−31.85 7.28	−13.91 6.12	38.26 −10.20	0.52 −5.35	−2.69 0.63	−30.86 8.39	−34.98 12.65
2	100 0	37.99 −8.53	−15.09 −1.14	−0.40 0.04	−31.85 7.28	−13.91 6.12	38.26 −10.20	0.52 −5.35	−2.69 0.63	−30.86 8.39	−34.98 12.65
3	−100 0	37.99 −8.53	−15.10 −1.14	−0.40 0.04	−31.85 7.28	−13.91 6.12	38.26 −10.20	0.52 −5.35	−2.69 0.63	−30.87 8.39	−34.99 12.65
4	200 0	37.99 −8.53	−15.09 −1.14	−0.40 0.04	−31.85 7.28	−13.91 6.12	38.27 −10.20	0.54 −5.35	−2.69 0.63	−30.86 8.39	−34.98 12.65
5	−200 0	37.99 −8.53	−15.09 −1.14	−0.40 0.04	−31.85 7.28	−13.91 6.12	38.25 −10.19	0.51 −5.35	−2.69 0.63	−30.87 8.39	−34.99 12.56
6	300 0	37.99 −8.53	−15.09 −1.14	−0.40 0.04	−31.84 7.28	−13.91 6.12	38.28 −10.20	0.54 −5.35	−2.69 0.63	−30.85 8.39	−34.98 12.65
7	−300 0	37.99 −8.53	−15.09 −1.14	−0.40 0.04	−31.86 7.28	−13.91 6.12	38.25 −10.19	0.50 −5.35	−2.69 0.63	−30.88 8.39	−34.99 12.56
8	0 −50	38.33 −8.60	−14.62 −1.24	−0.40 0.40	−31.54 7.21	−13.74 6.08	39.71 −10.57	2.36 −5.82	−2.66 0.62	−29.55 8.05	−34.46 12.52
9	0 50	37.65 −8.45	−15.57 −1.04	−0.39 0.04	−32.16 7.34	−14.08 6.16	36.81 −9.83	−1.32 −4.88	−2.71 0.63	−32.18 8.73	−35.51 12.79
10	0 −75	38.50 −8.64	−14.38 −1.80	−0.40 0.03	−31.39 7.18	−13.66 6.06	40.43 −10.75	3.28 −6.05	−2.65 0.62	−28.89 7.89	−34.19 12.45
11	0 75	37.48 −8.42	−15.81 −0.99	−0.40 0.04	−32.31 7.38	−14.16 6.17	36.09 −9.64	−2.23 −4.64	−2.72 0.64	−32.84 8.90	−35.78 12.86
12	100 −50	38.33 −8.60	−14.62 −1.25	−0.40 0.04	−31.54 7.21	−13.74 6.08	39.71 −10.57	2.37 −5.82	−2.67 0.62	−29.54 8.05	−34.46 12.52
13	−100 −50	38.33 −8.60	−14.62 −1.24	−0.40 0.03	−31.55 7.21	−13.74 6.08	39.70 −10.57	2.35 −5.82	−2.66 0.62	−29.55 8.06	−34.46 12.52
14	100 50	37.65 −8.45	−15.57 −1.04	−0.40 0.04	−32.15 7.34	−14.08 6.16	36.82 −9.83	−1.31 4.88	−2.71 0.63	−32.18 8.73	−35.51 12.79
15	−100 50	37.65 −8.45	−15.57 −1.04	−0.40 0.04	−32.16 7.34	−14.08 6.16	36.81 9.83	−1.32 −4.88	−2.71 0.63	−32.18 8.73	−35.52 12.79

Figure 4.4.2(iii) Percentage deviation of $\mathbf{b}^*_{2\sigma_1^2}$ from $\boldsymbol{\beta}$ if the prior intervals have lengths (200, 8)

Number of experiment	$(\beta_0-\beta)$ (in %)	$\mathbf{x}^{(1)}, \varepsilon^{(1)}$	$\mathbf{x}^{(1)}, \varepsilon^{(2)}$	$\mathbf{x}^{(1)}, \varepsilon^{(3)}$	$\mathbf{x}^{(1)}, \varepsilon^{(4)}$	$\mathbf{x}^{(1)}, \varepsilon^{(5)}$	$\mathbf{x}^{(2)}, \varepsilon^{(1)}$	$\mathbf{x}^{(2)}, \varepsilon^{(2)}$	$\mathbf{x}^{(2)}, \varepsilon^{(3)}$	$\mathbf{x}^{(2)}, \varepsilon^{(4)}$	$\mathbf{x}^{(2)}, \varepsilon^{(5)}$
$(\mathbf{b}_0-\beta)$ (in %)		38.05 -8.54	-15.08 -1.14	-0.40 0.04	-31.89 7.29	-13.93 6.12	38.56 -10.27	0.72 -5.40	-2.69 0.65	-31.09 8.45	-35.12 12.69
1	0 0	37.82 -8.49	-15.13 -1.14	-0.40 0.04	-31.72 7.25	-13.85 6.11	37.40 -9.98	-0.05 -5.20	2.69 0.53	-30.21 8.23	-34.59 12.55
2	100 0	37.82 -8.49	-15.13 -1.14	-0.40 0.04	-31.72 7.25	-13.85 6.11	37.40 -9.98	-0.04 -5.20	-2.69 0.63	-30.21 8.22	-34.58 12.55
3	-100 0	37.82 -8.49	-15.13 -1.14	-0.40 0.04	-31.72 7.25	-13.85 6.11	37.39 -9.97	0.06 -5.20	-2.69 0.63	-30.22 8.23	-34.59 12.55
4	200 0	37.82 -8.49	-15.12 -1.14	-0.40 0.04	-31.71 7.25	-13.85 6.11	37.41 -9.98	-0.04 -5.24	-2.69 0.63	-30.21 8.22	-34.58 12.59
5	-200 0	37.82 -8.49	-15.13 -1.13	-0.40 0.04	-31.72 7.25	-13.85 6.11	37.39 -9.97	0.06 -5.20	-2.69 0.63	-30.22 8.23	34.59 12.55
6	300 0	37.82 -8.49	-15.12 -1.14	-0.40 0.04	-31.71 7.25	-13.84 6.11	37.84 -9.98	-0.03 -5.21	-2.68 0.63	30.20 8.22	-34.58 12.55
7	-300 0	37.81 -8.49	-15.13 -1.13	-0.40 0.04	-31.72 7.25	-13.85 6.11	37.39 -9.97	-0.07 5.20	-2.69 0.63	-30.22 9.23	-34.59 12.55
8	0 -50	39.17 -8.78	-13.23 -1.54	-0.38 0.03	-30.49 6.99	-13.17 5.96	43.06 -11.42	-7.10 -7.03	-2.60 0.61	-25.05 6.91	-32.49 12.02
9	0 50	36.46 -8.20	-17.02 -0.73	-0.42 0.04	-32.94 7.51	-14.53 6.25	31.73 -8.53	-7.20 3.37	-2.77 0.65	-35.38 9.54	-36.68 13.09
10	0 -75	39.85 -8.93	-12.29 -1.75	-0.37 0.03	-29.88 6.85	-12.83 5.89	45.89 -12.15	10.57 -7.95	-0.56 0.59	-22.47 8.25	-31.4 11.75
11	0 75	35.78 -8.06	-17.97 -0.52	-0.43 -0.04	-33.55 7.65	-14.87 6.33	28.90 -7.80	-10.77 -2.46	-2.96 0.66	-37.96 10.20	37.77 13.3
12	100 -50	39.18 -8.78	-13.23 -1.54	-0.38 0.03	-30.49 6.99	-13.17 5.96	43.06 -11.42	17.11 -7.03	-2.60 0.61	-25.05 6.90	32.4 12.0
13	-100 -50	39.17 -8.78	-13.23 -1.54	-0.38 0.03	-30.50 6.99	-13.17 5.96	43.05 -11.42	7.09 -7.03	-2.60 0.61	-35.06 6.91	-32.49 12.02
14	100 50	36.46 -8.20	-17.02 -0.73	-0.42 0.04	-32.94 7.51	-14.53 6.25	31.74 8.53	-7.19 -3.38	-2.77 0.65	-35.37 9.54	-36.68 13.09
15	-100 50	36.46 -8.20	-17.02 -0.73	-0.42 0.04	-32.94 7.51	-14.53 6.25	-31.73 -8.53	-7.21 3.37	-2.77 0.65	-35.38 9.55	-36.69 13.09

Figure 4.4.3(iii) Percentage deviation of $\mathbf{b}^*_{2\sigma_1^2}$ from $\boldsymbol{\beta}$ if the prior intervals have lengths (200, 4)

Number of experiment	$(\beta_0-\beta)$ (in %)	$\mathbf{x}^{(1)}, \varepsilon^{(1)}$	$\mathbf{x}^{(1)}, \varepsilon^{(2)}$	$\mathbf{x}^{(1)}, \varepsilon^{(3)}$	$\mathbf{x}^{(1)}, \varepsilon^{(4)}$	$\mathbf{x}^{(1)}, \varepsilon^{(5)}$	$\mathbf{x}^{(2)}, \varepsilon^{(1)}$	$\mathbf{x}^{(2)}, \varepsilon^{(2)}$	$\mathbf{x}^{(2)}, \varepsilon^{(3)}$	$\mathbf{x}^{(2)}, \varepsilon^{(4)}$	$\mathbf{x}^{(2)}, \varepsilon^{(5)}$
$(\mathbf{b}_0-\boldsymbol{\beta})$	38.05	−15.08	−0.40	−31.89	−13.93	38.56	0.72	−2.69	−31.09	−35.12	
(in %)		−8.54	−1.14	0.04	7.29	6.12	−10.27	−5.40	0.63	8.45	12.69
1	0	37.82	−15.13	−0.40	−31.72	−13.85	37.40	−0.05	−2.69	−30.22	−34.59
	0	−8.49	−1.14	0.04	7.25	6.11	−9.98	−5.20	0.63	8.22	12.55
2	100	37.82	15.13	−0.40	−31.72	−13.85	37.40	−0.05	−2.69	−30.22	−34.59
	0	−8.49	−1.14	0.04	7.25	6.11	−9.98	−5.20	0.63	8.22	12.55
3	−100	37.82	−15.13	−0.40	−31.72	−13.85	37.40	−0.05	−2.69	−30.22	−34.59
	0	−8.49	−1.14	0.04	7.25	6.11	−9.98	−5.20	0.63	8.23	12.55
4	200	37.82	−15.13	−0.40	−31.72	−13.85	37.40	−0.05	−2.69	−30.21	−34.58
	0	−8.49	−1.14	0.04	7.25	6.11	−9.98	−5.20	0.63	8.22	12.55
5	−200	37.82	−15.13	−0.40	−31.72	−13.85	37.40	−0.05	−2.69	−30.22	−34.59
	0	−8.49	−1.13	0.04	7.25	6.10	−9.98	−5.20	0.63	8.23	12.55
6	300	37.82	−15.13	−0.40	−31.72	−13.85	37.40	−0.05	−2.68	−30.21	−34.58
	0	−8.49	−1.14	0.04	7.25	6.11	9.98	−5.20	0.63	8.22	12.55
7	−300	37.81	−15.13	−0.40	−31.72	−13.85	37.40	−0.06	−2.69	−30.22	−34.59
	0	−8.49	−1.14	0.04	7.25	6.10	−9.98	−5.20	0.63	8.23	12.55
8	0	39.17	−13.23	−0.38	−30.49	−13.17	43.06	7.10	−2.60	−25.05	−32.49
	−50	−8.78	−1.54	0.03	6.99	5.96	−11.42	−7.03	0.61	6.91	12.02
9	0	34.46	−17.02	−0.42	−32.94	−14.53	31.73	−7.20	−2.77	−35.38	−36.68
	50	−8.20	−0.73	0.04	7.51	6.25	8.53	−3.37	0.65	9.54	13.09
10	0	39.85	−12.28	−0.37	−29.88	−12.83	45.89	10.67	−2.56	−22.47	−31.44
	−75	−8.93	−1.75	0.03	6.85	5.88	−12.15	7.94	0.59	6.25	11.75
11	0	35.79	−17.97	−0.43	−33.55	−14.87	28.90	−10.78	−2.81	−37.96	−37.73
	75	−8.05	−0.52	0.04	7.65	6.33	−7.81	−2.46	0.66	10.20	13.36
12	100	39.17	−13.23	−0.38	−30.49	−13.17	43.06	7.10	−2.60	−25.05	−32.49
	−50	−8.78	−1.54	0.03	6.99	5.96	−11.42	−7.03	0.60	6.91	12.02
13	−100	39.17	−13.23	−0.38	−30.49	−13.17	43.06	7.10	−2.60	−25.05	−32.49
	−50	−8.78	−1.54	0.03	6.99	5.96	−11.42	−7.03	0.60	6.91	12.02
14	100	36.46	−17.02	−0.42	−32.94	−14.53	31.74	−7.20	−2.77	−35.38	−36.68
	50	−8.20	−0.73	0.04	7.51	6.25	8.53	3.38	0.65	9.54	13.09
15	−100	36.46	−17.02	−0.42	−32.94	−14.53	31.73	−7.20	−2.77	−35.38	−36.68
	50	−8.20	−0.73	0.04	7.51	6.25	−8.53	−3.37	0.65	9.54	13.09

Figure 4.4.4(iii) Percentage deviation of $\mathbf{b}^*_{2\sigma_1^2}$ from $\boldsymbol{\beta}$ if the prior intervals have lengths (400, 4)

Number of experiment	$(\beta_0-\beta)$ (in %)	$\mathbf{x}^{(1)}, \varepsilon^{(1)}$	$\mathbf{x}^{(1)}, \varepsilon^{(2)}$	$\mathbf{x}^{(1)}, \varepsilon^{(3)}$	$\mathbf{x}^{(1)}, \varepsilon^{(4)}$	$\mathbf{x}^{(1)}, \varepsilon^{(5)}$	$\mathbf{x}^{(2)}, \varepsilon^{(1)}$	$\mathbf{x}^{(2)}, \varepsilon^{(2)}$	$\mathbf{x}^{(2)}, \varepsilon^{(3)}$	$\mathbf{x}^{(2)}, \varepsilon^{(4)}$	$\mathbf{x}^{(2)}, \varepsilon^{(5)}$
$(\mathbf{b}_0-\beta)$ (in %)		38.05	−15.08	−0.40	−31.89	−13.93	38.56	0.72	−2.69	−31.09	−35.12
		−8.54	−1.14	0.04	7.29	6.12	−10.27	−5.40	0.63	8.45	12.69
1	0	37.81	−15.13	−0.40	−31.72	−13.85	37.39	−0.05	−2.69	−30.21	−34.58
	0	−8.49	−1.14	0.04	7.25	6.11	−9.97	−5.20	0.63	8.22	12.55
2	100	37.82	−15.12	−0.40	−31.71	−13.84	37.40	−0.03	−2.68	−30.19	−34.58
	0	−8.49	−1.14	0.04	7.25	6.10	−9.98	−5.21	0.63	8.22	12.55
3	−100	37.82	−15.13	−0.40	−31.72	−13.85	37.37	−0.07	−2.69	−30.23	−34.59
	0	−8.49	−1.13	0.04	7.25	6.11	−9.97	−5.20	0.63	8.23	12.56
8	0	39.17	−13.24	−0.38	−30.50	−13.17	43.05	7.10	−2.60	−25.05	−32.49
	−50	−8.78	−2.04	0.03	6.99	5.96	−11.42	−7.03	0.61	6.90	12.02

Figure 4.4.5(iii) Percentage deviation of $(\mathbf{b}^*_{2\sigma_1^2}$ from $\boldsymbol{\beta}$ if the prior intervals have lengths (100, 4)

Number of experiment	$(\beta_0-\beta)$ (in %)	$\mathbf{x}^{(1)}, \varepsilon^{(1)}$	$\mathbf{x}^{(1)}, \varepsilon^{(2)}$	$\mathbf{x}^{(1)}, \varepsilon^{(3)}$	$\mathbf{x}^{(1)}, \varepsilon^{(4)}$	$\mathbf{x}^{(1)}, \varepsilon^{(5)}$	$\mathbf{x}^{(2)}, \varepsilon^{(1)}$	$\mathbf{x}^{(2)}, \varepsilon^{(2)}$	$\mathbf{x}^{(2)}, \varepsilon^{(3)}$	$\mathbf{x}^{(2)}, \varepsilon^{(4)}$	$\mathbf{x}^{(2)}, \varepsilon^{(5)}$
$(\mathbf{b}_0-\beta)$ (in %)		38.05	−15.08	−0.40	−31.89	−13.93	38.56	0.72	−2.69	−31.09	−35.12
		−8.54	−1.14	0.04	7.29	6.12	−10.27	−5.40	0.63	8.45	12.69
1	0	37.99	−15.10	−0.40	−31.85	−13.91	38.26	0.52	−2.69	−30.87	−34.98
	0	−8.53	−1.14	0.04	7.28	6.12	−10.20	−5.35	0.63	8.39	12.65
2	100	37.99	−15.10	−0.40	−31.85	−13.91	38.26	0.53	−2.69	−30.86	−34.99
	0	−8.53	−1.14	0.04	7.28	6.12	−10.20	−5.35	0.63	8.39	12.65
3	−100	37.99	−15.10	−0.40	−31.85	−13.91	38.26	0.52	−2.69	−30.87	−34.98
	0	−8.53	1.14	0.04	7.28	6.12	−10.20	−5.35	0.63	8.39	12.65
8	0	38.33	−14.62	−0.40	−31.54	−13.74	39.71	2.36	−2.66	−29.55	−34.46
	−50	−8.60	−1.24	0.03	7.21	6.08	−10.57	−5.82	0.62	8.05	12.52

Figure 4.4.6(iii) Percentage deviation of $\mathbf{b}^*_{2\sigma_1^2}$ from $\boldsymbol{\beta}$ if the prior intervals have lengths (400, 8)

Length of prior intervals	$\mathbf{x}^{(1)}, \varepsilon^{(1)}$	$\mathbf{x}^{(1)}, \varepsilon^{(2)}$	$\mathbf{x}^{(1)}, \varepsilon^{(3)}$	$\mathbf{x}^{(1)}, \varepsilon^{(4)}$	$\mathbf{x}^{(1)}, \varepsilon^{(5)}$	$\mathbf{x}^{(2)}, \varepsilon^{(1)}$	$\mathbf{x}^{(2)}, \varepsilon^{(2)}$	$\mathbf{x}^{(2)}, \varepsilon^{(3)}$	$\mathbf{x}^{(2)}, \varepsilon^{(4)}$	$\mathbf{x}^{(2)}, \varepsilon^{(5)}$
$V(\mathbf{b}_0)_1$	1.3783					4.2578				
$V(\mathbf{b}_0)_2$	0.2350					1.0843				
400	0.27	0.38	0.0043	0.24	0.14	1.44	1.82	0.02	1.31	0.53
8	0.29	0.41	0.0051	0.26	0.15	1.48	1.87	0.02	1.34	0.54
200	0.27	0.38	0.0043	0.24	0.14	1.44	1.83	0.02	1.32	0.53
8	0.30	0.41	0.0051	0.27	0.15	1.48	1.88	0.02	1.35	0.54
400	1.07	1.49	0.02	0.96	0.54	5.56	6.99	0.08	5.08	2.08
4	1.16	1.62	0.02	1.05	0.59	5.71	7.18	0.09	5.21	2.13
200	1.07	1.50	0.02	0.97	0.54	5.57	7.00	0.08	5.08	2.08
4	1.17	1.63	0.02	1.05	0.59	5.72	7.19	0.09	5.22	2.14
100	1.08	1.51	0.02	0.97	0.54	5.59	7.03	0.08	5.10	2.09
4	1.17	1.64	0.02	1.06	0.59	5.74	7.22	0.09	5.24	2.15

Figure 4.4.7 Percentage decrease of variances (see (4.4.12))

Using (4.3.32) with the specifications $c = 2\sigma_1^2$, $W = I$ (therefore $S = X'X$), $k = 1$, T from (4.1.7), and replacing σ^{-2} by its estimator s^{-2}, we get

$$\sigma^{-2} \sup_{(\beta-\beta_0)' T(\beta-\beta_0) \leq 1} R\,(\mathbf{b}^*_{2\sigma_1^2}, \mathbf{aa}') = \mathbf{a}' D^{-1}_{2\sigma_1^2}(X'X + k^{-1}(2\sigma_1^2)^2 s^{-2}\, T)\, D^{-1}_{2\sigma_1^2}\, \mathbf{a}. \tag{4.4.14}$$

Both risks (13) and (14) are calculated for the ex-post predictors $\mathbf{a}'\mathbf{b}_0$ and $\mathbf{a}'\mathbf{b}^*_{2\sigma_1^2}$ where we use the regressor values of the last sample, i.e. $\mathbf{a} = (1, \mathbf{x}_T^{(i)})\,(i=1,2)$.

Now, we measure the percentage decrease of the minimax risk by

$$\Delta R = \frac{\sigma^{-2} R\,(\mathbf{b}_0, \mathbf{aa}') - \sigma^{-2} \sup_{(\beta-\beta_0)' T(\beta-\beta_0) \leq 1} R\,(\mathbf{b}^*_{2\sigma_1^2}, \mathbf{aa}')}{\sigma^{-2} R\,(\mathbf{b}_0, \mathbf{aa}')} \cdot \; 100\,\%. \tag{4.4.15}$$

These results are given in Figure 4.4.8.

4.4.3 Interpretation of the Simulation Experiment

Utilizing the numerical results of the simulation experiment (see Figures 4.4.1–4.4.8) we may come to the following conclusions.

(i) Interval Lengths Fixed but Centres Changed

The change of only one component of the centre $\boldsymbol{\beta}_0$ influences the values of all components of the minimax estimator, the extent of this alteration being dependent on the centre's change as well as on the lengths of all prior intervals $a_i \leq \beta_i \leq b_i$.

For the parameter of the constant term in (4.4.6) natural restraints are not known. Thus the corresponding prior interval is chosen with a wide range (the used lengths are 100, 200, and 400, respectively). If the centre of the prior intervals of the first component of $\boldsymbol{\beta}_0$ is changed by amounts which are not greater than 4 % of the length, then the minimax estimator is influenced to an inconsiderable degree.

As far as the length of the prior interval of the second component of $\boldsymbol{\beta}$ is concerned, a shrinkage to half the length leads to a considerable decrease of the variances as well as of the risks (see Figures 4.4.7 and 4.4.8). Clearly, that decrease is independent of both the centre of the prior rectangle and the centre of the prior ellipsoid.

(ii) Centres of Intervals Fixed but Lengths of Intervals Changed

The result (5) may be seen by comparing the headlines of the corresponding figures (which gives $\mathbf{b}_0$) and the first experiment, which has as centre $\boldsymbol{\beta}_0 = \boldsymbol{\beta}$ (see Figures 4.4.2(i)–4.4.6(i)).

The experiments with $(\mathbf{x}^{(1)}, \boldsymbol{\varepsilon}^{(3)})$ and $(\mathbf{x}^{(2)}, \boldsymbol{\varepsilon}^{(3)})$ yield OLSE's $\mathbf{b}_0$ as nearly the true parameter $\boldsymbol{\beta} = \begin{pmatrix} 1 \\ 2 \end{pmatrix}$. We may therefore conclude that a high level fitting of

Lengths of prior intervals	$\mathbf{x}^{(1)}, \varepsilon^{(1)}$	$\mathbf{x}^{(1)}, \varepsilon^{(2)}$	$\mathbf{x}^{(1)}, \varepsilon^{(3)}$	$\mathbf{x}^{(1)}, \varepsilon^{(4)}$	$\mathbf{x}^{(1)}, \varepsilon^{(5)}$	$\mathbf{x}^{(2)}, \varepsilon^{(1)}$	$\mathbf{x}^{(2)}, \varepsilon^{(2)}$	$\mathbf{x}^{(2)}, \varepsilon^{(3)}$	$\mathbf{x}^{(2)}, \varepsilon^{(4)}$	$\mathbf{x}^{(2)}, \varepsilon^{(5)}$
$\sigma^{-2}R(\mathbf{b}_0\mathbf{aa}')$	0.2922					0.4327				
400 8	0.10	0.13	0.004	0.08	0.05	0.58	0.73	0.01	0.50	0.24
200 8	0.10	0.13	0.004	0.08	0.05	0.58	0.73	0.01	0.51	0.24
400 4	0.38	0.53	0.006	0.33	0.19	2.27	2.82	0.03	1.98	0.97
200 4	0.38	0.53	0.006	0.33	0.19	2.27	2.82	0.03	1.98	0.97
100 4	0.38	0.53	0.006	0.34	0.19	2.28	2.83	0.03	1.99	0.97

Figure 4.4.8 Percentage decrease of minimax risk (see (4.4.15))

regression by ordinary least squares makes the minimax method superfluous in the sense that the decrease of variance and risk is negligible (see Figures 4.4.7 and 4.4.8).

Comparing the coefficients of determination (Figure 4.4.1) and the possible gain in efficiency by using the minimax estimator (Figure 4.4.8) gives the following result.

Let the coefficient of determination in the above simulation model be $R^2 \geq 0.90$; then the decrease in risk is not greater than 1%.

So we see that properly fitted regression models bring only an inconsiderable gain in efficiency by the minimax method.

(iii) A Fixed Prior Interval for the Constant's Parameter $\boldsymbol{\beta}_1$

If we change the arrangement and the length of the prior interval for $\boldsymbol{\beta}_2$ only, we see (Figure 4.4.9 and 4.4.10) that all estimators $\mathbf{b}^*_{2\sigma_1^2}$ are on the line

$$\beta_1 = \frac{2\sigma_1^2 \beta_{0(1)} t_1 + \sum_{t=1}^{T} y_t}{2\sigma_1^2 t_1 + T} - \frac{\sum_{t=1}^{T} X_t}{2\sigma_1^2 t_1 + T} \beta_2 \tag{4.4.16}$$

where $\beta_{0(1)} = 1/2\,(a_1 + b_1)$, $t_1 = (4/2)(b_1 - a_1)^{-2}$ (see (4.1.7)) as the prior interval is $a_1 \leq \beta_1 \leq b_1$.

If the prior interval for β_1 increases we see that the line (4.4.16) converges to $\beta_1 = \bar{y} - \bar{x}\beta_2$.

4.5 FURTHER ASPECTS OF MINIMAX ESTIMATION

4.5.1 Admissibility of Linear Estimators with Respect to a Restrained Parameter Set

As pointed out in Section 2.1.5, the concept of admissibility opens up the possibility of finding out estimators of $\boldsymbol{\beta}$ in the model (4.1.8) which are best with respect to a fixed risk and a well-defined class of estimators. Definition 2.3 described admissibility with respect to the MSE-risk where $\boldsymbol{\beta}$ could vary in the whole Euclidean space E^K.

Now we shall define admissibility with respect to a restrained parameter set of type

$$B = \{\boldsymbol{\beta} : \boldsymbol{\beta}' \boldsymbol{T} \boldsymbol{\beta} \leq \sigma^2\} \tag{4.5.1}$$

which is similar to B from (4.1.1).

Furthermore, let the class of homogeneous linear estimators be defined by

$$\mathscr{C} = \{\boldsymbol{C}'\mathbf{y},\ \boldsymbol{C}' \text{ a } K \times T\text{-matrix}\}. \tag{4.5.2}$$

Definition 4.2 $\hat{\boldsymbol{\beta}}$ is called admissible in $\mathscr{C}$ for $\boldsymbol{\beta} \in B$ with respect to the quadratic risk $R(\boldsymbol{\beta}, \boldsymbol{A}) = E(\boldsymbol{\beta} - \boldsymbol{\beta})'\boldsymbol{A}(\boldsymbol{\beta} - \boldsymbol{\beta})$ if $R(\hat{\boldsymbol{\beta}}, \boldsymbol{A}) \leq R(\tilde{\boldsymbol{\beta}}, \boldsymbol{A})$ for all $\hat{\boldsymbol{\beta}} \in \mathscr{C}$, $\boldsymbol{\beta} \in B$, and $R(\hat{\boldsymbol{\beta}}, \boldsymbol{A}) < R(\hat{\boldsymbol{\beta}}, \boldsymbol{A})$ for some $\boldsymbol{\beta}$.

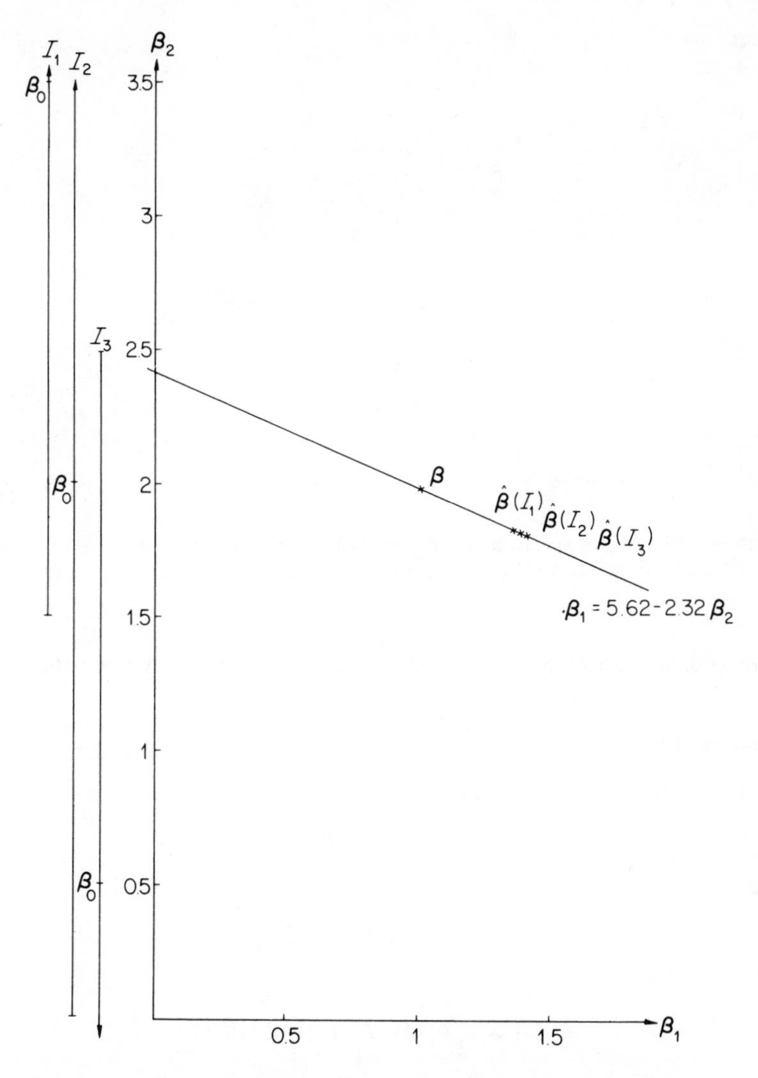

Figure 4.4.9 Minimax estimator for the model $\mathbf{y} = (\mathbf{i}, \mathbf{x}^{(1)})\boldsymbol{\beta} + \boldsymbol{\varepsilon}^{(1)}$ depending on the arrangement of the prior interval for $\boldsymbol{\beta}_2$ (length of this prior interval is const. = 4)

For convenience we abbreviate Definition 4.2 by saying that $\hat{\boldsymbol{\beta}}$ is B-admissible. We are interested in all linear estimators of $\mathscr{C}$ which are B-admissible.

We define the set of all nonnegative definite $K \times K$-matrices having eigenvalues in the interval $[0, 1]$ by

$$M = \{\boldsymbol{A} \geq \boldsymbol{0},\ (\boldsymbol{I} - \boldsymbol{A}) \geq \boldsymbol{0}\} \quad \text{(see Theorem A.14)} \tag{4.5.3}$$

and as a subset

$$\tilde{M} = \{\boldsymbol{A} \geq \boldsymbol{0},\ (\boldsymbol{I} - \boldsymbol{A}) > \boldsymbol{0}\}. \tag{4.5.4}$$

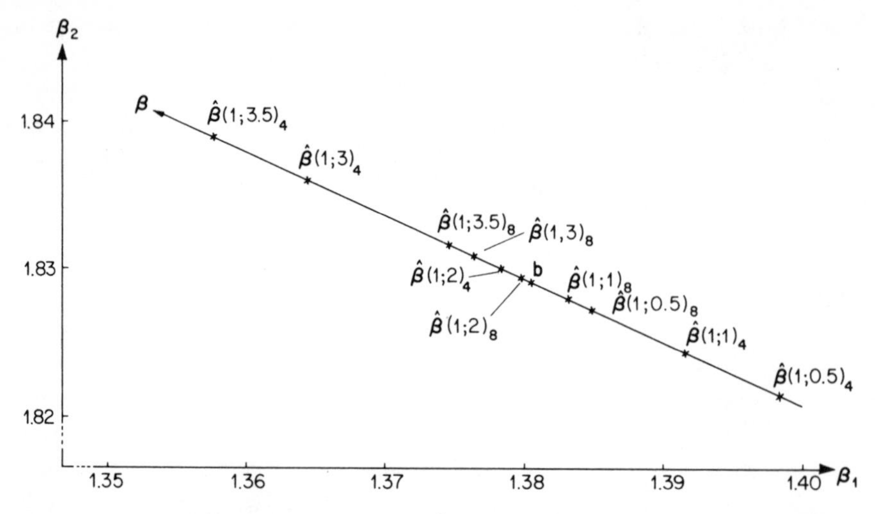

Figure 4.4.10 Minimax estimator $\mathbf{b}^*_{2\sigma_1^2} = \hat{\boldsymbol{\beta}}(\boldsymbol{\beta}_{0(1)}; \boldsymbol{\beta}_{0(2)})_{L_2}$ for the model $\mathbf{y} = (\mathbf{i}, \mathbf{x}^{(1)})\boldsymbol{\beta} + \boldsymbol{\varepsilon}^{(1)}$ which depends on the arrangement and length L_2 of the prior interval for $\boldsymbol{\beta}_2$

As before, we denote $\boldsymbol{S} = \boldsymbol{X}'\boldsymbol{W}^{-1}\boldsymbol{X}$. Then we have the following theorem.

Theorem 4.9

$\hat{\boldsymbol{\beta}} = \boldsymbol{C}'\mathbf{y}$ is B-admissible if and only if

$$\boldsymbol{C}' = \boldsymbol{L}\boldsymbol{X}'\boldsymbol{W}^{-1} \tag{4.5.5}$$

with

$$\boldsymbol{S}^{1/2}\boldsymbol{L}\boldsymbol{S}^{1/2} \in \tilde{M} \tag{4.5.6}$$

and

$$\operatorname{tr} \boldsymbol{S}^{-1}\boldsymbol{T}[(\boldsymbol{I}-\boldsymbol{L}\boldsymbol{S})^{-1} - \boldsymbol{I}] \leq 1. \tag{4.5.7}$$

Proof Let $\boldsymbol{P} = \boldsymbol{X}\boldsymbol{S}^{-1}\boldsymbol{X}'\boldsymbol{W}^{-1}$. Then we have $\boldsymbol{P}\boldsymbol{X} = \boldsymbol{X}$, $\boldsymbol{P}\boldsymbol{W}\boldsymbol{P}' = \boldsymbol{P}\boldsymbol{W}$ and therefore we get

$$\begin{aligned}
R(\boldsymbol{C}'\mathbf{y}, \boldsymbol{I}) &= E(\boldsymbol{C}'\mathbf{y} - \boldsymbol{\beta})'(\boldsymbol{C}'\mathbf{y} - \boldsymbol{\beta}) \\
&= \boldsymbol{\beta}'(\boldsymbol{C}'\boldsymbol{X} - \boldsymbol{I})'(\boldsymbol{C}'\boldsymbol{X} - \boldsymbol{I})\boldsymbol{\beta} + \sigma^2 \operatorname{tr} \boldsymbol{C}'\boldsymbol{W}\boldsymbol{C} \\
&= \boldsymbol{\beta}'(\boldsymbol{C}'\boldsymbol{P}\boldsymbol{X} - \boldsymbol{I})'(\boldsymbol{C}'\boldsymbol{P}\boldsymbol{X} - \boldsymbol{I})\boldsymbol{\beta} + \sigma^2 \operatorname{tr} \boldsymbol{C}'\boldsymbol{P}\boldsymbol{W}\boldsymbol{P}'\boldsymbol{C} \\
&\quad + \sigma^2 \operatorname{tr} \boldsymbol{C}'(\boldsymbol{I} - \boldsymbol{P})\boldsymbol{W}(\boldsymbol{I} - \boldsymbol{P})'\boldsymbol{C} \\
&\geq R(\boldsymbol{C}'\boldsymbol{P}\mathbf{y}, \boldsymbol{I})
\end{aligned}$$

and equality holds iff $\boldsymbol{C}'\boldsymbol{P} = \boldsymbol{C}'$. This is fulfilled for $\boldsymbol{C}' = \boldsymbol{L}\boldsymbol{X}'\boldsymbol{W}^{-1}$ and therefore we may confine ourselves to estimators $\hat{\boldsymbol{\beta}} = \boldsymbol{L}\boldsymbol{X}'\boldsymbol{W}^{-1}\mathbf{y}$. Now we transform the regression model $\mathbf{y} = \boldsymbol{X}\boldsymbol{\beta} + \boldsymbol{\varepsilon}$, $\boldsymbol{\varepsilon} \sim (\mathbf{0}, \sigma^2\boldsymbol{W})$ by premultiplying by the matrix

$S^{-1}X'W^{-1}$, which leads to the so-called location model

$$\tilde{y} = S^{-1}X'W^{-1}y = \beta + \tilde{\varepsilon}, \quad \tilde{\varepsilon} \sim (0, \sigma^2 S^{-1}). \tag{4.5.8}$$

It is easy to prove that $C'y$ is B-admissible for β in the regression model iff $LS\tilde{y}$ is B-admissible for β in the location model. Using Theorem 4.10 completes the proof.

Theorem 4.10

Let a location model be given by

$$y = \beta + \varepsilon, \quad \varepsilon \sim (0, \sigma^2 W) \tag{4.5.9}$$

and a prior ellipsoid by

$$B = \{\beta : \beta' T \beta \leq \sigma^2\}. \tag{4.5.10}$$

Then the estimator Ay is B-admissible iff

$$W^{1/2} A W^{1/2} \in M \text{ (see 4.5.3)}, \ I - A \text{ is regular, and}$$
$$\operatorname{tr} WT[(I-A)^{-1} - I] \leq 1$$

(for the complete proof the reader is referred to Hoffmann (1977), who has summarized as well as generalized results of Rao (1976) and Cohen (1966)).

As far as we are concerned in this book, we are interested in checking the B-admissibility of certain well-known linear estimators. Starting with the GLSE $\mathbf{b} = S^{-1}X'W^{-1}y$ we see from (5) that $L = S^{-1}$ and therefore condition (6) is not fulfilled:

$$S^{1/2} S^{-1} S^{1/2} = I \notin \tilde{M}.$$

Thus the GLSE $\mathbf{b}$ is not B-admissible. In other words, if $\beta \in B$ then there exist linear estimators which are better than $\mathbf{b}$. For that reason many attempts have been made to improve the GLSE by using the prior information $\beta \in B$.

A very important class of linear biased estimators is given by

$$\hat{\beta}(G) = (G+S)^{-1} X'W^{-1}y \tag{4.5.11}$$

where $G > 0$ is a known $K \times K$-matrix. If we set for G the following special matrices, we get

$G = kI$: the ridge estimator $\mathbf{b}(k)$ (see (4.2.28))
$G = k^{-1}\sigma^2 T$: the MILE $\mathbf{b}^*$ (4.2.7) under the prior information $\beta' T \beta \leq k$
$G = T$: the MILE $\mathbf{b}^* = (T+S)^{-1}X'W^{-1}y$ under the prior information $\beta' T \beta \leq \sigma^2$
$G = \rho S, \rho \geq 0$: a shrunken estimator $\hat{\beta}(\rho) = (\rho+1)^{-1}\mathbf{b}$ (see Mayer and Willke, 1973 and (2.3.40)).

The estimator $\hat{\beta}(G)$ defined in (11) is often called a general ridge estimator. As pointed out in Rao (1976), the symmetric matrix

$$S^{1/2}(G+S)^{-1}S^{1/2} \in \tilde{M}$$

(see (3)), so that $\hat{\boldsymbol{\beta}}(G)$ is admissible with respect to the whole parameter space E^K. In order to define the subclass of general ridge estimators which are also B-admissible we prove the following theorem.

Theorem 4.11

The general ridge estimator $\hat{\boldsymbol{\beta}}(G)$ is B-admissible iff

$$\text{tr } TG^{-1} \leq 1.$$

Proof We have to show that (7) holds, where $L = (G+S)^{-1}$. It is trivial to see that

$$I-(G+S)^{-1}S = (G^{-1}S+I)^{-1}$$

So (7) becomes

$$\text{tr } S^{-1}TG^{-1}S = \text{tr } TG^{-1} \leq 1.$$

As a conclusion of this theorem we see that the MILE $\mathbf{b}^*$ is not B-admissible. This surprising result becomes clearer if we remember that $\mathbf{b}^*$ (4.2.7) was derived as the optimal estimator with respect to the minimax risk $\sup_{\boldsymbol{\beta} \in B} R(\boldsymbol{\beta}, \mathbf{aa}')$ which has a loss matrix $A = \mathbf{aa}'$ of rank one.

If we look for the special case of a shrunken estimator

$$\hat{\boldsymbol{\beta}}(\rho) = (\rho+1)^{-1}\mathbf{b} \quad (\text{with } \mathbf{b} = S^{-1}X'W^{-1}\mathbf{y}) \tag{4.5.12}$$

we get from Theorem 4.11 that the shrunken estimator $\hat{\boldsymbol{\beta}}(\rho)$ is B-admissible iff

$$\rho \geq \text{tr } TS^{-1}. \tag{4.5.13}$$

Applying Theorem 2.10 for the constrained parameter set $B = \{\boldsymbol{\beta}: \boldsymbol{\beta}'T\boldsymbol{\beta} \leq \sigma^2\}$, we get

$$R(\mathbf{b}, A) - R(\hat{\boldsymbol{\beta}}(\rho), A) \geq 0 \quad \text{for all } \boldsymbol{\beta} \in B \tag{4.5.14}$$

if

$$\rho \leq 2\{\lambda_{\max}(T^{-1/2}AT^{-1/2})[\text{tr } S^{-1}A]^{-1} - 1\}^{-1}. \tag{4.5.15}$$

Combining the results (13) and (15) we have the following result (see also Hoffmann, 1977).

Theorem 4.12

The class of shrunken estimators $\hat{\boldsymbol{\beta}}(\rho)$ which are B-admissible, as well as being of smaller risk than the GLSE $\mathbf{b}$, may be characterized by all ρ with

$$\text{tr } TS^{-1} \leq \rho \leq 2 \text{ tr } AS^{-1}\{\lambda_{\max}(T^{-1/2}AT^{-1/2}) - \text{tr } AS^{-1}\}^{-1}. \tag{4.5.16}$$

4.5.2 Comparison of Mixed and Minimax Estimators

In situations of practical relevance certain prior information on the parameters β_i may be available in such a form that alternative mathematical formulations are possible. Let, for instance, prior restrictions of type (4.1.3) be given. As one possibility of reformulation of (4.1.3) we gave the construction of the ellipsoid (4.1.7). Another way is as follows (see Toutenburg and Roeder, 1978). To reformulate (4.1.3) as a linear restriction of type $\mathbf{r} = \boldsymbol{R\beta} + \mathbf{g} + \boldsymbol{\phi}$ (see (3.3.1)) we have to replace the certain events $\beta_i \in [a_i, b_i]$ by the weakened stochastic events: 'β_i is almost certainly an element of $[a_i, b_i]$'. 'Almost certainly' may be interpreted in different ways, e.g. in the sense of a 3σ-rule it may be applied to the error term $\boldsymbol{\phi}$ of the linear restriction. Writing

$$\tfrac{1}{2}(a_i + b_i) = \beta_i + g_i + \phi_i \quad (i = 1, \ldots, K) \tag{4.5.17}$$

gives

$$\mathbf{r} - \mathbf{g} = \frac{1}{2}\begin{pmatrix} a_1 + b_1 \\ \vdots \\ a_K + b_K \end{pmatrix}, \quad \boldsymbol{R} = \boldsymbol{I}_K \tag{4.5.18}$$

and $\boldsymbol{\phi} \sim (\mathbf{0}, \boldsymbol{V})$ with $\boldsymbol{V} = \operatorname{diag}(\sigma_1^2, \ldots, \sigma_K^2)$, where the 3σ-rule yields for any nonspecified distribution

$$P(|\mathbf{r} - \mathbf{g} - \boldsymbol{R\beta}|_i > 3\sigma_i) < \tfrac{1}{9} \quad (i = 1, \ldots, K).$$

For instance, we could choose $3\sigma_i = \frac{1}{2}(b_i - a_i)$ which gives for the covariance matrix of $\boldsymbol{\phi}$

$$\boldsymbol{V} = \tfrac{1}{36}\operatorname{diag}((b_1 - a_1)^2, \ldots, (b_K - a_K)^2). \tag{4.5.19}$$

Now, we have made all specifications of the linear restriction $\mathbf{r} = \boldsymbol{R\beta} + \mathbf{g} + \boldsymbol{\phi}$ so that the corresponding mixed estimator (3.3.9) is available.

This example of reformulating the cuboid (4.1.3) as weakened prior information in the form of an ellipsoid or as a linear restriction gives one argument for comparing the mixed and the minimax estimator with respect to a specified risk. This will be done in the following, where we leave the special case described above and handle the problem more generally (see Teräsvirta, 1980).

Without loss of generality we confine ourselves to the classical regression

$$\mathbf{y} = \boldsymbol{X\beta} + \boldsymbol{\varepsilon}, \quad \boldsymbol{\varepsilon} \sim (\mathbf{0}, \sigma^2\boldsymbol{I})$$

and to the special prior information

$$\boldsymbol{\beta} \in B = \{\boldsymbol{\beta} : \boldsymbol{\beta}'\boldsymbol{R}'\boldsymbol{R\beta} \le \sigma^2/c\}, \quad c > 0 \tag{4.5.20}$$

as well as

$$\mathbf{r} = \boldsymbol{R\beta} + \mathbf{g} + \boldsymbol{\phi}, \; \boldsymbol{\phi} \sim (\mathbf{0}, (\sigma^2/k)\boldsymbol{I}) \tag{4.5.21}$$

(see (3.3.8)). Such a situation (21) occurs when the prior opinion '$\boldsymbol{R\beta}$ is close to zero' is given a stochastic linear form. Nevertheless, $\boldsymbol{R\beta} \neq \mathbf{0}$ at probability one according to (20) (the term $\mathbf{g}$ is introduced to measure the deviation $\mathbf{r} - \boldsymbol{R\beta}$). Then the mixed estimator (see (3.3.9))

$$\mathbf{b}_R(k) = (\boldsymbol{X'X} + \boldsymbol{R'R})^{-1}(\boldsymbol{X}'\mathbf{y} + k\boldsymbol{R}'\mathbf{r}) \tag{4.5.22}$$

is biased because relation (21) is biased information due to

$$\mathbf{r} \sim (\boldsymbol{R\beta} + \mathbf{g}, (\sigma^2/k)\boldsymbol{I}).$$

The minimax linear estimator under the prior restriction (20) is deduced from (4.2.7) as

$$\mathbf{b}^* = (\boldsymbol{X'X} + c\boldsymbol{R'R})^{-1}\boldsymbol{X'Y} = \mathbf{b}_R^*(c), \text{ say.} \tag{4.5.23}$$

The stochastic properties of both estimators (22) and (23) are not identical because of the stochastic nature of $\mathbf{r}$. Since $\mathbf{r}$ and $\boldsymbol{\phi}$ are both stochastic, whereas $\boldsymbol{\beta}$ is not, it is not legitimate to claim that (23) is obtained from (22) by setting $\mathbf{r} = \mathbf{0}$, as Taylor (1974) does. This last operation is possible only if $k \to \infty$ in (21) so that at the limit $\boldsymbol{\phi} \equiv \mathbf{0}$. When $k \to \infty$ in (21) and $c \to \infty$ in (20) the two estimators do become identical, the limiting estimator being the restricted least squares estimator $\mathbf{b}_R$

Now, we shall compare the two estimators with respect to the risk $R(\hat{\boldsymbol{\beta}}, \boldsymbol{A}) = \operatorname{tr} \boldsymbol{A}E(\hat{\boldsymbol{\beta}} - \boldsymbol{\beta})(\hat{\boldsymbol{\beta}} - \boldsymbol{\beta})'$. As mentioned above, if rank $\boldsymbol{A} > 1$, then $\mathbf{b}_R^*(c)$ is no longer a minimax estimator. The risk of the mixed estimator is (see (3.3.15))

$$R(\mathbf{b}_R(k), \boldsymbol{A}) = \operatorname{tr} \boldsymbol{A}[\sigma^2\boldsymbol{U} - \sigma^2\boldsymbol{UR'S}_k\boldsymbol{RU} + \boldsymbol{UR'S}_k\mathbf{gg}'\boldsymbol{S}_k\boldsymbol{RU}] \tag{4.5.24}$$

where $\boldsymbol{U} = (\boldsymbol{X'X})^{-1}$ and $\boldsymbol{S}_k = (k^{-1}\boldsymbol{I} + \boldsymbol{RUR}')^{-1}$.

For the minimax estimator $\mathbf{b}_R^*(c)$ we have to specify the risk given in (4.2.15) by using $\boldsymbol{T} = \boldsymbol{R'R}$, $k = \sigma^2/c$, and $\boldsymbol{W} = \boldsymbol{I}$. If we denote

$$\boldsymbol{S}_c = (c^{-1}\boldsymbol{I} + \boldsymbol{RUR}')^{-1}$$

and use

$$\begin{aligned}(\boldsymbol{X'X} + c\boldsymbol{R'R})^{-1} &= \boldsymbol{U} - \boldsymbol{UR'S}_c\boldsymbol{RU} \\ \boldsymbol{S}_c\boldsymbol{RUR}' &= \boldsymbol{I} - c^{-1}\boldsymbol{S}_c\end{aligned}$$

and set $\mathbf{g} = -\boldsymbol{R\beta}$ (as $\mathbf{r} = \mathbf{0}$ in accordance with (20)) we get the risk (4.2.15) as

$$\begin{aligned}R(\mathbf{b}_R^*(c), \boldsymbol{A}) = \operatorname{tr} \boldsymbol{A}[&\sigma^2\boldsymbol{U} - \sigma^2\boldsymbol{UR'S}_c(2c^{-1}\boldsymbol{I} + \boldsymbol{RUR}')\boldsymbol{S}_c\boldsymbol{RU} \\ &+ \boldsymbol{UR'S}_c\mathbf{gg}'\boldsymbol{S}_c\boldsymbol{RU}].\end{aligned} \tag{4.5.25}$$

Before comparing the mixed and the minimax estimator we remember the well-known comparison of the two estimators with the OLSE $\mathbf{b}_0 = \boldsymbol{UX}'\mathbf{y}$. This gives with $R(\mathbf{b}_0, \boldsymbol{A}) = \sigma^2\boldsymbol{U}$ and (24)

$$R(\mathbf{b}_0, \boldsymbol{A}) - R(\mathbf{b}_R(k), \boldsymbol{A}) = \operatorname{tr} \boldsymbol{AUR'S}_k^{1/2}[\sigma^2\boldsymbol{I} - \boldsymbol{S}_k^{1/2}\mathbf{gg}'\boldsymbol{S}_k^{1/2}]\boldsymbol{S}_k^{1/2}\boldsymbol{RU} \geq 0$$

if and only if (see (3.3.17))

$$\sigma^{-2}\mathbf{g}'\boldsymbol{S}_k\mathbf{g} \leq 1. \tag{4.5.26}$$

Correspondingly, the MILE $\mathbf{b}_R^*(c)$ is better than $\mathbf{b}_0$ if and only if

$$\sigma^{-2}\mathbf{g}'(2c^{-1}\boldsymbol{I}+\boldsymbol{RUR}')^{-1}\mathbf{g} \leq 1 \tag{4.5.27}$$

(see (4.2.22) with $\boldsymbol{T} = \boldsymbol{R}'\boldsymbol{R}$ and $k = \sigma^2/c$, and also Swamy and Mehta (1977) and Farebrother (1978a,c)).

Choosing $c = 2k$, (26) and (27) become identical. If (26) is satisfied for a fixed k^*, then a sufficient condition for the minimax estimator to be better than the OLSE $\mathbf{b}_0$ is $c \leq 2k^*$.

Now, the minimax estimator is better than the mixed estimator with respect to the unconstrained risk when

$$R(\mathbf{b}_R(k), \boldsymbol{A}) - R(\mathbf{b}_R^*(c), \boldsymbol{A}) \geq 0 \quad \text{for all } \boldsymbol{A} \geq \boldsymbol{0}.$$

Using (24) and (25) we obtain for this difference

$$\operatorname{tr} \boldsymbol{AUR}'[(\sigma^2\boldsymbol{S}_c - \boldsymbol{S}_c\mathbf{gg}'\boldsymbol{S}_c) - (\sigma^2\boldsymbol{S}_k - \boldsymbol{S}_k\mathbf{gg}'\boldsymbol{S}_k) + \sigma^2c^{-1}\boldsymbol{S}_c^2]\boldsymbol{RU} \tag{4.5.28}$$

which is nonnegative iff the matrix in brackets $[\ \ ] \geq \boldsymbol{0}$.

(i) Let $\mathbf{g} = \mathbf{0}$

Then the matrix [ ] becomes

$$\boldsymbol{B} = \sigma^2(\boldsymbol{S}_c - \boldsymbol{S}_k + c^{-1}\boldsymbol{S}_c^2). \tag{4.5.29}$$

Using the spectral decomposition $\boldsymbol{C}_R\boldsymbol{\Lambda}_R\boldsymbol{C}'_R = \boldsymbol{RUR}'$ where $\boldsymbol{C}_R$ is the matrix of eigenvectors of $\boldsymbol{RUR}'$ and $\boldsymbol{\Lambda}_R$ the diagonal matrix of the corresponding eigenvalues $\lambda_1 \geq \ldots \geq \lambda_J$ we obtain

$$\sigma^{-2}\boldsymbol{B} = \boldsymbol{C}_R[(c^{-1}\boldsymbol{I}+\boldsymbol{\Lambda}_R)^{-1} - (k^{-1}\boldsymbol{I}+\boldsymbol{\Lambda}_R)^{-1} + c^{-1}(c^{-1}\boldsymbol{I}+\boldsymbol{\Lambda}_R)^{-2}]\boldsymbol{C}'_R.$$

If $\boldsymbol{B} \geq \boldsymbol{0}$, then all eigenvalues of the matrix in brackets [ ] have to be nonnegative, which gives

$$(c^{-1}+\lambda_j)^{-1} - (k^{-1}+\lambda_j)^{-1} + c^{-1}(c^{-1}+\lambda_j)^{-2} \geq 0, \quad (j = 1, \ldots, J)$$

or, equivalently,

$$c^{-2} - 2c^{-1}k^{-1} - k^{-1}\lambda_j \leq 0 \quad (j = 1, \ldots, J).$$

Solving the corresponding equality for c^{-1} gives

$$c^{-1} = k^{-1}(1 \pm (1+k\lambda_j)^{1/2}) \quad (j = 1, \ldots, J)$$

and since $c \geq 0$ must hold, we obtain as a necessary and sufficient condition for $\boldsymbol{B} \geq \boldsymbol{0}$

$$c \geq k(1+(1+k\lambda_J)^{1/2})^{-1}. \tag{4.5.30}$$

If $c \geq k/2$, then $\boldsymbol{B} \geq \boldsymbol{0}$ holds for all $\boldsymbol{X}$ and $\boldsymbol{R}$.

(ii) Let $c = k$

Then (28) becomes

$$\operatorname{tr} \boldsymbol{A}\boldsymbol{U}\boldsymbol{R}'(\sigma^2 k^{-1} \boldsymbol{S}_k^2)\boldsymbol{R}\boldsymbol{U} \geq 0 \tag{4.5.31}$$

for all $\boldsymbol{A} \geq \boldsymbol{0}$, independently of $\mathbf{g}$, $\boldsymbol{R}$, and $\boldsymbol{X}$. Thus, for all $\mathbf{g}$, prior information of type $\boldsymbol{\beta}'\boldsymbol{R}'\boldsymbol{R}\boldsymbol{\beta} = \sigma^2/k$ combined with sample information yields better estimates on the average than biased information $\mathbf{r} = \boldsymbol{R}\boldsymbol{\beta} + \mathbf{g} + \boldsymbol{\phi}$, $\boldsymbol{\phi} \sim (\mathbf{0}, (\sigma^2/k)\boldsymbol{I})$ and $\mathbf{r}$ observed as zero. This holds for all $\mathbf{g}$, i.e. for $\boldsymbol{R}\boldsymbol{\beta} = -\mathbf{g} = \mathbf{0}$ also, in which case $\mathbf{b}_{\boldsymbol{R}}(k)$ is unbiased.

(iii) Let $c \neq k$ and $\mathbf{g} \neq \mathbf{0}$

This is the most general case. The difference (28) is nonnegative for all $\boldsymbol{A} \geq \boldsymbol{0}$ if and only if

$$\sigma^2 \boldsymbol{B} - (\boldsymbol{S}_c \mathbf{g}\mathbf{g}' \boldsymbol{S}_c - \boldsymbol{S}_k \mathbf{g}\mathbf{g}' \boldsymbol{S}_k) \geq \boldsymbol{0} \tag{4.5.32}$$

Now, by a lemma of Teräsvirta (Theorem A.16), the second matrix in (32) is indefinite for $c \neq k$ provided it is not a scalar. Thus $\boldsymbol{B} > \boldsymbol{0}$ is a necessary condition for (32) to hold. Assuming $\boldsymbol{B} > \boldsymbol{0}$ and setting $\boldsymbol{B} = \boldsymbol{B}^{1/2}\boldsymbol{B}^{1/2}$ where $\boldsymbol{B}^{1/2}$ is regular, (32) is fulfilled if

$$\sigma^2 \boldsymbol{I} - (\boldsymbol{B}^{-1/2}\boldsymbol{S}_c \mathbf{g}\mathbf{g}' \boldsymbol{S}_c \boldsymbol{B}^{-1/2} - \boldsymbol{B}^{-1/2}\boldsymbol{S}_k \mathbf{g}\mathbf{g}' \boldsymbol{S}_k \boldsymbol{B}^{-1/2}) \geq \boldsymbol{0} \tag{4.5.33}$$

Applying Theorem A.16, with $\mathbf{h} = \boldsymbol{B}^{-1/2}\boldsymbol{S}_k \mathbf{g}$ and $\tilde{\mathbf{g}} = \boldsymbol{B}^{-1/2}\boldsymbol{S}_c \mathbf{g}$, yields the result that (33) holds if and only if

$$\begin{aligned} &\mathbf{g}'\boldsymbol{S}_k \boldsymbol{B}^{-1}\boldsymbol{S}_k \mathbf{g} - \mathbf{g}'\boldsymbol{S}_c \boldsymbol{B}^{-1}\boldsymbol{S}_c \mathbf{g} \\ &\quad + [(\mathbf{g}'\boldsymbol{S}_k \boldsymbol{B}^{-1}\boldsymbol{S}_k \mathbf{g} - \mathbf{g}'\mathrm{S}_c \boldsymbol{B}^{-1}\boldsymbol{S}_c \mathbf{g})^2 \\ &\quad + \mathbf{g}'\mathrm{S}_c \boldsymbol{B}^{-1}\boldsymbol{S}_c \mathbf{g}\mathbf{g}'\boldsymbol{S}_k \boldsymbol{B}^{-1}\boldsymbol{S}_k \mathbf{g} \\ &\quad - (\mathbf{g}'\boldsymbol{S}_c \boldsymbol{B}^{-1}\boldsymbol{S}_k \mathbf{g})^2]^{1/2} \leq \sigma^2. \end{aligned} \tag{4.5.34}$$

It is difficult to draw very specific conclusions from this general result. As may be proved (Teräsvirta, 1980), $c > k$ is a necessary condition for (34) to hold for all $\mathbf{g}$. (In the single regressor case $c > k$ is necessary and sufficient for the minimax estimator to be better than the mixed estimator.)

Superiority of the Mixed over the Minimax Estimator

As may be seen from the above, the use of the special types of prior information (20) and (21) leads to superiority of the minimax over the mixed estimator

(i) if $c \geq k/2$, when $\mathbf{g} = \mathbf{0}$ (see (30))
(ii) in any case, when $c = k$ (see (31))
(iii) if $c > k$, in the general case (see (34)).

Thus, in cases (i) and (ii) there exist situations where the mixed estimator will dominate the minimax estimator. In the following we will give an important example of this fact (Toutenburg and Roeder, 1978), where despite the above investigations we will use the minimax risk to measure the estimator's efficiency.

Let prior information of type (4.1.3) be given, that is all components β_i of the vector $\boldsymbol{\beta}$ are prior known to lie in known intervals

$$a_i \leq \beta_i \leq b_i \quad (i = 1, \ldots, K). \tag{4.5.35}$$

Using the reformulation of the cuboid (35) proposed in Section 4.1.1 gives the weaker restriction $\boldsymbol{\beta} \in B = \{\boldsymbol{\beta} : (\boldsymbol{\beta} - \boldsymbol{\beta}_0)' \boldsymbol{T} (\boldsymbol{\beta} - \boldsymbol{\beta}_0) \leq 1\}$ with $\boldsymbol{\beta}_0$ and $\boldsymbol{T}$ from (4.1.7). The corresponding MILE becomes (see (4.2.11))

$$\mathbf{b}^* (\boldsymbol{\beta}_0) = \boldsymbol{\beta}_0 + \boldsymbol{D}^{-1} \boldsymbol{X}' \boldsymbol{W}^{-1} (\mathbf{y} - \boldsymbol{X}\boldsymbol{\beta}_0)$$

where $\boldsymbol{\beta}_0' = \frac{1}{2}(a_1 + b_1, \ldots, a_K + b_K)$

and

$$\boldsymbol{T} = \operatorname{diag}(4/K)((b_1 - a_1)^{-2}, \ldots, (b_K - a_K)^{-2}). \tag{4.5.36}$$

This estimator has the minimax risk (see (4.2.14))

$$\sup_{(\boldsymbol{\beta} - \boldsymbol{\beta}_0)' \boldsymbol{T} (\boldsymbol{\beta} - \boldsymbol{\beta}_0) \leq 1} R[\mathbf{b}^* (\boldsymbol{\beta}_0), \mathbf{aa}'] = \sigma^2 \mathbf{a}' \boldsymbol{D}^{-1} \mathbf{a}. \tag{4.5.37}$$

Now, using the reformulation of (35) as proposed in (17) and confining ourselves to the assumption that the mean of the distribution of the fictitious observations lies in the centre of the cuboid (in which case $\mathbf{g} = \mathbf{0}$) leads to the stochastic linear prior information

$$\left.\begin{aligned} &\mathbf{r} = \boldsymbol{R}\boldsymbol{\beta} + \boldsymbol{\phi}, \ \boldsymbol{\phi} \sim (\mathbf{0}, \boldsymbol{V}) \\ &\mathbf{r} = \tfrac{1}{2} \begin{pmatrix} a_1 + b_1 \\ \vdots \\ a_K + b_K \end{pmatrix}, \quad \boldsymbol{R} = \boldsymbol{I}_K, \\ &\boldsymbol{V} = \tfrac{1}{36} \operatorname{diag}((b_1 - a_1)^2, \ldots, (b_K - a_K)^2) \end{aligned}\right\}. \tag{4.5.38}$$

The mixed estimator corresponding to this information is (see (3.1.9)) $\hat{\boldsymbol{\beta}}(\sigma^2)$ which is unbiased (due to $\mathbf{g} = \mathbf{0}$) and has the risk (see (3.1.10) and (4.2.16))

$$\begin{aligned} \sup_{(\boldsymbol{\beta} - \boldsymbol{\beta}_0)' \boldsymbol{T} (\boldsymbol{\beta} - \boldsymbol{\beta}_0) \leq 1} R(\hat{\boldsymbol{\beta}}(\sigma^2), \mathbf{aa}') &= R(\hat{\boldsymbol{\beta}}(\sigma^2), \mathbf{aa}') \\ &= \mathbf{a}' \boldsymbol{V}[\hat{\boldsymbol{\beta}}(\sigma^2)] \mathbf{a} \\ &= \mathbf{a}' (\sigma^{-2} \boldsymbol{S} + \boldsymbol{R}' \boldsymbol{V}^{-1} \boldsymbol{R})^{-1} \mathbf{a}. \end{aligned} \tag{4.5.39}$$

Thus, the mixed estimator $\hat{\boldsymbol{\beta}}(\sigma^2)$ is superior to the MILE $\mathbf{b}^* (\boldsymbol{\beta}_0)$ if

$$\Delta^k (\mathbf{b}^* (\boldsymbol{\beta}_0), \hat{\boldsymbol{\beta}}(\sigma^2)) = \mathbf{a}' [\sigma^2 \boldsymbol{D}^{-1} - (\sigma^{-2} \boldsymbol{S} + \boldsymbol{R}' \boldsymbol{V}^{-1} \boldsymbol{R})^{-1}] \mathbf{a} \geq 0 \tag{4.5.40}$$

or, equivalently (Theorem A.12), if

$$R'V^{-1}R - k^{-1}T = C \geq 0. \tag{4.5.41}$$

Inserting the corresponding matrices from (38) and (36) gives

$$C = \operatorname{diag}\left(\frac{36K-4}{K(b_i-a_i)^2}\right) \geq 0. \tag{4.5.42}$$

Note As a situation of greater practical relevance we will handle the case of restricting only a subset of coefficients, that is

$$a_i \leq \beta_i \leq b_i \quad (i = 1, \ldots, J < K). \tag{4.5.43}$$

We get in a similar way:

$$\mathbf{r} = R\beta + \phi \text{ with } \mathbf{r} = \frac{1}{2}\begin{pmatrix} a_1 + b_1 \\ \vdots \\ a_J + b_J \end{pmatrix}$$

$$R = (I_J, 0_{K-J})$$

$$V = \tfrac{1}{36}\operatorname{diag}\left((b_1-a_1)^2, \ldots, (b_J-a_J)^2\right)$$

and, therefore,

$$R'V^{-1}R = 36 \operatorname{diag}\left((b_1-a_1)^{-2}, \ldots, (b_J-a_J)^{-2}, 0, \ldots, 0\right).$$

As the elements β_i ($i = J+1, \ldots, K$) are not constrained by (43) we may rewrite this as a trivial restriction: $\bar{a}_i \leq \beta_i \leq \bar{b}_i$ ($i = J+1, \ldots, K$) where $\bar{a}_i \to -\infty$ and $\bar{b}_i \to \infty$. Therefore, the matrix C in the limit will take the form

$$C \underset{\substack{\bar{a}_i \to \infty \\ \bar{b}_i \to \infty}}{\longrightarrow} \operatorname{diag}\left(\frac{36K-4}{K(b_1-a_1)^2}, \ldots, \frac{36K-4}{K(b_J-a_J)^2}, 0, \ldots, 0\right) \geq 0. \tag{4.5.44}$$

Thus, if the information $a_i \leq \beta_i \leq b_i$ is weakened as proposed above and if $\mathbf{g} = \mathbf{0}$ can be assumed, then the mixed estimator dominates the MILE with respect to the minimax risk. Clearly, the assumption $\mathbf{g} = \mathbf{0}$ has to be justified.

Summarizing the results of this section leads to the conclusion that superiority of the mixed estimator over the minimax estimator (and conversely) depends on the correctness of the linear prior information $\mathbf{r} = R\beta + \mathbf{g} + \phi$, that is the relationship between both estimators depends on $\mathbf{g}$.

4.5.3 On the Combination of Equality and Inequality Restrictions

In Section 3.5.1 the estimator $\mathbf{b}_R(V)$ was proposed, which enabled us to mix exact and stochastic restrictions. Similarly, we will give in the following a method for using equality and inequality restrictions simultaneously on β (see Toutenburg, 1980b). This is done by combining the ideas of mixed estimation and the minimax estimation.

Example

Let us assume linear equality restrictions on the subvector $\boldsymbol{\beta}_1$ of $(\boldsymbol{\beta}_1, \boldsymbol{\beta}_2)$ of type

$$\mathbf{r} = \boldsymbol{R\beta} \text{ with } \boldsymbol{R} = (\boldsymbol{I}, \boldsymbol{0}) \tag{4.5.45}$$

so that $\boldsymbol{\beta}_2$ is not influenced by (45). Furthermore, it may be that we have prior information on the remaining subvector $\boldsymbol{\beta}_2$ in the form of inequality constraints such as

$$\boldsymbol{\beta}'_2 \boldsymbol{T}_2 \boldsymbol{\beta}_2 \leq k. \tag{4.5.46}$$

This can be given the equivalent form

$$(\boldsymbol{\beta}'_1, \boldsymbol{\beta}'_2) \begin{pmatrix} \boldsymbol{0} & \boldsymbol{0} \\ \boldsymbol{0} & \boldsymbol{T}_2 \end{pmatrix} \begin{pmatrix} \boldsymbol{\beta}_1 \\ \boldsymbol{\beta}_2 \end{pmatrix} \leq k$$

i.e. $\boldsymbol{\beta}' \boldsymbol{T\beta} \leq k$.

Thus we have an important example of the situation in which equality and inequality restrictions are imposed on $\boldsymbol{\beta}$ simultaneously.

We now return to the question of combining restrictions as

$$\mathbf{r} = \boldsymbol{R\beta} \quad \text{and} \quad \boldsymbol{\beta}'\boldsymbol{T\beta} \leq k.$$

Based on the reasoning of Section 4.2 a direct solution of the optimization problem

$$\min_{\hat{\boldsymbol{\beta}}} \sup_{\boldsymbol{\beta}'\boldsymbol{T\beta} \leq k} \{R(\hat{\boldsymbol{\beta}}, \mathbf{aa}')/\mathbf{r} = \boldsymbol{R\beta}\}$$

is not possible, as the restriction $\mathbf{r} = \boldsymbol{R\beta}$ would be neglected (Roeder, 1978). We will set up an estimator using an heuristic approach. We choose the type of the conditional restricted least squares (see (2.3.71))

$$\tilde{\mathbf{b}}_R = \mathbf{b} + \boldsymbol{S}^{-1}\boldsymbol{R}'(\boldsymbol{RS}^{-1}\boldsymbol{R}')^{-1}(\mathbf{r} - \boldsymbol{R}\mathbf{b}) \tag{4.5.47}$$

but instead of $\mathbf{b}$ we will use the minimax estimator $\mathbf{b}^* = \boldsymbol{D}^{-1}\boldsymbol{X}'\boldsymbol{W}^{-1}\mathbf{y}$ (4.2.7). Comparing the minimax risks of both estimators (i.e. (4.2.10) and (4.2.18), respectively) we see that the matrix $\boldsymbol{S}$ would correspond to $\boldsymbol{D}$ (4.2.6). Replacing $\mathbf{b}$ in (47) by $\mathbf{b}^*$ and, correspondingly, $\boldsymbol{S}$ by $\boldsymbol{D}$ gives the, so-to-speak, restricted minimax linear estimator

$$\mathbf{b}^*_R = \mathbf{b}^* + \boldsymbol{D}^{-1}\boldsymbol{R}'(\boldsymbol{RD}^{-1}\boldsymbol{R}')^{-1}(\mathbf{r} - \boldsymbol{R}\mathbf{b}^*). \tag{4.5.48}$$

This estimator is biased:

$$\begin{aligned} \text{bias } \mathbf{b}^*_R &= k^{-1}\sigma^2(-\boldsymbol{I} + \boldsymbol{D}^{-1}\boldsymbol{R}'(\boldsymbol{RD}^{-1}\boldsymbol{R}')^{-1}\boldsymbol{R})\boldsymbol{D}^{-1}\boldsymbol{T\beta} \\ &= k^{-1}\sigma^2 \boldsymbol{ND}^{-1}\boldsymbol{T\beta}, \text{ say} \end{aligned}$$

where

$$\boldsymbol{N} = -\boldsymbol{I} + \boldsymbol{D}^{-1}\boldsymbol{R}'(\boldsymbol{RD}^{-1}\boldsymbol{R}')^{-1}\boldsymbol{R}. \tag{4.5.49}$$

This gives

$$\sup_{\beta' T\beta \le k} \mathbf{a}'(\text{bias } \mathbf{b}_R^*)(\text{bias } \mathbf{b}_R^*)'\mathbf{a} = k^{-1}\sigma^4 \mathbf{a}' N D^{-1} T D^{-1} N' \mathbf{a}. \tag{4.5.50}$$

As far as the dispersion matrix of $\mathbf{b}_R^*$ is concerned, we evaluate

$$\begin{aligned}\sigma^{-2} V(\mathbf{b}_R^*) = {} & D^{-1}SD^{-1} + D^{-1}R'(RD^{-1}R')^{-1}RD^{-1}SD^{-1}R'(RD^{-1}R')^{-1} \\ & - D^{-1}R'(RD^{-1}R')^{-1}RD^{-1}SD^{-1} \\ & - D^{-1}SD^{-1}R'(RD^{-1}R')^{-1}RD^{-1}.\end{aligned} \tag{4.5.51}$$

Thus, adding (50) and (51) we get the minimax risk

$$\sup_{\beta' T\beta \le k} \mathbf{a}'R(\mathbf{b}_R^*)\mathbf{a} = \sigma^2 \mathbf{a}'[D^{-1} - D^{-1}R'(RD^{-1}R')^{-1}RD^{-1}]\mathbf{a}. \tag{4.5.54}$$

If we compare the MILE $\mathbf{b}^*$ (4.2.7) and the restricted MILE $\mathbf{b}_R^*$, we can then use their minimax risks to check if $\mathbf{b}_R^*$ is superior to $\mathbf{b}^*$. This gives (see (4.2.25) and (4.2.10))

$$\mathbf{a}'\Delta^k(\mathbf{b}^*, \mathbf{b}_R^*)\mathbf{a} = \sigma^2 \mathbf{a}' D^{-1}R'(RD^{-1}R')^{-1}RD^{-1}\mathbf{a} \ge 0. \tag{4.5.53}$$

Thus the proposed estimator $\mathbf{b}_R^*$ which uses equality and inequality restrictions, simultaneously, dominates the standard MILE $\mathbf{b}^*$ which is based on inequality restrictions only.

Note (i) The quadratic form

$$\mathbf{a}'\Delta^k(\mathbf{b}^*, \mathbf{b}_R^*)\mathbf{a}$$

is nonnegative definite regardless of the concrete background imposed by the restrictions $\mathbf{r} = R\boldsymbol{\beta}$ and $\boldsymbol{\beta}' T \boldsymbol{\beta} \le k$. This result should not always lead to an application of $\mathbf{b}_R^*$ in practice. As in the above example, we may ensure that there is no overlapping in restricting any component β_i of $\boldsymbol{\beta}$, i.e. we can give β_i either an equality restriction or an inequality restriction. If both types of restrictions on a certain β_i are available, we may decide to take the stronger one in the sense that it is more informative on β_i.

Note (ii) If the linear restrictions are stochastic, rather than exact, i.e. if $\mathbf{r} = R\boldsymbol{\beta} + \boldsymbol{\phi}$, then analogously to the above estimation procedure we would propose to take the estimator

$$\mathbf{b}_R^*(V) = \mathbf{b}^* + D^{-1}R'(\sigma^{-2}V + RD^{-1}R')^{-1}(\mathbf{r} - R\mathbf{b}^*) \tag{4.5.54}$$

which is derived from the generalized RLSE $\mathbf{b}_R(V)$ (3.5.6) by replacing $\mathbf{b}$ by the MILE $\mathbf{b}^*$. Clearly we have

$$\mathbf{b}_R^*(V) \xrightarrow[V \to 0]{} \mathbf{b}_R^*.$$

Furthermore, we derive that $\mathbf{b}_R^*(V)$ has the minimax risk

$$\sup_{\beta' T\beta \leq k} \mathbf{a}'R[\mathbf{b}_R^*(V)]\mathbf{a} = \sigma^2 \mathbf{a}'[D^{-1} - D^{-1}R'(\sigma^{-2}V + RD^{-1}R')^{-1}RD^{-1}]\mathbf{a} \tag{4.5.55}$$

so that

$$\mathbf{a}'\Delta^k[\mathbf{b}^*, \mathbf{b}_R^*(V)]\mathbf{a} = \sigma^2 \mathbf{a}'D^{-1}R'(\sigma^{-2}V + RD^{-1}R')^{-1}RD^{-1}\mathbf{a} \geq 0.$$

That is, the estimator $\mathbf{b}_R^*(V)$ is superior to the MILE $\mathbf{b}^*$.

V Problems in Model Choice

5.1 ON PREDICTION

5.1.1 Introduction

Man has always sought to forecast the future. Initially, this was done by subjective estimates and verbal descriptions of various phenomena. In other words, the first basis of forecasting was purely empirical. Later, simple mathematical rules were used, especially if the interesting phenomenon was a physical one. Every time, the sight into the future was given a philosophical aspect: '*Time present and time past are both perhaps present in time future, and time future contained in time past.*' (T. S. Eliot: *Burnt Norton*)

During the present century, the problem of forecasting has become more and more interesting. The understanding of this problem focused on methods based on series of observations and rules which allow the extrapolation of laws or experience of the past. Thus forecasting methods have been developed which are of a statistical nature.

Clearly, no method of prediction can be perfect, but in a fixed situation (perhaps described by a mathematical model) one method may be better than some alternatives.

As far as linear regression as the underlying model is concerned, the monograph of Bibby and Toutenburg (1978) was devoted to the problem of best predictors as well as to investigation of regions where one predictor improves another solution. One of the main aims of the cited book was to examine the conditions under which biased predictors can lead to an improvement over conventional unbiased procedures. Thus, for a full discussion of these problems we can refer to it and to Toutenburg (1970c,e). In the following we will concentrate on results which are connected to the problem of model choice.

Conventional notation Within this book we will regard 'prediction' and 'forecasting' as synonymous.

5.1.2 Some Simple Linear Models

To demonstrate the development of statistical prediction in regression we will first give some illustrative examples of linear models.

(i) The Constant Mean Model

The most simple 'regression' may be described by

$$y_t = \mu + \varepsilon_t \qquad (t = 1, \ldots, T) \tag{5.1.1}$$

where $\boldsymbol{\varepsilon} = (\varepsilon_1, \ldots, \varepsilon_T)' \sim (\mathbf{0}, \sigma^2 \boldsymbol{I})$ and μ is a scalar constant. T denotes the index (time) of last observation of the random process $\{y_t\}$. We assume that a prediction of a future observation $y_{T+\tau}$ is required. Extrapolating model (1) gives

$$y_{T+\tau} = \mu + \varepsilon_{T+\tau}. \tag{5.1.2}$$

One would expect to estimate $y_{T+\tau}$ by adding the estimators of μ and $\varepsilon_{T+\tau}$. The actual value of the random variable $\varepsilon_{T+\tau}$ cannot be predicted; thus we simply forecast $\varepsilon_{T+\tau}$ by its expected value, i.e. $E\varepsilon_{T+\tau} = 0$. The quantity μ is constant over time, so its estimate from the past will give a predictor for the future.

Thus we are led to the predictor

$$\hat{y}_{T+\tau} = T^{-1} \sum_{t=1}^{T} y_t = \bar{y} \tag{5.1.3}$$

which is unbiased:

$$E\hat{y}_{T+\tau} = E\bar{y} = \mu \tag{5.1.4}$$

and has variance

$$\operatorname{var}(\hat{y}_{T+\tau}) = \sigma^2/T. \tag{5.1.5}$$

An increasing period of observations (i.e. an increasing size T) will improve the precision of the predictor.

A more realistic model arises if the constancy of the mean over all indices t is replaced by the assumption that μ is only locally constant. That is, μ is regarded as a slowly varying quantity. If, for example, t denotes the months and if one can believe in a quarterly constancy of the mean, then μ is estimated for a three-month period:

$$\hat{\mu}(t, t+2) = (y_t + y_{t+1} + y_{t+2})/3.$$

This is called a moving average and the model is given the name local constant mean model. Then $\hat{\mu}(T-2, T)$ gives the predictor of $y_{T+\tau}$.

(ii) The Linear Trend Model

Assume that a structure which has a linear trend gives the model

$$y_t = \alpha + \beta t + \varepsilon_t \qquad (t = 1, \ldots, T) \tag{5.1.6}$$

(see Figure 5.1.1) where α is the expectation of y_0, β is the slope, and $\{\varepsilon_t\}$ is the added random variation.

The predictor of any future value $y_{T+\tau}$ is simply obtained by

$$\hat{y}_{T+\tau} = \hat{\alpha} + \hat{\beta}(T + \tau) \tag{5.1.7}$$

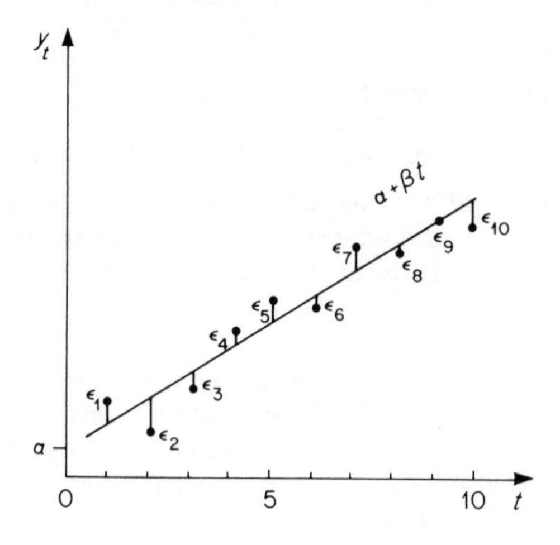

Figure 5.1.1 A linear trend model

where $\hat{\alpha}$, $\hat{\beta}$ are the unbiased ordinary least squares estimates of α, β (see Section 2.1.8):

$$\left.\begin{aligned} \hat{\alpha} = \bar{y}, \ \hat{\beta} &= \frac{\sum y_t t}{\sum t^2}, \\ \text{var } \hat{\alpha} = \sigma^2/T, \qquad \text{var } \hat{\beta} &= \sigma^2 / \sum_{t=1}^{T} t^2. \end{aligned}\right\} \tag{5.1.8}$$

Denoting the forecast error by $e_{T+\tau}$ gives

$$\begin{aligned} e_{T+\tau} &= y_{T+\tau} - \hat{y}_{T+\tau} \\ &= [\alpha + \beta(T+\tau) + \varepsilon_{T+\tau}] - [\hat{\alpha} + \hat{\beta}(T+\tau)] \\ &= (\alpha - \hat{\alpha}) + (\beta - \hat{\beta})(T+\tau) + \varepsilon_{T+\tau}. \end{aligned}$$

Hence, $Ee_{T+\tau} = 0$ and the predictor $\hat{y}_{T+\tau}$ is unbiased. This leads to the mean square error

$$\begin{aligned} \text{MSE}(\hat{y}_{T+\tau}) &= E(e_{T+\tau})^2 \\ &= \text{var }(\hat{\alpha}) + \text{var }(\hat{\beta}) + \sigma^2 \\ &= \sigma^2\left(\frac{1}{T} + \frac{(T+\tau)^2}{\sum t^2} + 1\right). \end{aligned} \tag{5.1.9}$$

From (9) it will be seen that increasing the predictor's horizon (i.e. τ) will decrease the expected precision of the forecast. As pointed out in Gilchrist (1976, p. 64), the linear trend will be the only stable aspect of the data that one can reasonably expect to extrapolate into the future, at least for a short time.

(iii) Polynomial Models

The polynomial trend model of order K is of the form

$$y_t = \alpha + \beta_1 t + \beta_2 t^2 + \ldots + \beta_K t^K + \varepsilon_t \qquad (5.1.10)$$

and its forecast again is based on the OLSE of $\alpha, \beta_1, \ldots, \beta_K$:

$$\hat{y}_{T+\tau} = \hat{\alpha} + \hat{\beta}_1 (T+\tau) + \ldots + \hat{\beta}_K (T+\tau)^K. \qquad (5.1.11)$$

Before using quadratic and higher-order trends, the background of that model's choice has to be investigated. An increase of the polynomial's degree does not lead to an immediate growth of stability of the model. For $T = 11$ and $\tau = 1$ (i.e. a *one-step-ahead forecast*) the variance of the forecast error is $1.647\sigma^2$ for a linear trend model, $2.098\sigma^2$ for a quadratic model, and $2.771\sigma^2$ for a cubic model (see Gilchrist, 1976, p. 76). If we set

$$\mathbf{x}_1 = \mathbf{i},\ \mathbf{x}_2 = t\mathbf{i}, \ldots, \mathbf{x}_{K+1} = t^K \mathbf{i}$$

we are led to the form of a regression model

$$\begin{aligned} \mathbf{y} &= (\mathbf{x}_1, \ldots, \mathbf{x}_{K+1}) \begin{pmatrix} \alpha \\ \beta_1 \\ \vdots \\ \beta_K \end{pmatrix} + \boldsymbol{\varepsilon} \\ &= \boldsymbol{X\beta} + \boldsymbol{\varepsilon}. \end{aligned}$$

Thus predicting in regression gives the above results as special cases.

5.1.3 Classical Prediction in Regression

Suppose that we have the general linear model $\mathbf{y} = \boldsymbol{X\beta} + \boldsymbol{\varepsilon}$, $\boldsymbol{\varepsilon} \sim (\mathbf{0}, \sigma^2 \boldsymbol{W})$. Suppose also that we have a known matrix $\boldsymbol{X}_*$ consisting of further 'observations' on the set of regressors $\boldsymbol{X}_1, \ldots, \boldsymbol{X}_K$ which form the independent variables in the sample of the linear model. That is, we assume the model $\mathbf{y} = \boldsymbol{X\beta} + \boldsymbol{\varepsilon}$ (at indices $t = 1, \ldots, T$) to be extrapolated to a future period. Denoting this period by the set of indices $\mathcal{T} = \{T+1, \ldots, T+n\}$ gives

$$y_\tau = \mathbf{x}'_\tau \boldsymbol{\beta} + \varepsilon_\tau \quad (\tau \in \mathcal{T}) \qquad (5.1.12)$$

or in its matrix form

$$\mathbf{y}_* = \boldsymbol{X}_* \boldsymbol{\beta} + \boldsymbol{\varepsilon}_*. \qquad (5.1.13)$$

(The asterisk indicates the future period.)

Here

- $\mathbf{y}_*$ is a $n \times 1$ vector of (unknown) future realizations of the dependent variable Y,
- $\boldsymbol{X}_*$ is the $n \times K$ matrix of (known) future values of the regressors,
- $\boldsymbol{\varepsilon}_*$ is the $n \times 1$ vector of disturbances having

$$\boldsymbol{\varepsilon}_* \sim (\mathbf{0}, \sigma^2 \boldsymbol{W}_*). \qquad (5.1.14)$$

As the classical set-up of a predictor of $\mathbf{y}_*$ we denote the estimation of the conditional expectation $E(\mathbf{y}_*/X_*) = X_*\boldsymbol{\beta}$. By Theorem 2.4 and (2.1.49) we get immediately

$$\hat{\mathbf{y}}_* = X_*\mathbf{b} = X_*\boldsymbol{\beta} + S^{-1}X'W^{-1}\boldsymbol{\varepsilon}. \tag{5.1.15}$$

Since $E\hat{\mathbf{y}}_* = X_*\boldsymbol{\beta}$, this equals the expectation of $\mathbf{y}_*$ given by (13). Thus we may say that the classical predictor $\hat{\mathbf{y}}_*$ is unbiased for $E(\mathbf{y}_*)$. Its dispersion matrix is

$$V(\hat{\mathbf{y}}_*) = X_* V(\mathbf{b})X'_* = \sigma^2 X_* S^{-1} X'_*. \tag{5.1.16}$$

This measures the MSE of $\hat{\mathbf{y}}_*$ from its mean $X_*\boldsymbol{\beta}$. However, a more useful measure of the suitability of $\hat{\mathbf{y}}_*$ is given by its mean square error of prediction (MSEP):

Definition 5.1 Let a predictor $\mathbf{p}$ of $\mathbf{y}_*$ be given. Then the matrix-valued mean square error of prediction (MSEP) of $\mathbf{p}$ is defined by

$$\text{MSEP}(\mathbf{p}) = E(\mathbf{p} - \mathbf{y}_*)(\mathbf{p} - \mathbf{y}_*)'. \tag{5.1.17}$$

Using this measure we get for the classical predictor $\hat{\mathbf{y}}_*$

$$\text{MSEP}(\hat{\mathbf{y}}_*) = V(\hat{\mathbf{y}}_*) + V(\mathbf{y}_*) - C - C' \tag{5.1.18}$$

where

$$C = E(\hat{\mathbf{y}}_* - X_*\boldsymbol{\beta})(\mathbf{y}_* - X_*\boldsymbol{\beta})'. \tag{5.1.19}$$

If the vector of future disturbances $\boldsymbol{\varepsilon}_*$ is uncorrelated with the vector of 'sample' disturbances $\boldsymbol{\varepsilon}$, then C in (18) is a zero matrix. Hence in this case

$$\text{MSEP}(\hat{\mathbf{y}}_*) = V(\hat{\mathbf{y}}_*) + V(\mathbf{y}_*) = \sigma^2 X_* S^{-1} X'_* + \sigma^2 W_*. \tag{5.1.20}$$

Comparing this result with the MSE of the estimator $X_*\mathbf{b}$ of $X_*\boldsymbol{\beta}$, we see that the prediction of $\mathbf{y}_* = X_*\boldsymbol{\beta} + \boldsymbol{\varepsilon}_*$ leads to a larger MSE by the uncertainty contained in $\boldsymbol{\varepsilon}_*$.

In measuring the efficiency of estimators, in Chapter II we defined the quadratic risk $R(\hat{\boldsymbol{\beta}}, A)$ (2.1.27) which is connected with the MSE of $\hat{\boldsymbol{\beta}}$ (see Theorem 2.1). Analogously, we may define a quadratic risk for predictors.

Definition 5.2 Suppose that we have a predictor $\mathbf{p}$ of $\mathbf{y}_*$. Then its risk is defined by

$$r(\mathbf{p}, A) = E(\mathbf{p} - \mathbf{y}_*)' A (\mathbf{p} - \mathbf{y}_*) \tag{5.1.21}$$

where $A > \boldsymbol{0}$ is a known $n \times n$ matrix.

Using Definition 5.1 gives

$$r(\mathbf{p}, A) = \operatorname{tr} A \ \text{MSEP}(\mathbf{p}). \tag{5.1.22}$$

Thus minimizing $r(\mathbf{p}, A)$ is equivalent to minimizing MSEP$(\mathbf{p})$. (The proof of this follows the proof of Theorem 2.1.)

5.1.4 Best Linear Unbiased Prediction

We consider the regression model $\mathbf{y} = \boldsymbol{X\beta} + \boldsymbol{\varepsilon}, \boldsymbol{\varepsilon} \sim (\mathbf{0}, \sigma^2 \boldsymbol{W})$ and the corresponding future model (13) with specification (14). Now in view of the interdependence of disturbances in the sample as well as in the future (i.e. $E\boldsymbol{\varepsilon\varepsilon}' = \sigma^2 \boldsymbol{W} \neq \sigma^2 \boldsymbol{I}$ and $E\boldsymbol{\varepsilon}_* \boldsymbol{\varepsilon}_*' = \sigma^2 \boldsymbol{W}_* \neq \sigma^2 \boldsymbol{I}$, in general) it is reasonable to assume that the future disturbances are dependent on the sample disturbances. Therefore, we will assume

$$E\boldsymbol{\varepsilon\varepsilon}_*' = \sigma^2 \boldsymbol{W}_0 \qquad (5.1.23)$$

where $\sigma^2 \boldsymbol{W}_0$ is the $T \times n$ matrix of covariances of $\boldsymbol{\varepsilon}$ with $\boldsymbol{\varepsilon}_*$. Goldberger (1962) was the first to use this additional information which brought a movement to prediction in regression. Using the covariance structure (23), the matrix $\boldsymbol{C}$ (19) becomes

$$\boldsymbol{C} = E\boldsymbol{X}_* \boldsymbol{S}^{-1} \boldsymbol{X}' \boldsymbol{W}^{-1} \boldsymbol{\varepsilon\varepsilon}_*' = \sigma^2 \boldsymbol{X}_* \boldsymbol{S}^{-1} \boldsymbol{X}' \boldsymbol{W}^{-1} \boldsymbol{W}_0. \qquad (5.1.24)$$

Then the classical predictor $\hat{\mathbf{y}}_*$ has

$$\begin{aligned} \text{MSEP}(\hat{\mathbf{y}}_*) = \sigma^2 \boldsymbol{X}_* \boldsymbol{S}^{-1} \boldsymbol{X}_*' + \sigma^2 \boldsymbol{W}_* - \sigma^2 \boldsymbol{X}_* \boldsymbol{S}^{-1} \boldsymbol{X}' \boldsymbol{W}^{-1} \boldsymbol{W}_0 \\ - \sigma^2 \boldsymbol{W}_0' \boldsymbol{W}^{-1} \boldsymbol{X} \boldsymbol{S}^{-1} \boldsymbol{X}_*'. \end{aligned} \qquad (5.1.25)$$

To derive an r-optimal unbiased predictor we consider the general linear estimator of $\mathbf{y}_*$ defined by

$$\mathbf{p} = \boldsymbol{D}\mathbf{y} + \mathbf{d} \qquad (5.1.26)$$

where $\boldsymbol{D}$ is $n \times T$ and $\mathbf{d}$ is $n \times 1$. This has expectation

$$E\mathbf{p} = \boldsymbol{DX\beta} + \mathbf{d}. \qquad (5.1.27)$$

Then the condition for lack of bias becomes

$$E(\mathbf{p} - \mathbf{y}_*) = (\boldsymbol{DX} - \boldsymbol{X}_*)\boldsymbol{\beta} + \mathbf{d} = \mathbf{0} \text{ for all } \boldsymbol{\beta}.$$

Thus, choosing $\boldsymbol{\beta} = \mathbf{0}$, we have $\mathbf{d} = \mathbf{0}$ and therefore $\mathbf{p}$ is unbiased if

$$\boldsymbol{DX} = \boldsymbol{X}_*. \qquad (5.1.28)$$

From (26) with $\mathbf{d} = \mathbf{0}$ and (28) we have

$$\mathbf{p} - \mathbf{y}_* = \boldsymbol{D\varepsilon} - \boldsymbol{\varepsilon}_*,$$

$$\text{MSEP}(\mathbf{p}) = \sigma^2 \boldsymbol{DWD}' + \sigma^2 \boldsymbol{W}_* - \sigma^2 \boldsymbol{DW}_0 - \sigma^2 \boldsymbol{W}_0' \boldsymbol{D}' \qquad (5.1.29)$$

and therefore the predictor's risk is of the form

$$r(\mathbf{p}, \boldsymbol{A}) = \sigma^2 \operatorname{tr} \boldsymbol{A}(\boldsymbol{DWD}' + \boldsymbol{W}_* - 2\boldsymbol{DW}_0). \qquad (5.1.30)$$

Then the r-optimal unbiased predictor of $\mathbf{y}_*$ is the solution of

$$\min_{\boldsymbol{D}, \boldsymbol{\lambda}} \left\{ r(\mathbf{p}, \boldsymbol{A}) - 2 \sum_{\tau=1}^{n} \boldsymbol{\lambda}_\tau' (\boldsymbol{DX} - \boldsymbol{X}_*)_\tau' \right\} \qquad (5.1.31)$$

where $\boldsymbol{\lambda} = (\boldsymbol{\lambda}_1, \ldots, \boldsymbol{\lambda}_n)$ is a matrix of Lagrangian multipliers and $(\ )_\tau$ denotes the

τth column of the matrix in brackets. Differentiating (31) with respect to $\boldsymbol{D}$ and $\boldsymbol{\lambda}$ and equating to zero leads to the normal equations (see Goldberger, 1962, for $\tau = 1$, and Toutenburg, 1968, and 1975a, p. 72)

$$\boldsymbol{A}\boldsymbol{D}\boldsymbol{W} - \boldsymbol{A}\boldsymbol{W}_0' - \sigma^{-2}\boldsymbol{\lambda}\boldsymbol{X}' = \boldsymbol{0}, \tag{5.1.32}$$

$$\boldsymbol{D}\boldsymbol{X} - \boldsymbol{X}_* = \boldsymbol{0}. \tag{5.1.33}$$

Premultiplying (32) by $\boldsymbol{A}^{-1}$ and postmultiplying by $\boldsymbol{W}^{-1}\boldsymbol{X}$ gives

$$\boldsymbol{D} = \boldsymbol{W}_0'\boldsymbol{W}^{-1} + \sigma^{-2}\boldsymbol{\lambda}\boldsymbol{X}'\boldsymbol{W}^{-1}. \tag{5.1.34}$$

Thus

$$\boldsymbol{D}\boldsymbol{X} = \boldsymbol{W}_0'\boldsymbol{W}^{-1}\boldsymbol{X} + \sigma^{-2}\boldsymbol{\lambda}\boldsymbol{S} = \boldsymbol{X}_* \quad \text{(see (33))}$$

and therefore,

$$\sigma^{-2}\hat{\boldsymbol{\lambda}} = \boldsymbol{X}_*\boldsymbol{S}^{-1} - \boldsymbol{W}_0'\boldsymbol{W}^{-1}\boldsymbol{X}\boldsymbol{S}^{-1}. \tag{5.1.35}$$

Inserting (35) in equation (34) leads to the optimal matrix $\boldsymbol{D}_{\text{opt}} = \hat{\boldsymbol{D}}$

$$\hat{\boldsymbol{D}} = \boldsymbol{X}_*\boldsymbol{S}^{-1}\boldsymbol{X}'\boldsymbol{W}^{-1} + \boldsymbol{W}_0'\boldsymbol{W}^{-1}(\boldsymbol{I} - \boldsymbol{X}\boldsymbol{S}^{-1}\boldsymbol{X}'\boldsymbol{W}^{-1}). \tag{5.1.36}$$

Then the *r-optimal unbiased predictor* of $\mathbf{y}_*$ is of the form

$$\begin{aligned}\hat{\mathbf{p}} = \hat{\boldsymbol{D}}\mathbf{y} &= \boldsymbol{X}_*\mathbf{b} + \boldsymbol{W}_0'\boldsymbol{W}^{-1}(\mathbf{y} - \boldsymbol{X}\mathbf{b}) \\ &= \boldsymbol{X}_*\mathbf{b} + \boldsymbol{W}_0'\boldsymbol{W}^{-1}\hat{\boldsymbol{\varepsilon}}\end{aligned}$$

where $\mathbf{b}$ is the GLSE of $\boldsymbol{\beta}$ and $\hat{\boldsymbol{\varepsilon}}$ is the GLS residual vector. Thus, the first element $\boldsymbol{X}_*\mathbf{b}$ of $\hat{\mathbf{p}}$ estimates the expectation $\boldsymbol{X}_*\boldsymbol{\beta}$ of $\mathbf{y}_*$, and the second element $\boldsymbol{W}_0'\boldsymbol{W}^{-1}\hat{\boldsymbol{\varepsilon}}$ is an estimator of the future disturbance vector $\boldsymbol{\varepsilon}_*$.

Theorem 5.1

Assume the generalized regression model $\mathbf{y} = \boldsymbol{X}\boldsymbol{\beta} + \boldsymbol{\varepsilon}$, $\boldsymbol{\varepsilon} \sim (\boldsymbol{0}, \sigma^2\boldsymbol{W})$ and the future model (13) with specifications (14) and (23). Then the best linear unbiased predictor of $\mathbf{y}_*$ under risk $r(\mathbf{p}, \boldsymbol{A})$ (21) is

$$\hat{\mathbf{p}} = \boldsymbol{X}_*\mathbf{b} + \boldsymbol{W}_0'\boldsymbol{W}^{-1}(\mathbf{y} - \boldsymbol{X}\mathbf{b}) \tag{5.1.37}$$

with

$$\begin{aligned}r(\hat{\mathbf{p}}, \boldsymbol{A}) &= \sigma^2 \operatorname{tr} \boldsymbol{A}(\boldsymbol{W}_* - \boldsymbol{W}_0'\boldsymbol{W}^{-1}\boldsymbol{W}_0) \\ &\quad + \sigma^2 \operatorname{tr} \boldsymbol{A}(\boldsymbol{X}_* - \boldsymbol{W}_0'\boldsymbol{W}^{-1}\boldsymbol{X})\boldsymbol{S}^{-1}(\boldsymbol{X}_* - \boldsymbol{W}_0'\boldsymbol{W}^{-1}\boldsymbol{X})'.\end{aligned} \tag{5.1.38}$$

In practice, when there is no auxiliary information available, $\hat{\mathbf{p}}$(37) may be regarded as the most sensible predictor to use. This becomes clearer if we seek biased solutions of r-optimal prediction (see next section).

As an important special case we refer to the one-step-ahead predictor, that is, to the case $\tau = 1$ (Goldberger, 1962). Using model (12) gives

$$y_* = y_{T+1} = \mathbf{x}_{T+1}'\boldsymbol{\beta} + \varepsilon_{T+1} \tag{5.1.39}$$

with $\varepsilon_{T+1} \sim (\mathbf{0}, \sigma^2 \mathrm{w}_*) = (\mathbf{0}, \sigma_*^2)$, say. Then the vector of correlation of $\boldsymbol{\varepsilon}$ with the scalar disturbance ε_{T+1} is the first column of $\sigma^2 \boldsymbol{W}_0$ (23):

$$E\boldsymbol{\varepsilon}\varepsilon_{T+1} = \sigma^2 \mathbf{w}. \tag{5.1.40}$$

The best linear unbiased predictor specializes to

$$\hat{p} = \mathbf{x}'_{T+1}\mathbf{b} + \mathbf{w}' \boldsymbol{W}^{-1} (\mathbf{y} - \boldsymbol{X}\mathbf{b}) \tag{5.1.41}$$

with

$$r(\hat{p}) = \sigma_*^2 - \sigma^2 \mathbf{w}' \boldsymbol{W}^{-1} \mathbf{W} + \sigma^2 (\mathbf{x}'_{T+1} - \mathbf{w}' \boldsymbol{W}^{-1} \boldsymbol{X}) \boldsymbol{S}^{-1} (\mathbf{x}'_{T+1} - \mathbf{w}' \boldsymbol{W}^{-1} \boldsymbol{X})'.$$

Applying this result to the one-step-ahead predictor of y_* if the disturbances $\boldsymbol{\varepsilon}$ and $\boldsymbol{\varepsilon}_*$ follow a first-order autoregression leads to the specification $\boldsymbol{W}$ from (2.2.23) and to

$$\mathbf{w} = E\boldsymbol{\varepsilon}\varepsilon_{T+1} = \sigma^2 \begin{pmatrix} \rho^T \\ \rho^{T-1} \\ \vdots \\ \rho \end{pmatrix}.$$

Then $\mathbf{w}' \boldsymbol{W}^{-1}$ reduces to (see $\boldsymbol{W}^{-1}$ from (2.2.24))

$$\mathbf{w}' \boldsymbol{W}^{-1} = \rho(0, \ldots, 0, 1)$$

and the predictor (41) becomes

$$\hat{p} = \mathbf{x}'_{T+1}\mathbf{b} + \rho \hat{\varepsilon}_T \tag{5.1.42}$$

(see also (2.2.32)).

Gain in Efficiency

To indicate the gain in efficiency if the r-optimal unbiased predictor $\hat{\mathbf{p}}$ (37) is used instead of the classical predictor $\hat{\mathbf{y}}_*$ (15) we have to analyse the reduction of risk. With (25) and (38) we arrive at

$$r(\hat{\mathbf{y}}_*, \boldsymbol{A}) - r(\hat{\mathbf{p}}, \boldsymbol{A}) = \sigma^2 \operatorname{tr} \boldsymbol{A} \boldsymbol{W}_0' (\boldsymbol{W}^{-1} - \boldsymbol{W}^{-1} \boldsymbol{X} \boldsymbol{S}^{-1} \boldsymbol{X}' \boldsymbol{W}^{-1}) \boldsymbol{W}_0. \tag{5.1.43}$$

As $\boldsymbol{W}^{-1} = \boldsymbol{W}^{-1/2} \boldsymbol{W}^{-1/2}$ (Theorem A.2), the inner matrix may be written as the product

$$(\boldsymbol{W}^{-1/2} - \boldsymbol{W}^{-1} \boldsymbol{X} \boldsymbol{S}^{-1} \boldsymbol{X}' \boldsymbol{W}^{-1/2})(\boldsymbol{W}^{-1/2} - \boldsymbol{W}^{-1/2} \boldsymbol{X} \boldsymbol{S}^{-1} \boldsymbol{X}' \boldsymbol{W}^{-1})$$

which is nonnegative definite (Theorem A.10). Thus the difference (43) is nonnegative, which is not surprising if we notice that the optimal predictor $\hat{\mathbf{p}}$ uses the additional information contained in the correlation matrix $E\boldsymbol{\varepsilon}\boldsymbol{\varepsilon}'_* = \sigma^2 \boldsymbol{W}_0$.

Clearly, in the case that $\boldsymbol{\varepsilon}$ is uncorrelated with $\boldsymbol{\varepsilon}_*$ (i.e. if $\boldsymbol{W}_0 = \boldsymbol{0}$), then classical and optimal prediction coincide.

5.1.5 Further Optimal Predictors

Corresponding to R-optimal estimation, we may also seek r-optimal heterogeneous and homogeneous predictors. As may be expected, the solutions will be of the same quality in the sense that they contain the unknown parameters of the model.

(i) r-optimal Heterogeneous Prediction

Using the set-up (26) of a linear predictor and inserting this in the MSE-risk (21) gives

$$r(\mathbf{p}, \mathbf{A}) = \operatorname{tr} \boldsymbol{A}[(\boldsymbol{D}\boldsymbol{X}-\boldsymbol{X}_*)\boldsymbol{\beta}+\mathbf{d}][(\boldsymbol{D}\boldsymbol{X}-\boldsymbol{X}_*)\boldsymbol{\beta}+\mathbf{d}]' + \sigma^2 \operatorname{tr} \boldsymbol{A}[\boldsymbol{D}\boldsymbol{W}\boldsymbol{D}' + \boldsymbol{W}_* - 2\boldsymbol{D}\boldsymbol{W}_0]. \tag{5.1.44}$$

Due to the fact that the second term is independent of $\mathbf{d}$, the risk is minimized for

$$\mathbf{d} = -(\boldsymbol{D}\boldsymbol{X}-\boldsymbol{X}_*)\boldsymbol{\beta}. \tag{5.1.45}$$

Differentiating the second term with respect to $\boldsymbol{D}$ leads to

$$\boldsymbol{A}\boldsymbol{D}\boldsymbol{W} = \boldsymbol{A}\boldsymbol{W}_0'.$$

Substituting the solution $\hat{\boldsymbol{D}}_1 = \boldsymbol{W}_0'\boldsymbol{W}^{-1}$ in (45) gives

$$\hat{\mathbf{d}} = -(\boldsymbol{W}_0'\boldsymbol{W}^{-1}\boldsymbol{X}-\boldsymbol{X}_*)\boldsymbol{\beta}$$

and the r-optimal heterogeneous predictor becomes

$$\hat{\mathbf{p}}_1 = \hat{\boldsymbol{D}}_1\mathbf{y}+\hat{\mathbf{d}} = \boldsymbol{X}_*\boldsymbol{\beta}+\boldsymbol{W}_0'\boldsymbol{W}^{-1}(\mathbf{y}-\boldsymbol{X}\boldsymbol{\beta}). \tag{5.1.46}$$

Note that $\hat{\mathbf{p}}_1$ is unbiased in the sense $E(\hat{\mathbf{p}}_1 - \boldsymbol{X}_*\boldsymbol{\beta}) = \mathbf{0}$ and has

$$r(\hat{\mathbf{p}}_1, \boldsymbol{A}) = \sigma^2 \operatorname{tr} \boldsymbol{A}(\boldsymbol{W}_* - \boldsymbol{W}_0'\boldsymbol{W}^{-1}\boldsymbol{W}_0). \tag{5.1.47}$$

Thus we have found a lower bound for the risk of a linear predictor. Comparing with (38) gives

$$r(\hat{\mathbf{p}}_1, \boldsymbol{A}) \le r(\hat{\mathbf{p}}, \boldsymbol{A}).$$

(ii) r-optimal Homogeneous Predictor

The homogeneous set-up of a linear predictor is

$$\mathbf{p} = \boldsymbol{D}\mathbf{y}.$$

Its risk is given in (44) when $\mathbf{d} = \mathbf{0}$. Differentiating with respect to $\boldsymbol{D}$ and equating to zero leads to the optimal value of $\boldsymbol{D}$:

$$\boldsymbol{D}_{\text{opt}} = \hat{\boldsymbol{D}}_2 = (\boldsymbol{W}_0' + \sigma^2\boldsymbol{X}_*\boldsymbol{\beta}\boldsymbol{\beta}'\boldsymbol{X}')(\sigma^{-2}\boldsymbol{X}\boldsymbol{\beta}\boldsymbol{\beta}'\boldsymbol{X}' + \boldsymbol{W})^{-1}$$

and, therefore, to the r-optimal homogeneous predictor $\hat{\mathbf{p}}_2 = \hat{\boldsymbol{D}}_2\mathbf{y}$, i.e.

$$\hat{\mathbf{p}}_2 = \boldsymbol{X}_*\hat{\boldsymbol{\beta}}_2 + \boldsymbol{W}_0'(\sigma^{-2}\boldsymbol{X}\boldsymbol{\beta}\boldsymbol{\beta}'\boldsymbol{X}' + \boldsymbol{W})^{-1}\mathbf{y} \tag{5.1.48}$$

where $\hat{\boldsymbol{\beta}}_2$ is the R-optimal homogeneous estimator of $\boldsymbol{\beta}$ (see (2.4.6)). The risk is found by inserting $\hat{\boldsymbol{D}}_2$ in (44), where $\mathbf{d} = \mathbf{0}$. Furthermore, in accordance with (2.4.20) we have the relation

$$r(\hat{\mathbf{p}}_1, A) \le r(\hat{\mathbf{p}}_2, A) \le r(\hat{\mathbf{p}}, A).$$

Thus the best linear unbiased predictor $\hat{\mathbf{p}}$ (37) has the greatest risk, but—on the other hand—up to now, $\hat{\mathbf{p}}$ is the only practicable solution among r-optimal predictors when no other than the sample information is available. This property of $\hat{\mathbf{p}}$ corresponds to the optimality of the GLSE $\mathbf{b}$ (see Section 2.4.3).

Note To overcome the lack of immediate practicability of $\hat{\mathbf{p}}_2$, strong prior information on the direction as well as on the size of $\sigma^{-1}\boldsymbol{\beta}$ has to be given. For a full discussion of this problem the reader is referred to Toutenburg (1968) and Bibby and Toutenburg, 1978, pp. 90–93.

5.1.6 Relationship Between Unbiased Estimation and Unbiased Prediction

Let an unbiased estimator $\hat{\boldsymbol{\beta}} = C\mathbf{y}$ with dispersion matrix $V(\hat{\boldsymbol{\beta}})$ be given. As $CX = I$ we have $\hat{\boldsymbol{\beta}} - \boldsymbol{\beta} = C\boldsymbol{\varepsilon}$ and, therefore,

$$E(\hat{\boldsymbol{\beta}} - \boldsymbol{\beta})(W_0'W^{-1}\boldsymbol{\varepsilon} - \boldsymbol{\varepsilon}_*)' = \mathbf{0}. \tag{5.1.49}$$

Let us look for the structure of the best linear unbiased predictor $\hat{\mathbf{p}}$ (37) which consists of two components, the first of which estimates the expectation $X_*\boldsymbol{\beta}$ of $\mathbf{y}_*$ and the second estimates $\boldsymbol{\varepsilon}_*$ by $W_0'W^{-1}\hat{\boldsymbol{\varepsilon}}$. Instead of the GLSE $\mathbf{b}$ we will now use another unbiased estimator $\hat{\boldsymbol{\beta}}$. This is done by choosing the linear predictor

$$\mathbf{p}(\hat{\boldsymbol{\beta}}) = X_*\hat{\boldsymbol{\beta}} + W_0'W^{-1}(\mathbf{y} - X\hat{\boldsymbol{\beta}}). \tag{5.1.50}$$

This predictor is unbiased in the familiar sense, i.e.

$$E\mathbf{p}(\hat{\boldsymbol{\beta}}) = X_*\boldsymbol{\beta}.$$

Using the representation

$$\mathbf{p}(\hat{\boldsymbol{\beta}}) - \mathbf{y}_* = (X_* - W_0'W^{-1}X)(\hat{\boldsymbol{\beta}} - \boldsymbol{\beta}) + (W_0'W^{-1}\boldsymbol{\varepsilon} - \boldsymbol{\varepsilon}_*)$$

gives (see (49) and (47))

$$r[\mathbf{p}(\hat{\boldsymbol{\beta}}), A] = \operatorname{tr} A(X_* - W_0'W^{-1}X)V(\hat{\boldsymbol{\beta}})(X_* - W_0'W^{-1}X)' + r(\hat{\mathbf{p}}_1, A). \tag{5.1.51}$$

Thus, given two estimators $\hat{\boldsymbol{\beta}}_1$ and $\hat{\boldsymbol{\beta}}_2$ with $V(\hat{\boldsymbol{\beta}}_2) - V(\hat{\boldsymbol{\beta}}_1) \ge \mathbf{0}$ leads to the relation of the corresponding predictors:

$$\begin{aligned} &r[\mathbf{p}(\hat{\boldsymbol{\beta}}_2), A] - r[\mathbf{p}(\hat{\boldsymbol{\beta}}_1), A] \\ &\quad = \operatorname{tr} A(X_* - W_0'W^{-1}X)[V(\hat{\boldsymbol{\beta}}_2) - V(\hat{\boldsymbol{\beta}}_1)](X_* - W_0'W^{-1}X)' \ge \mathbf{0}. \end{aligned} \tag{5.1.52}$$

In other words, *a better unbiased estimator $\hat{\boldsymbol{\beta}}$ will give a better unbiased predictor* $\mathbf{p}(\hat{\boldsymbol{\beta}})$.

Note (i) This result may be generalized to unbiased estimators and their corresponding predictors (see Bunke, 1977).

(ii) Using (52), it can be seen that all relations between restricted estimators as found in the preceding chapters are valid for the corresponding predictors. Moreover, a direct application of r-optimality to the various restricted models would lead to solutions of type $\mathbf{p}(\hat{\beta})$ (50) where $\hat{\beta}$ is the R-optimal estimator in the restricted model under consideration (see Bibby and Toutenburg, 1978; Toutenburg 1971, 1973).

This result will be demonstrated in the following examples.

Example (i)

Let the general stochastic prior information $\mathbf{r} = R\beta + \phi$ (3.1.2) be given. Then the r-optimal unbiased predictor becomes

$$\mathbf{p}[\hat{\beta}(\sigma^2)] = X_* \hat{\beta}(\sigma^2) + W_0' W^{-1}(\mathbf{y} - X\hat{\beta}(\sigma^2)) \tag{5.1.53}$$

with $\hat{\beta}(\sigma^2)$ the mixed estimator defined in (3.1.9). This predictor has the risk

$$r\{\mathbf{p}[\hat{\beta}(\sigma^2)]\} = \operatorname{tr} A Z_* V[\hat{\beta}(\sigma^2)] Z_*' + r(\hat{\mathbf{p}}_1, A) \tag{5.1.54}$$

where $Z_* = X_* - W_0' W^{-1} X$ and $V[\hat{\beta}(\sigma^2)]$ is as in (3.1.10). The relation between the estimators $\mathbf{b}$ and $\hat{\beta}(\sigma^2)$, i.e.

$$V(\mathbf{b}) - V[\hat{\beta}(\sigma^2)] \geq 0 \quad \text{(see (3.1.11))}$$

immediately leads to

$$r(\hat{\mathbf{p}}) - r\{\mathbf{p}[\hat{\beta}(\sigma^2)]\} \geq 0. \tag{5.1.55}$$

Example (ii)

In the exact restriction case (2.3.66) the best estimator was given by $\tilde{\mathbf{b}}_R$ (2.3.71), the conditional restricted LSE. The corresponding (conditional unbiased) predictor becomes

$$\mathbf{p}(\tilde{\mathbf{b}}_R) = X_* \tilde{\mathbf{b}}_R + W_0' W^{-1}(\mathbf{y} - X\tilde{\mathbf{b}}_R) \tag{5.1.56}$$

having

$$r(\mathbf{p}(\tilde{\mathbf{b}}_R)) = \operatorname{tr} A Z_* V(\tilde{\mathbf{b}}_R) Z_*' + r(\hat{\mathbf{p}}_1, A) \tag{5.1.57}$$

where $V(\tilde{b}_R)$ was given in (2.3.73). Using relation (2.3.74), i.e. $V(\mathbf{b}) - V(\tilde{\mathbf{b}}_R) \geq 0$, leads to

$$r(\hat{\mathbf{p}}) - r(\mathbf{p}(\tilde{\mathbf{b}}_R)) \geq 0. \tag{5.1.58}$$

5.1.7 Prediction Ellipsoids

Assuming normality of the disturbances gives

$$\mathbf{y}_* \sim N_n(X_* \beta, \sigma^2 W_*) \tag{5.1.59}$$

and

$$\mathbf{y} \sim N_T(\boldsymbol{X\beta}, \sigma^2 \boldsymbol{W}). \tag{5.1.60}$$

If we confine ourselves to the case $E\boldsymbol{\varepsilon\varepsilon}'_* = \boldsymbol{0}$ (so that perhaps $\boldsymbol{W} = \boldsymbol{I}$ and $\boldsymbol{W}_* = \boldsymbol{I}$), then $\mathbf{p}(\boldsymbol{\beta}) = \boldsymbol{X}_*\boldsymbol{\beta}$ where $\boldsymbol{\beta}$ is an unbiased estimator of $\boldsymbol{\beta}$. Thus (see Theorem A.44) we are led to

$$\begin{aligned} Q(\hat{\boldsymbol{\beta}}) &= n^{-1}(\mathbf{y}_* - \boldsymbol{X}_*\hat{\boldsymbol{\beta}})'(\sigma^{-2}\boldsymbol{X}_* \boldsymbol{V}(\hat{\boldsymbol{\beta}})\boldsymbol{X}'_* + \boldsymbol{W}_*)^{-1}(\mathbf{y}_* - \boldsymbol{X}_*\boldsymbol{\beta}) \\ &\sim \sigma^2 n^{-1}\chi^2_n. \end{aligned} \tag{5.1.61}$$

Using s^2 as an estimate for σ^2 (see (2.2.9)) gives

$$Q(\hat{\boldsymbol{\beta}})s^{-2} \sim F_{n,T-K}. \tag{5.1.62}$$

Then the set

$$\begin{aligned} B(\hat{\boldsymbol{\beta}}) = \{\mathbf{y}_*: &(\mathbf{y}_* - \boldsymbol{X}_*\hat{\boldsymbol{\beta}})'(\sigma^{-2}\boldsymbol{X}_* \boldsymbol{V}(\hat{\boldsymbol{\beta}})\boldsymbol{X}'_* + \boldsymbol{W}_*)^{-1}(\mathbf{y}_* - \boldsymbol{X}_*\hat{\boldsymbol{\beta}}) \\ &< s^2 n F_{n,T-K,1-q}\} \end{aligned} \tag{5.1.63}$$

defines a prediction region for $\mathbf{y}_*$. $B(\hat{\boldsymbol{\beta}})$ is an ellipsoid centred around $\boldsymbol{X}_*\hat{\boldsymbol{\beta}}$ and has the *expected coverage* q:

$$E_{\mathbf{y}} P_{\mathbf{y}_*}\{\mathbf{y}_* \in B(\hat{\boldsymbol{\beta}})\} = q.$$

If q is fixed, the efficiency of the region $B(\hat{\boldsymbol{\beta}})$ may be measured by its expected volume (see Bibby and Toutenburg, 1978, p. 124):

$$\bar{V}[B(\hat{\boldsymbol{\beta}})] = V_n \sqrt{nF_{n,T-K,1-q}}\, E(s)[\sigma^{-2}\boldsymbol{X}_* \boldsymbol{V}(\hat{\boldsymbol{\beta}})\boldsymbol{X}'_* + \boldsymbol{W}_*]^{1/2} \tag{5.1.64}$$

where V_n is the volume of the n-dimensional unit sphere. Thus the efficiency of the prediction ellipsoid $B(\hat{\boldsymbol{\beta}})$ depends only on the dispersion matrix $\boldsymbol{V}(\hat{\boldsymbol{\beta}})$ of the chosen unbiased estimator. In other words, the better the estimator, the smaller is the prediction ellipsoid. (For a full discussion of problems in interval prediction the reader is referred to Chapter 8 of Bibby and Toutenburg (1978) and to Aitchison and Dunsmore (1968), Guttman (1970), and Liebermann and Miller (1963).)

5.2 PREDICTION AND MODEL CHOICE

5.2.1 Reduced Regressor Set

In many practical situtations some explanatory variables may be costly or difficult to record. Hence one must often work with a reduced model such as

$$\mathbf{y} = \boldsymbol{X}_2\boldsymbol{\beta}_2 + \boldsymbol{\varepsilon}, \tag{5.2.1}$$

Here the variables $\boldsymbol{X}_1$ of $\boldsymbol{X} = (\boldsymbol{X}_1, \boldsymbol{X}_2)$ are omitted (see also Hocking, 1976). As the whole model $\mathbf{y} = \boldsymbol{X\beta} + \boldsymbol{\varepsilon}$ is assumed to be the true one, estimation in the reduced model (1) will lead to solutions which are neither unbiased nor optimal. On the other hand, in the light of our previous results one can expect that these estimators will possess other desirable characteristics, as, for instance, a small mean square error.

The step from the full model to the reduced model (1) can be expressed equivalently by adding a linear restriction

$$\mathbf{r} = \boldsymbol{R\beta}, \quad \text{with } \mathbf{r} = \mathbf{0}, \boldsymbol{R} = (\boldsymbol{I}, \boldsymbol{0}) \tag{5.2.2}$$

to the full model $\mathbf{y} = \boldsymbol{X\beta} + \boldsymbol{\varepsilon}$. Here the dimension of $\boldsymbol{I}$ corresponds to the number of columns of $\boldsymbol{X}_1$. It can be seen that model reduction is a special case of regression under restrictions. Moreover, as the full model is assumed to be true, the restrictions are misspecified.

5.2.2 Misspecified Linear Restrictions

More generally, we will seek to compare the unrestricted model

$$\mathbf{y} = \boldsymbol{X\beta} + \boldsymbol{\varepsilon}, \quad \boldsymbol{\varepsilon} \sim (\mathbf{0}, \sigma^2 \boldsymbol{W}) \tag{5.2.3}$$

with the model under linear restrictions

$$\mathbf{y} = \boldsymbol{X\beta} + \boldsymbol{\varepsilon}, \quad \mathbf{r} = \boldsymbol{R\beta}, \quad \boldsymbol{\varepsilon} \sim (\mathbf{0}, \sigma^2 \boldsymbol{W}) \tag{5.2.4}$$

where the unrestricted model is the true one. The optimal estimators are the GLSE $\mathbf{b}$ and the conditional RLSE $\tilde{\mathbf{b}}_R$ (defined in (2.3.71)), respectively.

To get scalar variables, instead of the two estimators we will use the one-step-ahead predictors $p(\mathbf{b}) = \hat{p}(5.1.41)$ and

$$p(\tilde{\mathbf{b}}_R) = \mathbf{x}'_{T+1} \tilde{\mathbf{b}}_R + \mathbf{w}' \boldsymbol{W}^{-1} (\mathbf{y} - \boldsymbol{X} \tilde{\mathbf{b}}_R). \tag{5.2.5}$$

Let us denote $\mathbf{x}_{T+1}$ by $\mathbf{x}_*$.
Under the standard assumptions of the previous section we may conclude that

$$\hat{p} = \mathbf{x}'_* \mathbf{b} + \mathbf{w}' \boldsymbol{W}^{-1} (\mathbf{y} - \boldsymbol{X}\mathbf{b}) \tag{5.2.6}$$

has mean

$$\mu = \mathbf{x}'_* \boldsymbol{\beta} \tag{5.2.7}$$

and variance

$$\sigma_p^2 = \sigma^2 (\mathbf{x}'_* \boldsymbol{S}^{-1} \mathbf{x}_* + \mathbf{w}' \boldsymbol{W}^{-1} \mathbf{w} - \mathbf{w}' \boldsymbol{W}^{-1} \boldsymbol{X} \boldsymbol{S}^{-1} \boldsymbol{X}' \boldsymbol{W}^{-1} \mathbf{w}). \tag{5.2.8}$$

We may note that

$$p(\tilde{\mathbf{b}}_R) = \hat{p} + \hat{p}_R \tag{5.2.9}$$

where we define

$$\begin{aligned} \hat{p}_R &= (\mathbf{x}_* - \boldsymbol{X}' \boldsymbol{W}^{-1} \mathbf{w})' (\tilde{\mathbf{b}}_R - \mathbf{b}) \\ &= (\mathbf{x}_* - \boldsymbol{X}' \boldsymbol{W}^{-1} \mathbf{w})' \boldsymbol{S}^{-1} \boldsymbol{R}' (\boldsymbol{R} \boldsymbol{S}^{-1} \boldsymbol{R}')^{-1} (\mathbf{r} - \boldsymbol{R}\mathbf{b}) \\ &= \mathbf{z}' (\mathbf{r} - \boldsymbol{R}\mathbf{b}), \text{ say.} \end{aligned} \tag{5.2.10}$$

If we put

$$\gamma = \mathbf{z}' (\mathbf{r} - \boldsymbol{R\beta}) \tag{5.2.11}$$

then

$$E(\hat{p}_R) = \gamma. \tag{5.2.12}$$

As the unrestricted model (3) was assumed to be true, we may expect in general that $\mathbf{r} \neq \boldsymbol{R\beta}$ and, therefore, $\gamma \neq 0$. Furthermore, we may deduce that

$$\text{var}(\hat{p}_R) = \sigma^2 \mathbf{z}' \boldsymbol{R S}^{-1} \boldsymbol{R}' \mathbf{z} = \sigma_R^2, \text{ say.} \tag{5.2.13}$$

The covariance between $\hat{p}$ and $\hat{p}_R$ is

$$E(\hat{p} - \mu)(\hat{p}_R - \gamma) = -\tfrac{1}{2}(\sigma_R^2 + h) \tag{5.2.14}$$

where

$$h = 2\sigma^2 \mathbf{w}' \boldsymbol{W}^{-1} \boldsymbol{X S}^{-1} \boldsymbol{R}' \mathbf{z} + \sigma_R^2. \tag{5.2.15}$$

Thus $p(\tilde{\mathbf{b}}_R)$ (9) has mean $(\mu + \gamma)$ and variance

$$\sigma_p^2 + \sigma_R^2 - (\sigma_R^2 + h) = \sigma_p^2 - h. \tag{5.2.16}$$

Hence the biased predictor $p(\tilde{\mathbf{b}}_R)$ has a smaller variance than the unbiased predictor p if and only if $h > 0$. This is true, for example, when $\mathbf{w} = \mathbf{0}$. Then we have $h = \sigma_R^2$.

5.2.3 MSE-Criterion for Model Choice

If in general a parameter μ has to be estimated we say that an estimator $\hat{\mu}_1$ is MSE-better than another estimator $\hat{\mu}_2$ if

$$E(\hat{\mu}_1 - \mu)^2 < E(\hat{\mu}_2 - \mu)^2. \tag{5.2.17}$$

Applying this criterion to the predictors $p(\tilde{\mathbf{b}}_R)$ and $\hat{p}$ gives the result that the restricted predictor $p(\mathbf{b}_R)$ (and thus the restricted model (4)) can be preferred if

$$E[p(\tilde{\mathbf{b}}_R) - \mathbf{x}_*' \boldsymbol{\beta}]^2 < E[\hat{p} - \mathbf{x}_*' \boldsymbol{\beta}]^2. \tag{5.2.18}$$

Using (16) and (8), this condition is equivalent to

$$\sigma_p^2 - h + \gamma^2 < \sigma_p^2$$

or

$$\gamma^2 \sigma_R^{-2} < h \sigma_R^{-2} = \delta_0, \text{ say.} \tag{5.2.19}$$

If we define the test statistic

$$F = \left(\frac{\hat{p}_R}{s}\right)^2 \tag{5.2.20}$$

with s^2 from (2.2.9) we may show (Toutenburg, 1970a, 1973) that F follows a noncentral F when $\boldsymbol{\varepsilon}$ is normally distributed. That is

$$F \sim F_{1,T-K}(\gamma^2 \sigma_R^{-2}). \tag{5.2.21}$$

Due to relation (19) the noncentrality parameter $\gamma^2\sigma_R^{-2}$ provides a basis for deciding whether to use $\hat{p}$ or $p(\tilde{\mathbf{b}}_R)$ (that is, whether to use the unrestricted model (3) or the restricted model (4) regardless of the falsity of the restriction).

Hence we may choose as our null hypothesis

$$H_0:\gamma^2\sigma_R^{-2} < \delta_0 \tag{5.2.22}$$

against

$$H_1:\gamma^2\sigma_R^{-2} \geq \delta_0.$$

For any given type I error level α we may calculate a critical value F_α such that

$$P(F \geq F_\alpha | \gamma^2\sigma_R^{-2} = \delta_0) = \alpha. \tag{5.2.23}$$

If H_0 is accepted, then $p(\tilde{\mathbf{b}}_R)$ is MSE-better than $\hat{p}$ and, moreover, the restricted model will be used. In the special case of model reduction we then would take the reduced model (1). Clearly, in the case of model reduction, the estimator $\tilde{\mathbf{b}}_R$ specializes to $\mathbf{b}_R(0)$ (3.5.28).

Note The connection of the above results to the MSE-criterion I for comparing the estimators $\mathbf{b}$ and $\tilde{\mathbf{b}}_R$ is as follows. Let $\mathbf{w} = \mathbf{0}$ so that $\hat{p} = \mathbf{x}_*'\mathbf{b}$ and $p(\tilde{\mathbf{b}}_R) = \mathbf{x}_*'\tilde{\mathbf{b}}_R$. Then $p(\tilde{\mathbf{b}}_R)$ is MSE-better than $\hat{p}$ if (see (19))

$$h - \gamma^2 = \sigma^2\mathbf{z}'\{(RS^{-1}R') - \sigma^{-2}(\mathbf{r} - R\boldsymbol{\beta})(\mathbf{r} - R\boldsymbol{\beta})'\}\mathbf{z} \geq 0 \tag{5.2.24}$$

If we require (24) to hold for all $\mathbf{z} = \mathbf{x}_*'S^{-1}R'(RS^{-1}R')^{-1}$, we are led to condition (3.4.15). (For alternative criteria of model choice see Bibby and Toutenburg, 1978, pp. 113–115.)

5.3 TWO-STAGE PROCEDURES FOR ESTIMATION AND MODEL CHOICE

5.3.1 Models with Replications of the Input Matrix

Consider the following set of classical linear models

$$\left.\begin{aligned} &\mathbf{y}_j = X\boldsymbol{\beta} + \boldsymbol{\varepsilon}_j, j = 1, \ldots, r \\ &\boldsymbol{\varepsilon}_j \sim N(\mathbf{0}, \sigma^2 I), E\boldsymbol{\varepsilon}_j\boldsymbol{\varepsilon}_{j'}' = \mathbf{0} \quad (j \neq j') \end{aligned}\right\} \tag{5.3.1}$$

That is, model (1) is of multivariate regression type (see (6.1.14)). From Theorem 6.2 it follows immediately that the optimal unbiased estimator of $\boldsymbol{\beta}$ in model (1) is

$$\hat{\boldsymbol{\beta}}_r = r^{-1}\sum_{j=1}^{r}(X'X)^{-1}X'\mathbf{y}_j \tag{5.3.2}$$

which has the dispersion matrix

$$V(\hat{\boldsymbol{\beta}}_r) = r^{-1}\sigma^2(X'X)^{-1} = r^{-1}\sigma^2 U. \tag{5.3.3}$$

In the context of this chapter we consider (1) as a model which replies (at the maximum) r times the same input matrix $\boldsymbol{X}$ to get an efficient estimator of $\boldsymbol{\beta}$. In practice it is often sufficient to use only some of the r replications of the model, say $r_1 < r = r_1 + r_2$. The corresponding unbiased estimator of $\boldsymbol{\beta}$ is then

$$\hat{\boldsymbol{\beta}}_{r_1} = r_1^{-1} \sum_{j=1}^{r_1} (\boldsymbol{X}'\boldsymbol{X})^{-1}\boldsymbol{X}'\mathbf{y}_j \tag{5.3.4}$$

with dispersion

$$V(\hat{\boldsymbol{\beta}}_{r_1}) = r_1^{-1}\sigma^2 \boldsymbol{U}. \tag{5.3.5}$$

Thus the components of $\hat{\boldsymbol{\beta}}_r$ are of smaller variance than the components of $\hat{\boldsymbol{\beta}}_{r_1}$ when $r_2 > 0$. On the other hand we notice the cost-saving benefit of $\hat{\boldsymbol{\beta}}_{r_1}$, for $\hat{\boldsymbol{\beta}}_{r_1}$ needs less replications of the model than $\hat{\boldsymbol{\beta}}_r$.

Let us assume that we will prefer the less accurate estimator $\hat{\boldsymbol{\beta}}_{r_1}$ if it belongs to a certain acceptance region $B(\boldsymbol{\beta}, \sigma^2) = B \subset E^K$ (see Mayer, Singh, and Willke, 1974). Thus we are led to an estimator which depends on the (up to now unknown) region B:

$$\hat{\boldsymbol{\beta}}(B) = \begin{cases} \hat{\boldsymbol{\beta}}_{r_1} & \text{if } \hat{\boldsymbol{\beta}}_{r_1} \in B \\ \hat{\boldsymbol{\beta}}_r & \text{otherwise.} \end{cases} \tag{5.3.6}$$

The region B will be determined by minimizing the MSE-risk

$$R(B/\boldsymbol{\beta}, \sigma^2) = \operatorname{tr} \boldsymbol{A}E[\hat{\boldsymbol{\beta}}(B) - \boldsymbol{\beta}][\hat{\boldsymbol{\beta}}(B) - \boldsymbol{\beta}]' \tag{5.3.7}$$

with $\boldsymbol{A} > \boldsymbol{0}$ a given matrix.

If we define

$$\hat{\boldsymbol{\beta}}_{r_2} = r_2^{-1} \sum_{j=r_1+1}^{r} (\boldsymbol{X}'\boldsymbol{X})^{-1}\boldsymbol{X}'\mathbf{y}_j \tag{5.3.8}$$

the estimator $\hat{\boldsymbol{\beta}}_r$ (2) can be written

$$\hat{\boldsymbol{\beta}}_r = \frac{r_1}{r}\hat{\boldsymbol{\beta}}_{r_1} + \frac{r_2}{r}\hat{\boldsymbol{\beta}}_{r_2}. \tag{5.3.9}$$

Let $F_1(.)$ and $F_2(.)$ denote the distribution function of $\hat{\boldsymbol{\beta}}_{r_1}$ and $\hat{\boldsymbol{\beta}}_{r_2}$, respectively. From the above definitions of the estimators and the normality of $\boldsymbol{\varepsilon}$ it follows that

$$\hat{\boldsymbol{\beta}}_{r_1} \sim N(\boldsymbol{\beta}, r_1^{-1}\sigma^2 \boldsymbol{U}), \quad \hat{\boldsymbol{\beta}}_{r_2} \sim N(\boldsymbol{\beta}, r_2^{-1}\sigma^2 \boldsymbol{U}). \tag{5.3.10}$$

Moreover, according to the assumed independence of $\boldsymbol{\varepsilon}_j$ and $\boldsymbol{\varepsilon}_{j'}$ $(j \neq j')$ the components $\hat{\boldsymbol{\beta}}_{r_1}$ and $\hat{\boldsymbol{\beta}}_{r_2}$ of $\hat{\boldsymbol{\beta}}_r$ are independent. The density function of $\hat{\boldsymbol{\beta}}_r$ is therefore

$$\mathrm{d}F = \mathrm{d}F_1 \mathrm{d}F_2. \tag{5.3.11}$$

Then the following Theorem holds (Mayer *et al.*, 1974).

Theorem 5.2

If

$$B^*(\boldsymbol{\beta}, \sigma^2) = \{\tilde{\boldsymbol{\beta}}: \operatorname{tr} \boldsymbol{A}(\tilde{\boldsymbol{\beta}} - \boldsymbol{\beta})(\tilde{\boldsymbol{\beta}} - \boldsymbol{\beta})' \leq \sigma^2 (r_1 + r)^{-1} \operatorname{tr} \boldsymbol{A}\boldsymbol{U}\} \quad (5.3.12)$$

then

$$\min_B R(B/\boldsymbol{\beta}, \sigma^2) = R(B^*/\boldsymbol{\beta}, \sigma^2).$$

Proof Let $\overline{B} = E^K - B$ denote the complement of B. According to (6), (7), and (11) we have

$$R(B/\boldsymbol{\beta}, \sigma^2) = \int_{\hat{\boldsymbol{\beta}}_{r_1} \in B} \int_{\hat{\boldsymbol{\beta}}_{r_2}} \operatorname{tr} \boldsymbol{A}(\hat{\boldsymbol{\beta}}_{r_1} - \boldsymbol{\beta})(\hat{\boldsymbol{\beta}}_{r_1} - \boldsymbol{\beta})' \, \mathrm{d}F_1 \, \mathrm{d}F_2$$

$$+ \int_{\hat{\boldsymbol{\beta}}_{r_1} \in \overline{B}} \int_{\hat{\boldsymbol{\beta}}_{r_2}} \operatorname{tr} \boldsymbol{A}(\hat{\boldsymbol{\beta}}_r - \boldsymbol{\beta})(\hat{\boldsymbol{\beta}}_r - \boldsymbol{\beta})' \, \mathrm{d}F_1 \, \mathrm{d}F_2$$

$$= \int_{\hat{\boldsymbol{\beta}}_{r_1} \in B} \int_{\hat{\boldsymbol{\beta}}_{r_2}} r^{-2}(r^2 - r_1^2) \operatorname{tr} \boldsymbol{A}(\hat{\boldsymbol{\beta}}_{r_1} - \boldsymbol{\beta})(\hat{\boldsymbol{\beta}}_{r_1} - \boldsymbol{\beta})' \mathrm{d}F_1 \, \mathrm{d}F_2$$

$$+ \int_{\hat{\boldsymbol{\beta}}_{r_1} \in E^K} \int_{\hat{\boldsymbol{\beta}}_{r_2}} r^{-2} r_1^2 \operatorname{tr} \boldsymbol{A}(\hat{\boldsymbol{\beta}}_{r_1} - \boldsymbol{\beta})(\hat{\boldsymbol{\beta}}_{r_1} - \boldsymbol{\beta})' \, \mathrm{d}F_1 \, \mathrm{d}F_2$$

$$+ \int_{\hat{\boldsymbol{\beta}}_{r_1} \in \overline{B}} \int_{\hat{\boldsymbol{\beta}}_{r_2}} r^{-2} r_2^2 \operatorname{tr} \boldsymbol{A}(\hat{\boldsymbol{\beta}}_{r_2} - \boldsymbol{\beta})(\hat{\boldsymbol{\beta}}_{r_2} - \boldsymbol{\beta})' \, \mathrm{d}F_1 \, \mathrm{d}F_2.$$

The second term can be verified to equal

$$r^{-2} r_1 \sigma^2 \operatorname{tr} \boldsymbol{A}\boldsymbol{U}.$$

The last term equals

$$r^{-2} r_2 \sigma^2 \operatorname{tr} \boldsymbol{A}\boldsymbol{U} - r^{-2} r_2^2 \int_{\hat{\boldsymbol{\beta}}_{r_1} \in B} \int_{\hat{\boldsymbol{\beta}}_{r_2}} \operatorname{tr} \boldsymbol{A}(\hat{\boldsymbol{\beta}}_{r_2} - \boldsymbol{\beta})(\hat{\boldsymbol{\beta}}_{r_2} - \boldsymbol{\beta})' \, \mathrm{d}F_1 \, \mathrm{d}F_2$$

$$= r^{-2} r_2 \sigma^2 \operatorname{tr} \boldsymbol{A}\boldsymbol{U} - r^{-2} r_2 \sigma^2 \operatorname{tr} \boldsymbol{A}\boldsymbol{U} \int_{\hat{\boldsymbol{\beta}}_r \in B} \mathrm{d}F_1.$$

So we get after simple manipulation

$$R(B/\boldsymbol{\beta}, \sigma^2) = r^{-2} \int_{\hat{\boldsymbol{\beta}}_{r_1} \in B} \left[(r^2 - r_1^2) \operatorname{tr} \boldsymbol{A}(\hat{\boldsymbol{\beta}}_{r_1} - \boldsymbol{\beta})(\hat{\boldsymbol{\beta}}_{r_1} - \boldsymbol{\beta})' - r_2 \sigma^2 \operatorname{tr} \boldsymbol{A}\boldsymbol{U} \right] \mathrm{d}F_1$$

$$+ r^{-1} \sigma^2 \operatorname{tr} \boldsymbol{A}\boldsymbol{U}. \quad (5.3.13)$$

$R(B/\boldsymbol{\beta},\sigma^2)$ is minimized if $\hat{\boldsymbol{\beta}}_{r_1} \in B^*(\boldsymbol{\beta},\sigma^2)$ where $B^*(\boldsymbol{\beta},\sigma^2)$ was defined in Theorem 5.2.

Corollary

The estimator

$$\hat{\boldsymbol{\beta}}(B^*) = \begin{cases} \hat{\boldsymbol{\beta}}_{r_1} & \text{if } \hat{\boldsymbol{\beta}}_{r_1} \in B^*(\boldsymbol{\beta},\sigma^2) \\ \hat{\boldsymbol{\beta}}_r & \text{otherwise} \end{cases} \tag{5.3.14}$$

(with $\hat{\boldsymbol{\beta}}_{r_1}$ from (4) and $\hat{\boldsymbol{\beta}}_r$ the BLUE) has smaller MSE-risk than the unrestricted BLUE $\hat{\boldsymbol{\beta}}_r$.

Proof From (13) it follows that

$$\begin{aligned} \operatorname{tr} \mathrm{MSE}(\hat{\boldsymbol{\beta}}(B^*)) &= R(B^*/\boldsymbol{\beta},\sigma^2) \\ &\leq r^{-1}\sigma^2 \operatorname{tr} \boldsymbol{AU} = \operatorname{tr} \mathrm{MSE}(\hat{\boldsymbol{\beta}}_r) = R(\hat{\boldsymbol{\beta}}_r). \end{aligned}$$

To make the criterion practicable we may choose an initial estimator $\boldsymbol{\beta}_0$ of $\boldsymbol{\beta}$ and an initial estimator c^2 or σ^2 and decide to take the smaller but less efficient estimator $\hat{\boldsymbol{\beta}}_{r_1}$ if $\hat{\boldsymbol{\beta}}_{r_1} \in B^*(\boldsymbol{\beta}_0, c^2)$. Then one needs only $(r_1 r^{-1}) \times 100$ per cent of the costs which would be necessary for the whole estimator $\hat{\boldsymbol{\beta}}_r$.

Example

Let a sample $(y_1,\ldots,y_r)$ of a one-dimensional normal distributed variable $y \sim N(\mu,\sigma^2)$ be given. We may estimate μ by the whole sample mean

$$\bar{y} = r^{-1} \sum_{i=1}^{r} y_i \sim N(\mu,\sigma^2 r^{-1})$$

or by the partial sample mean

$$\bar{y}_1 = r_1^{-1} \sum_{i=1}^{r_1} y_i \sim N(\mu,\sigma^2 r_1^{-1}).$$

Both estimators are unbiased, but $\bar{y}$ needs full costs and yields a smaller variance.

According to the previous theorem we decide to take $\bar{y}_1$ if

$$(\bar{y}_1 - \mu)^2 \leq \sigma^2 (r_1 + r)^{-1} = \sigma^2 \frac{r_2}{r^2 - r_1^2}.$$

If $r_1 \to r$ this inequality becomes $(\bar{y} - \mu)^2 < \infty$ so that the acceptance region for $\bar{y}$ tends to the whole space E^1.

5.3.2 Two-Stage Restricted Estimator

The idea of the preceding subsection was to save costs by decreasing the number of replications of the measurements. Applying this idea to the reduction of the regressor set (see Section 5.2.1) will lead us to a two-stage procedure which allows us to choose between a full model

$$\mathbf{y} = \boldsymbol{X\beta} + \boldsymbol{\varepsilon}, \quad \boldsymbol{\varepsilon} \sim N(\mathbf{0}, \sigma^2 \boldsymbol{W}) \tag{5.3.15}$$

and the reduced model (5.2.1).

As pointed out in Section 5.2.2, this can be included in the more general problem of deciding between model (15) and the restricted model

$$\mathbf{y} = \boldsymbol{X\beta} + \boldsymbol{\varepsilon}, \quad \mathbf{r} = \boldsymbol{R\beta}, \quad \boldsymbol{\varepsilon} \sim N(\mathbf{0}, \sigma^2 \boldsymbol{W}) \tag{5.3.16}$$

where in general these restrictions are misspecified. Using the notation of Section 3.4.2 we have

$$\mathbf{r} - \boldsymbol{R\beta} = \mathbf{g} \neq \mathbf{0}. \tag{5.3.17}$$

With the abbreviation (see (3.4.9))

$$\boldsymbol{D} = \boldsymbol{S}^{-1} \boldsymbol{R}' (\boldsymbol{R S}^{-1} \boldsymbol{R}')^{-1}$$

the conditional RLSE $\tilde{\mathbf{b}}_R$(2.3.71) can be written

$$\tilde{\mathbf{b}}_R = \mathbf{b} + \boldsymbol{D}(\mathbf{r} - \boldsymbol{R}\mathbf{b}) = \mathbf{b} - \hat{\boldsymbol{\beta}}_*, \ \textit{say} \tag{5.3.18}$$

where $\hat{\boldsymbol{\beta}}_*$ is an 'auxiliary estimator' which allows us to apply an analysis similar to that introduced by (5.3.9). We have (see (3.4.10) and (3.4.11))

$$\tilde{\mathbf{b}}_R \sim N_K(\boldsymbol{\beta} + \boldsymbol{D}\mathbf{g}, \boldsymbol{V}(\tilde{\mathbf{b}}_R)) \tag{5.3.19}$$

with

$$\boldsymbol{V}(\tilde{\mathbf{b}}_R) = \sigma^2 \boldsymbol{S}^{-1} - \sigma^2 \boldsymbol{D}(\boldsymbol{R S}^{-1} \boldsymbol{R}') \boldsymbol{D}' \leq \boldsymbol{V}(\mathbf{b}) = \sigma^2 \boldsymbol{S}^{-1},$$
$$\text{bias } \tilde{\mathbf{b}}_R = \boldsymbol{D}\mathbf{g},$$

and

$$\hat{\boldsymbol{\beta}}_* \sim N_K(-\boldsymbol{D}\mathbf{g}, \sigma^2 \boldsymbol{D}(\boldsymbol{R S}^{-1} \boldsymbol{R}')\boldsymbol{D}'). \tag{5.3.20}$$

Moreover, $\tilde{\mathbf{b}}_R$ and $\hat{\boldsymbol{\beta}}_*$ are independent:

$$E(\tilde{\mathbf{b}}_R - E\tilde{\mathbf{b}}_R)(\hat{\boldsymbol{\beta}}_* - E\hat{\boldsymbol{\beta}}_*)' = \mathbf{0}. \tag{5.3.21}$$

Clearly,

$$\mathbf{b} \sim N_K(\boldsymbol{\beta}, \boldsymbol{V}(\mathbf{b})). \tag{5.3.22}$$

Furthermore, we may derive that

$$E(\tilde{\mathbf{b}}_R - \boldsymbol{\beta})\hat{\boldsymbol{\beta}}'_* = -(\text{bias } \tilde{\mathbf{b}}_R)(\text{bias } \tilde{\mathbf{b}}_R)' = -\boldsymbol{D}\mathbf{g}\mathbf{g}'\boldsymbol{D}' \tag{5.3.23}$$

and

$$E\hat{\boldsymbol{\beta}}_*\hat{\boldsymbol{\beta}}'_* = \boldsymbol{V}(\mathbf{b}) - \mathbf{V}(\tilde{\mathbf{b}}_R) + (\text{bias } \tilde{\mathbf{b}}_R)(\text{bias } \tilde{\mathbf{b}}_R)'. \tag{5.3.24}$$

Now we will define a two-stage estimator $\hat{\boldsymbol{\beta}}(B)$ which depends on an acceptance region $B \subseteq E^K$ and which allows us to take into account the prior weight given to the restricted (perhaps the reduced) model:

$$\hat{\boldsymbol{\beta}}(B) = \begin{cases} \tilde{\mathbf{b}}_R & \text{if } \tilde{\mathbf{b}}_R \in B \\ \mathbf{b} & \text{otherwise.} \end{cases} \tag{5.3.25}$$

Using (18), (23), and (24) gives

$$\begin{aligned} \mathrm{MSE}(\mathbf{b}) = E(\mathbf{b}-\boldsymbol{\beta})(\mathbf{b}-\boldsymbol{\beta})' &= E(\tilde{\mathbf{b}}_R - \boldsymbol{\beta} + \hat{\boldsymbol{\beta}}_*)(\tilde{\mathbf{b}}_R - \boldsymbol{\beta} + \hat{\boldsymbol{\beta}}_*)' \\ &= \mathrm{MSE}(\tilde{\mathbf{b}}_R) + 2E(\tilde{\mathbf{b}}_R - \boldsymbol{\beta})\hat{\boldsymbol{\beta}}'_* + E\hat{\boldsymbol{\beta}}_*\hat{\boldsymbol{\beta}}'_* \\ &= [V(\tilde{\mathbf{b}}_R) + (\text{bias } \tilde{\mathbf{b}}_R)(\text{bias } \tilde{\mathbf{b}}_R)'] \qquad (5.3.26) \\ &\quad + [-2(\text{bias } \tilde{\mathbf{b}}_R)(\text{bias } \tilde{\mathbf{b}}_R)'] \\ &\quad + [V(\mathbf{b}) - V(\tilde{\mathbf{b}}_R) + (\text{bias } \tilde{\mathbf{b}}_R)(\text{bias } \tilde{\mathbf{b}}_R)']. \end{aligned}$$

Let us denote by $F_1(.)$ and $F_2(.)$ the distribution functions of $\tilde{\mathbf{b}}_R$ and $\hat{\boldsymbol{\beta}}_*$, respectively. If $F(.)$ denotes the common distribution function of $(\tilde{\mathbf{b}}_R, \hat{\boldsymbol{\beta}}_*)$, we have $\mathrm{d}F = \mathrm{d}F_1\,\mathrm{d}F_2$ according to (21). Then the following theorem holds (Toutenburg, 1977b).

Theorem 5.3

If

$$\begin{aligned} B^*(\boldsymbol{\beta}, \sigma^2) = B^* &= [\tilde{\mathbf{b}}_R : 2\mathrm{tr}\, A(\tilde{\mathbf{b}}_R - \boldsymbol{\beta})(\tilde{\mathbf{b}}_R - \boldsymbol{\beta})' \le \mathrm{tr}\, A\{V(\tilde{\mathbf{b}}_R) + V(\mathbf{b}) \\ &\quad + (\text{bias } \tilde{\mathbf{b}}_R)(\text{bias } \tilde{\mathbf{b}}_R)'\}] \end{aligned} \tag{5.3.27}$$

then $\min_B R(B/\boldsymbol{\beta}, \sigma^2) = R(B^*/\boldsymbol{\beta}, \sigma^2)$.

Proof According to (25) and (26) we have

$$\begin{aligned} R(B/\boldsymbol{\beta}, \sigma^2) &= \int_{\tilde{\mathbf{b}}_R \in B} \int_{\hat{\boldsymbol{\beta}}_*} \mathrm{tr}\, A(\tilde{\mathbf{b}}_R - \boldsymbol{\beta})(\tilde{\mathbf{b}}_R - \boldsymbol{\beta})'\,\mathrm{d}F_1\,\mathrm{d}F_2 \\ &\quad + \int_{\tilde{\mathbf{b}}_R \in \bar{B}} \int_{\hat{\boldsymbol{\beta}}_*} \mathrm{tr}\, A(\mathbf{b} - \boldsymbol{\beta})(\mathbf{b} - \boldsymbol{\beta})'\,\mathrm{d}F_1\,\mathrm{d}F_2 \\ &= [V(\tilde{\mathbf{b}}_R) + (\text{bias } \tilde{\mathbf{b}}_R)(\text{bias } \tilde{\mathbf{b}}_R)'] \\ &\quad + 2\int_{\tilde{\mathbf{b}}_R \in \bar{B}} \int_{\hat{\boldsymbol{\beta}}_*} \mathrm{tr}\, A(\tilde{\mathbf{b}}_R - \boldsymbol{\beta})\hat{\boldsymbol{\beta}}'_*\,\mathrm{d}F_1\,\mathrm{d}F_2 \\ &\quad + \int_{\tilde{\mathbf{b}}_R \in \bar{B}} \int_{\hat{\boldsymbol{\beta}}_*} \mathrm{tr}\, A\hat{\boldsymbol{\beta}}_*\hat{\boldsymbol{\beta}}'_*\,\mathrm{d}F_1\,\mathrm{d}F_2. \end{aligned}$$

If we collect under 'const.' all terms which are independent of the region B we get after simple manipulation

$$R(B/\boldsymbol{\beta},\sigma^2)-\text{const.} = -\int_{\tilde{\mathbf{b}}_R\in B}\int_{\hat{\boldsymbol{\beta}}_*}\text{tr}\, \boldsymbol{A}[2(\tilde{\mathbf{b}}_R-\boldsymbol{\beta})\hat{\boldsymbol{\beta}}'_* + \hat{\boldsymbol{\beta}}_*\hat{\boldsymbol{\beta}}'_*]\,\mathrm{d}F_1\,\mathrm{d}F_2$$

$$= \int_{\tilde{\mathbf{b}}_R\in B} \text{tr}\, \boldsymbol{A}\{2(\tilde{\mathbf{b}}_R-\boldsymbol{\beta})(\tilde{\mathbf{b}}_R-\boldsymbol{\beta})'$$

$$-(\boldsymbol{V}(\tilde{\mathbf{b}}_R)+\boldsymbol{V}(\mathbf{b})+(\text{bias}\,\tilde{\mathbf{b}}_R)(\text{bias}\,\tilde{\mathbf{b}}_R)')\}\,\mathrm{d}F_1.$$

The integral is minimal if the integrand is less or equal to zero, that is if $\tilde{\mathbf{b}}_R\in B^*$ defined in (27). This completes the proof.

If we choose as risk matrix $\boldsymbol{A} = \boldsymbol{I}_K$ and if $\mathbf{r} = \boldsymbol{R\beta}$ holds, then bias $\tilde{\mathbf{b}}_R = \mathbf{0}$ and $B^* = E^K$ follows. This corresponds to the superiority of $\tilde{\mathbf{b}}_R$ over $\mathbf{b}$ so long as the restriction is true.

In a similar way as for the corollary to Theorem 5.2 we may show that for $\mathbf{r} \neq \boldsymbol{R\beta}$

$$\text{MSE}[\hat{\boldsymbol{\beta}}(B^*)] = R(B^*/\boldsymbol{\beta},\sigma^2) \leq \text{MSE}(\mathbf{b}). \tag{5.3.28}$$

Let again $\mathbf{r} \neq \boldsymbol{R\beta}$. Then as a case of special interest let us choose $\boldsymbol{A} = \sigma^2\boldsymbol{V}^{-1}(\mathbf{b}) = \boldsymbol{S}$ where this choice may be motivated as in Definition 3.3. Then the region $B^*(\boldsymbol{\beta},\sigma^2)$ can be shown to have the form of a concentration ellipsoid

$$B^*(\boldsymbol{\beta},\sigma^2) = \{\tilde{\mathbf{b}}_R: (\tilde{\mathbf{b}}_R-\boldsymbol{\beta})'\boldsymbol{S}(\tilde{\mathbf{b}}_R-\boldsymbol{\beta}) \leq \sigma^2(2K-J+\lambda)/2\} \tag{5.3.29}$$

with $\lambda = \sigma^{-2}\mathbf{g}'(\boldsymbol{RS}^{-1}\boldsymbol{R}')^{-1}\mathbf{g}$ (see (3.4.15)). This leads to a comparison with the MSE-III criterion (see Section 3.4.2). $\tilde{\mathbf{b}}_R$ is MSE-III better than $\mathbf{b}$ if

$$E(\tilde{\mathbf{b}}_R-\boldsymbol{\beta})'\boldsymbol{S}(\tilde{\mathbf{b}}_R-\boldsymbol{\beta}) \leq \sigma^2 K \tag{5.3.30}$$

or, equivalently, if

$$-J+\lambda \leq 0 \quad \text{(see (3.4.23))}. \tag{5.3.31}$$

That is, if $\tilde{\mathbf{b}}_R$ is MSE-better than $\mathbf{b}$ the region B^* (29) becomes

$$B^*(\boldsymbol{\beta},\sigma^2) = \{\tilde{\mathbf{b}}_R: (\tilde{\mathbf{b}}_R-\boldsymbol{\beta})'\boldsymbol{S}(\tilde{\mathbf{b}}_R-\boldsymbol{\beta}) \leq \sigma^2 K\}. \tag{5.3.32}$$

Thus the two-stage estimator

$$\hat{\boldsymbol{\beta}}(B^*) = \begin{cases} \tilde{\mathbf{b}}_R & \text{if } \tilde{\mathbf{b}}_R\in B^* \\ \mathbf{b} & \text{otherwise} \end{cases}$$

leads to the estimator $\tilde{\mathbf{b}}_R$ only for concrete realizations of the model which fulfil (32). In other words, the two-stage estimation procedure pre-tests the concrete observation matrix before deciding which model to take.

Note As may be seen from the region B^* (27), the estimator $\hat{\beta}(B^*)$ (25) depends on the unknown parameters β and σ. To make this estimator practicable for model choice we have to replace β and σ by certain initial estimators β_0 and σ_0. If then $\tilde{\mathbf{b}}_R \in B^*(\beta_0, \sigma_0^2)$ one would decide to prefer the restricted model (16). The question arises, how to find appropriate initial estimates β_0, σ_0. Thus we are led to similar questions of stability as in the context of the mixed estimator and the minimax estimation procedure.

5.3.3 Further Results on Pre-Test Estimators

Confining ourselves to uncorrelated errors, i.e. $W = I$, gives the test statistic (3.4.25) for testing the linear hypothesis $\mathbf{r} = R\beta$ as

$$F = \frac{(\mathbf{r} - R\mathbf{b}_0)'(RUR')^{-1}(\mathbf{r} - R\mathbf{b}_0)}{(\mathbf{y} - X\mathbf{b}_0)'(\mathbf{y} - X\mathbf{b}_0)} \frac{T-K}{J}. \tag{5.3.33}$$

Under the null hypothesis, $\mathbf{r} = R\beta$, F has a central $F_{J,T-K}$-distribution. Under the alternative hypothesis, $\mathbf{g} = \mathbf{r} - R\beta \neq \mathbf{0}$, F is distributed as noncentral $F_{J,T-K}(\lambda)$ with the noncentrality parameter $\lambda = \sigma^{-2}\mathbf{g}'(RUR')^{-1}\mathbf{g}$ (see (3.4.15)).

Brook (1976) defines a *preliminary test estimator* for β due to

$$\beta^* = \begin{cases} \mathbf{b}_0 & \text{if} \quad F \geq c \\ \tilde{\mathbf{b}}_R & \text{if} \quad F < c \end{cases} \tag{5.3.34}$$

where $\tilde{\mathbf{b}}_R$ is the conditional RLSE in the case $W = I$, i.e.

$$\tilde{\mathbf{b}}_R = \mathbf{b}_0 + UR'(RUR')^{-1}(\mathbf{r} - R\mathbf{b}_0). \tag{5.3.35}$$

c can be taken to be the critical value of the F-distribution at the level α. On the other hand, if one is interested in an optimal value of c one has to calculate the risk of the estimator β^* (34) and to minimize it with respect to c. Instead of β^* one can consider a pre-test estimator for $E(Y|X)$:

$$X\beta^* = \begin{cases} X\mathbf{b}_0 & \text{if} \quad F \geq c \\ X\tilde{\mathbf{b}}_R & \text{if} \quad F < c. \end{cases} \tag{5.3.36}$$

As risk function of this estimator we choose the familiar MSE-function of the conditional forecast:

$$\begin{aligned} R(X\beta^*) &= E(X\beta^* - X\beta)'(X\beta^* - X\beta) \qquad (5.5.37) \\ &= E(\beta^* - \beta)'X'X(\beta^* - \beta). \end{aligned}$$

Comparing this with (3.4.22) we note that the risk $R(X\beta^*)$ is just the MSE-III risk of the estimator β^*. Thus the risk becomes a function only of degrees of freedom, noncentrality parameter, and the critical value c (Toyoda and Wallace, 1976). As T, K, and J are fixed in a given model, we may write $R(X\beta^*) = R(X\beta^*, c, \lambda)$. Note that for $c = 0$ one would always prefer the OLSE $\mathbf{b}_0$, and, for $c = \infty$, $\boldsymbol{X\beta}^*$ becomes $\boldsymbol{X}\tilde{\mathbf{b}}_R$ with probability one. The corresponding values of the risk function are

$$R(\boldsymbol{X\beta}^*, 0, \lambda) = \text{MSE III}(\mathbf{b}_0) = \sigma^2 K \tag{5.3.38}$$

and

$$R(\boldsymbol{X\beta}^*, \infty, \lambda) = \text{MSE III}(\tilde{\mathbf{b}}_R) = \sigma^2 (K - J + \lambda). \tag{5.3.39}$$

For $0 < c < \infty$ the risk is

$$R(\boldsymbol{X\beta}^*, c, \lambda) = \sigma^2 \{K - J + J f_1(c, \lambda) + 2\lambda [1 - 2f_1(c, \lambda) + f_2(c, \lambda)]\} \tag{5.3.40}$$

where

$$f_1(c, \lambda) = P\left\{\psi(J+2, T-K, \lambda) \geq \frac{Jc}{J+2}\right\}$$

and

$$f_2(c, \lambda) = P\left\{\psi(J+4, T-K, \lambda) \geq \frac{Jc}{J+4}\right\}$$

(ψ is the density of the noncentral F distribution: see Brook, 1976, and Figure 5.3.1).

If we define the relative risk of $\boldsymbol{X\beta}^*$ with respect to the best either unrestricted or restricted estimator as

$$G(c) = \sigma^{-2} \{R(\boldsymbol{X\beta}^*) - \min_{\lambda} [R(\boldsymbol{X}\mathbf{b}_0), R(\boldsymbol{X}\tilde{\mathbf{b}}_R)]\} \tag{5.3.41}$$

and minimize $G(c)$ with respect to c then we get the result

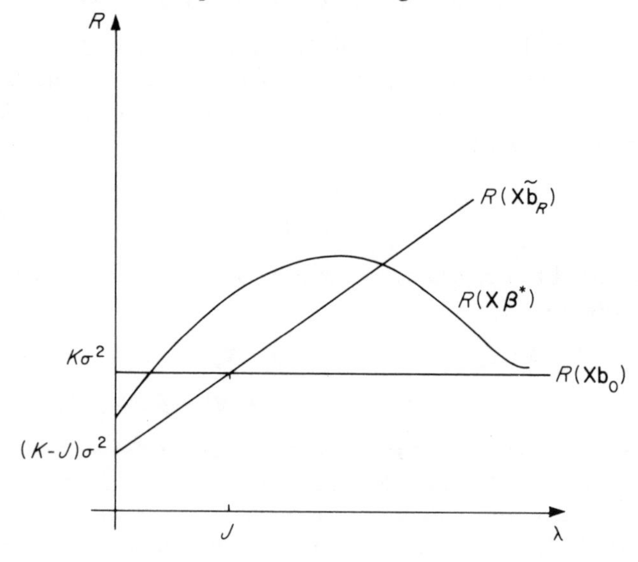

Figure 5.3.1 Risk-function of the pre-test estimator

(i) if $J \leq 4$ then the optimal value of c is $c = 0$ (that is, $\boldsymbol{X\beta}^* = \boldsymbol{X}\mathbf{b}_0$ so long as the number of restrictions is less than or equal to 4),
(ii) if $J \geq 5$ then the optimal value of c is nonzero. (Note that these values of c are tabulated in Toyoda and Wallace, 1976.)

For a more detailed discussion of problems related to pre-testing the reader is referred to Judge and Bock (1978).

VI Prior Information in Econometric Models

6.1 MULTIVARIATE AND STOCHASTIC REGRESSION

6.1.1 Introduction

The preceding chapter was concerned with methods for estimating parameters in single-equation relations, where various types of prior information could be used. Economic models, however, frequently involve a set of relationships designed to explain the connection between, and the behaviour of, a set of exogenous variables as well as a set of endogenous variables. In such models the estimation of parameters has special features that are not present when a model has only a single relation. In particular, when a regression model is a part of an econometric system, *some exogenous variables (regressors) are stochastic* and thus correlated with the disturbances. Moreover, the various relations of the econometric system are interdependent in the sense that variables in one relation may be 'on the right hand' (i.e. regressors) whereas in other relations they are in a 'left-hand position' (i.e. regressands). On the other hand, the interdependence of the systems equations is caused by the multivariate distribution of the disturbances of all equations. The estimation procedures of a single relation therefore have to be generalized in two directions:

(i) from single-equation models with fixed regressors to single-equation models with stochastic regressors which may be correlated with the disturbance term;
(ii) from single-equation models to simultaneous-equation models (with fixed as well as with stochastic regressors).

In other words, *the simultaneous equation model may be called multivariate stochastic regression.*

From this definition it follows that the statistical methods for these models would be greatly different from those for the single-equation procedure.

6.1.2 Prior Information and the Simultaneous-Equation Model

In building a highly complicated simultaneous-equation model to describe the behaviour of interdependent economic relationships we use prior information (or at least hypotheses) on

—the common distribution of the disturbance terms
—the statistical and/or causal dependence of variables
—the values of parameters, or at least on subregions, where certain parameters may vary.

The most compact information for the construction of econometric models has to be deduced from macro-economic theory. Macro-economic variables are thought of as determined by a complete system of equations, i.e. by a model which contains as many equations as endogenous variables. The following types of equations are possible: equations of economic behaviour, institutional laws (e.g. definition of equilibrium relations), and technological laws (transformation rules, definitions, identities).

In order that there should be practical relevance, systems of structural equations must not be composed entirely on the basis of economic theory. The system which is deduced from economic laws and observed rules of economic behaviour has to be combined with systematically collected data for the relevant variables for a given period and an economic unit (e.g. a country). Koopmans (1953) describes this situation as follows:

> 'Where statistical data are used as one of the foundation stones on which the equation system is erected, the modern methods of statistical inference are an indispensable instrument. However, without economic theory as another foundation stone, it is impossible to make such statistical inference apply directly to the equations of economic behaviour which are most relevant to analysis and to policy discussion. Statistical inference unsupported by economic theory applies to whatever statistical regularities and stable relationships can be discerned in the data.'

Data in economics can be found only in a non-experimental way. From this it follows that there is no possibility in macro-economics of finding the exact probability distributions (see Menges, 1959, for a full discussion of the relation of econometrics to sciences such as physics, chemistry, and so on, which live by experiments). This fact is one of the reasons for the high relevance of various types of prior information on the concrete structure of the equations, both on the relations between parameters as well as on the values of single parameters. The use of constraints involving parameters within a single relation is well known from the identification problem. On the other hand, there is a growing interest in cross-equation constraints (Kelly, 1975).

The occurrence of restrictions between the parameters of several structural equations cannot be excluded *a priori*. It shall be the main intention of this chapter to handle the use of prior information in the form of constraints on the parameters. In the preceding chapters there has been developed a wide range of methods for estimating and predicting single regression equations whose parameters are under linear and nonlinear restrictions. The aim of these methods is to overcome the restrained usefulness of the OLSE $\mathbf{b}_0$ or of Aitken's GLSE $\mathbf{b}$.

Methods such as the ridge principle, the restricted least squares, the mixed estimator, or the minimax estimator have given estimators which are better in the sense of a smaller quadratic loss compared with the unrestricted OLSE or GLSE.

But up to now these methods have been applied to structural equation estimation and prediction only to an inconsiderable degree. Rothenberg (1973) gave a survey of the problems of prior information in econometrics. Kakwani and Court (1970) developed a special method for using exact linear restrictions and gave it the name 'restricted two-stage least squares'. Basmann (1957) gives a generalization of classical estimators which leads to estimators possessing optimal properties equivalent to those of limited-information estimators. As far as a theory of optimal estimation such as that for single-equation regression is concerned, there are only attempts to solve this problem in econometrics. Apart from the application of the maximum likelihood idea in normal distributed models, most of the econometric estimation procedures were developed more from the point of practicability than from the point of mathematical optimality.

A new idea to overcome this was given by Dhrymes (1974). His set-up allows the interpretation of the two-stage least squares estimator as a GLSE in a transformed model, so that the application of known regression procedures may be possible (see 6.3.2).

6.1.3 Single-Equation Model with Stochastic Regressors

In preparing the estimation of parameters in simultaneous-equation models we at first consider the well-known regression model, which we generalize by the assumption that the exogenous variables (regressors) are generated by a stochastic process. From this fact we deduce that in general the exogenous variables are time-dependent and are correlated with the disturbances process. Based on this stochastic dependence a classification of stochastic regression arises.

(i) Independent Stochastic Regression

Let the regression model be of the form $\mathbf{y} = \boldsymbol{X\beta} + \boldsymbol{\varepsilon}$, as above. The basic assumption of the stochastic regression model says that

$$\boldsymbol{X} \text{ is stochastic, but independent of } \boldsymbol{\varepsilon}. \tag{6.1.1}$$

From (1) we may deduce that

$$E(\boldsymbol{\varepsilon}|\boldsymbol{X}) = E\boldsymbol{\varepsilon} = \mathbf{0}$$

and, therefore, we have the unbiasedness of the OLSE $\mathbf{b}_0$:

$$\begin{aligned} E(\mathbf{b}_0|\boldsymbol{X}) &= \boldsymbol{\beta} + E((\boldsymbol{X}'\boldsymbol{X})^{-1}\boldsymbol{X}'\boldsymbol{\varepsilon}|\boldsymbol{X}) \\ &= \boldsymbol{\beta}. \end{aligned}$$

So, the OLSE $\mathbf{b}_0$ is unbiased in the stochastic regression model but it is no longer the best *linear* unbiased estimator of $\boldsymbol{\beta}$ in general, as the process generating $\boldsymbol{X}$ is now a stochastic one. Nevertheless, certain optimality properties of $\mathbf{b}_0$ are valid. So $\mathbf{b}_0$ is the maximum-likelihood estimator if ε is normally distributed.

Remark As far as the optimal prediction in this model is concerned, the reader is referred to Bibby and Toutenburg, 1978, pp. 135–137.

(ii) Contemporaneously Uncorrelated Regression

'How much further can we weaken the assumption of independence of regressors and disturbances without destroying the consistency of the classical least-squares estimators?' This question of Goldberger (1964, p. 278) may be answered as follows.

Definition 6.1 An estimator $\hat{\theta}$ of a parameter θ is called consistent if

$$\operatorname{p\,lim} \hat{\theta} = \theta$$

(where p lim denotes the probability limit, i.e. $\operatorname{p\,lim} \hat{\theta} = \theta$ if $\lim_{n \to \infty} P\{|\hat{\theta} - \theta| < \delta\} = 1$ for every $\delta > 0$; see Definition A.56).

If we write the OLSE $\mathbf{b}_0$ as

$$\mathbf{b}_0 = \boldsymbol{\beta} + (X'X)^{-1} X'\boldsymbol{\varepsilon}$$

we have by the well-known theorem of Slutsky (Theorem A.57)

$$\operatorname{p\,lim} \mathbf{b}_0 = \boldsymbol{\beta} + (\operatorname{p\,lim} T^{-1} X'X)^{-1} (\operatorname{p\,lim} T^{-1} X'\boldsymbol{\varepsilon}).$$

The critical requirement to ensure the consistency of $\mathbf{b}_0$ is therefore

$$\operatorname{p\,lim} T^{-1} X'\boldsymbol{\varepsilon} = \mathbf{0} \tag{6.1.2}$$

as far as the existence and regularity of $\operatorname{p\,lim} T^{-1} X'X$ can be ensured.

Now, the contemporaneously uncorrelated linear regression is defined by the assumption that every row vector $\mathbf{x}_t'$ of

$$X = \begin{pmatrix} \mathbf{x}_1' \\ \vdots \\ \mathbf{x}_T' \end{pmatrix}$$

is distributed independently of $\varepsilon_t, \varepsilon_{t+1}, \ldots, \varepsilon_T$ whereas $\mathbf{x}_t'$ may be dependent on $\varepsilon_1, \ldots, \varepsilon_{t-1}$, i.e. we have

$$E\mathbf{x}_t' E\varepsilon_{t+\tau} = E\mathbf{x}_t'\varepsilon_{t+\tau} = \mathbf{0} \qquad (\tau = 1, 2, \ldots). \tag{6.1.3}$$

We now check the condition (2) for the contemporaneously uncorrelated process $\{\varepsilon_t\}$. Let the jth component of the vector $T^{-1} X'\boldsymbol{\varepsilon}$ be

$$c_j = T^{-1} \sum_{t=1}^{T} x_{tj} \varepsilon_t.$$

By (3) we have

$$Ec_j = T^{-1} \sum E x_{tj} E\varepsilon_t = 0 \tag{6.1.4}$$

and, moreover, the following relations hold:

$$Ec_j^2 = T^{-2} E\left(\sum x_{tj}\varepsilon_t\right)^2 = T^{-2} \sum_t \sum_s E(x_{tj} x_{sj} \varepsilon_t \varepsilon_s)$$

$$E(x_{tj} x_{sj} \varepsilon_t \varepsilon_s) = \begin{cases} E x_{tj}^2 \varepsilon_t^2 = E x_{tj}^2 E\varepsilon_t^2 & \text{for } s = t \\ 0 & \text{for } s \neq t \end{cases}$$

Let us assume that the process X_t is stationary, i.e. we have $E x_{tj}^2 = \sigma_j^2$ for $t = 1, \ldots, T$. Thus we get

$$Ec_j^2 = T^{-2} \sigma_j^2 E\varepsilon_t^2 = T^{-2} \sigma_j^2 \operatorname{var}(\varepsilon_t)$$

and, therefore,

$$\lim_{T \to \infty} Ec_j^2 = 0. \tag{6.1.5}$$

The combination of (4) and (5) (for all $j = 1, \ldots, K$) and Theorem A.61 give the result

$$\text{l.i.m.}\, T^{-1} X'\boldsymbol{\varepsilon} = \mathbf{0} \tag{6.1.6}$$

and by Theorem A.59 we conclude that $\operatorname{plim} T^{-1} X'\boldsymbol{\varepsilon} = \mathbf{0}$. This completes the proof.

Summarizing, we can say for the contemporaneously uncorrelated linear regression that the OLSE $\mathbf{b}_0 = (X'X)^{-1} X'\mathbf{y}$ is no longer unbiased but is at least a *consistent* estimator of $\boldsymbol{\beta}$.

(iii) General Stochastic Regression

If the critical condition (2) is not fulfilled, a special estimation procedure is available which nevertheless results in a consistent estimator. This method is known as the *instrumental variable estimator*.

We assume that in addition to our observations $\mathbf{y}$ and X there are also available observations on K 'instrumental variables' $Z_1, \ldots, Z_K$ which are collected in the $T \times K$-matrix Z. Furthermore, we suppose on Z

$$\operatorname{plim} T^{-1} Z'\boldsymbol{\varepsilon} = \mathbf{0} \tag{6.1.7}$$

(i.e. Z and $\boldsymbol{\varepsilon}$ shall be contemporaneously uncorrelated) and

$$\operatorname{plim} T^{-1} Z'X = \boldsymbol{\Sigma}_{ZX} \tag{6.1.8}$$

exists and is regular.

Then the IVE (*I*nstrumental *V*ariable *E*stimator) is defined as

$$\mathbf{b}_{\text{IVE}} = (Z'X)^{-1} Z'\mathbf{y}. \tag{6.1.9}$$

Theorem 6.1

The IVE is consistent:

$$\begin{aligned} \operatorname{plim} \mathbf{b}_{\mathrm{IVE}} &= \boldsymbol{\beta} + (\operatorname{plim} T^{-1} \mathbf{Z}'\mathbf{X})^{-1} (\operatorname{plim} T^{-1} \mathbf{Z}'\boldsymbol{\varepsilon}) \\ &= \boldsymbol{\beta} + \boldsymbol{\Sigma}_{ZX}^{-1} . 0 = \boldsymbol{\beta}. \end{aligned} \tag{6.1.10}$$

If we define the asymptotic covariance matrix of an estimator $\hat{\boldsymbol{\beta}}$ of $\boldsymbol{\beta}$ by

$$\bar{V}(\hat{\boldsymbol{\beta}}) = T^{-1} \operatorname{plim}[\sqrt{T}(\hat{\boldsymbol{\beta}} - \boldsymbol{\beta}) \sqrt{T}(\hat{\boldsymbol{\beta}} - \boldsymbol{\beta})'] \tag{6.1.11}$$

we get

$$\begin{aligned} \bar{V}(\mathbf{b}_{\mathrm{IVE}}) &= T^{-1} \boldsymbol{\Sigma}_{ZX}^{-1} \operatorname{plim} (T^{-1} \mathbf{Z}'\boldsymbol{\varepsilon}\boldsymbol{\varepsilon}'\mathbf{Z}) \boldsymbol{\Sigma}_{ZX}^{-1} \\ &= T^{-1} \sigma^2 \boldsymbol{\Sigma}_{ZX}^{-1} \boldsymbol{\Sigma}_{ZZ} \boldsymbol{\Sigma}_{ZX}^{-1} \end{aligned} \tag{6.1.12}$$

where the notation $\boldsymbol{\Sigma}_{ZZ} = \operatorname{plim} T^{-1} \mathbf{Z}'\mathbf{Z}$ is used (a proof of the relation $\operatorname{plim} T^{-1} \mathbf{Z}'\boldsymbol{\varepsilon}\boldsymbol{\varepsilon}'\mathbf{Z} = \sigma^2 \boldsymbol{\Sigma}_{ZZ}$ may be found in Goldberger, 1964, p. 271).

In practice it seems to be highly difficult to find a set of instrumental variables with which to construct the IVE (9), so this method has its main relevance in proving the consistency of other estimators. If a given estimator can be shown to have the form of an IVE we have proved its consistency and, moreover, by (12) we have its asymptotic covariance matrix. By this the expense in technical calculations can be decreased.

6.1.4 General Multivariate Regression

As another point of preparing the simultaneous equation model we now regard the multivariate regression model which is composed of M interrelated single-equation regression models

$$\mathbf{y}_m = \mathbf{X}\tilde{\boldsymbol{\beta}}_m + \boldsymbol{\varepsilon}_m \qquad (m = 1, \ldots, M). \tag{6.1.13}$$

The multivariate regression model may also be written as

$$\begin{pmatrix} \mathbf{y}_1 \\ \mathbf{y}_2 \\ \vdots \\ \mathbf{y}_M \end{pmatrix} = \begin{pmatrix} \mathbf{X} & 0 & \ldots & 0 \\ 0 & \mathbf{X} & \ldots & 0 \\ \vdots & \vdots & & \vdots \\ 0 & 0 & \ldots & \mathbf{X} \end{pmatrix} \begin{pmatrix} \tilde{\boldsymbol{\beta}}_1 \\ \tilde{\boldsymbol{\beta}}_2 \\ \vdots \\ \tilde{\boldsymbol{\beta}}_M \end{pmatrix} + \begin{pmatrix} \boldsymbol{\varepsilon}_1 \\ \boldsymbol{\varepsilon}_2 \\ \vdots \\ \boldsymbol{\varepsilon}_M \end{pmatrix}. \tag{6.1.14}$$

Relation (14) may be understood as the multivariate regression in its non-identified form which is characterized by the occurrence of all K regressors $\boldsymbol{X}_1, \ldots, \boldsymbol{X}_K$ in each of the M equations. But in practice there is prior information on the equationwise occurrence of regressors which are relevant for any of the M regressands $\boldsymbol{Y}_1, \ldots, \boldsymbol{Y}_M$. This information is used in such a way that in every equation the nonrelevant regressors $\boldsymbol{X}_K$ are left out. The corresponding parameters $\tilde{\beta}_{km}$ are given the value zero. This procedure leads to reduced regressor matrices X_m (of type $T \times K_m$) and corresponding reduced parameter vectors β_m. So after using this identification information the mth equation has the form

$$y_m = X_m \beta_m + \varepsilon_m \qquad (m = 1, \ldots, M) \tag{6.1.15}$$

and the (identified) multivariate regression model becomes

$$\begin{pmatrix} y_1 \\ \vdots \\ y_M \end{pmatrix} = \begin{pmatrix} X_1 & & 0 \\ & \ddots & \\ 0 & & X_M \end{pmatrix} \begin{pmatrix} \beta_1 \\ \vdots \\ \beta_M \end{pmatrix} + \begin{pmatrix} \varepsilon_1 \\ \vdots \\ \varepsilon_M \end{pmatrix} \tag{6.1.16}$$

or in matrix notation

$$\underset{MT \times 1}{y} = \underset{MT \times \tilde{K}}{Z} \; \underset{\tilde{K} \times 1}{\beta} + \underset{MT \times 1}{\varepsilon}. \tag{6.1.17}$$

We assume

$$\text{rank } X_m = K_m \quad (m = 1, \ldots, M),$$

$$\text{rank } Z = \tilde{K} = \sum_{m=1}^{M} K_m$$

and, moreover, we restrict ourselves to the case of nonstochastic regressors. As far as the disturbances are concerned, the following assumptions are made:

$$E\varepsilon_{mt} = 0, \; E\varepsilon_{mt}\varepsilon_{m't'} = w_{mm'}(t, t') \tag{6.1.18}$$

which results in

$$E\varepsilon_m = \mathbf{0}, \quad E\varepsilon = \mathbf{0}, \tag{6.1.19}$$

$$E\varepsilon_m \varepsilon'_{m'} = \underset{T \times T}{W_{mm'}} = \begin{pmatrix} w_{mm'}(1,1) & \ldots & w_{mm'}(1,T) \\ \vdots & & \vdots \\ w_{mm'}(T,1) & \ldots & w_{mm'}(T,T) \end{pmatrix}. \tag{6.1.20}$$

The covariance matrix of the systems disturbances vector ε is denoted by ϕ:

$$E\varepsilon\varepsilon' = \underset{MT \times MT}{\phi} = \begin{pmatrix} W_{11} & \ldots & W_{1M} \\ \vdots & & \vdots \\ W_{M1} & \ldots & W_{MM} \end{pmatrix}. \tag{6.1.21}$$

The structure of ϕ fully determines the type of multivariate regression as well as the choice between single-equation or system estimators for the parameters.

Single-equation Estimator

Ignoring the relation of one of the equations (15) to the system (16) the single-equation estimators of $\boldsymbol{\beta}_m$ $(m = 1, \ldots, M)$ are

$$\hat{\boldsymbol{\beta}}_m = (X'_m W_{mm}^{-1} X_m)^{-1} X'_m W_{mm}^{-1} \mathbf{y}_m \tag{6.1.22}$$

which are unbiased and have the covariance matrix

$$V(\hat{\boldsymbol{\beta}}_m) = (X'_m W_{mm}^{-1} X_m)^{-1}. \tag{6.1.23}$$

As these estimators $\hat{\boldsymbol{\beta}}_m$ do not reflect the systems information, they cannot be expected to be *best* linear unbiased.

System Estimator

Following the compact set-up (17) we see that it is of the form of a generalized linear regression model. Thus Theorem 2.5 gives the BLUE of $\boldsymbol{\beta}$ (and of all components $\boldsymbol{\beta}_m$) as the Aitken estimator

$$\mathbf{b} = (Z'\phi^{-1}Z)^{-1} Z'\phi^{-1}\mathbf{y} \tag{6.1.24}$$

which has covariance matrix

$$V(\mathbf{b}) = (Z'\phi^{-1}Z)^{-1}. \tag{6.1.25}$$

Theorem 6.2

In the multivariate regression model, the system estimator $\mathbf{b}$ (25) has smaller variance (and, moreover, smaller MSE-risk) than the estimator $\hat{\boldsymbol{\beta}}' = (\hat{\boldsymbol{\beta}}'_1, \ldots, \hat{\boldsymbol{\beta}}'_M)$ whose components are set up from the single-equation estimators $\hat{\boldsymbol{\beta}}_m$ (22) and which does not use the system information contained in the systems covariance matrix ϕ. If $X_1 = \ldots = X_M$ (i.e. the model is unidentified) or if

$$\phi = \begin{pmatrix} W_{11} & & 0 \\ & \ddots & \\ 0 & & W_{MM} \end{pmatrix}$$

(i.e. the M equations are pairwise uncorrelated), the single-equation estimator $\hat{\boldsymbol{\beta}}$ and the system estimator coincide.

For a proof of this theorem the reader is referred to Bibby and Toutenburg, 1978, pp. 138–141 and to Toutenburg, 1970b.

6.1.5 Multivariate Contemporaneously Independent Regression

In the general multivariate model (17) the regressors were assumed to be nonstochastic, i.e. Z was fixed in repeated sampling. Now we leave this restriction and work with stochastic regressors which shall be weakly correlated with the disturbances such that the consistency of estimators is ensured (see condition (2)). We define the multivariate (classical) contemporaneously uncorrelated regression model by the following equations:

$$\mathbf{y}_m = \boldsymbol{X}\boldsymbol{\beta}_m + \boldsymbol{\varepsilon}_m \quad (m = 1, \ldots, M) \tag{6.1.26}$$

$$E\boldsymbol{\varepsilon}_m = \mathbf{0}, E\boldsymbol{\varepsilon}_m\boldsymbol{\varepsilon}'_{m'} = w_{mm'}\boldsymbol{I} \quad (m, m' = 1, \ldots, M) \tag{6.1.27}$$

$$\operatorname{p\,lim} T^{-1}\boldsymbol{\varepsilon}'_m\boldsymbol{\varepsilon}_m = w_{mm'} \tag{6.1.28}$$

$$E\mathbf{x}(t)\mathbf{x}'(t) = \boldsymbol{\Sigma}_{\mathbf{xx}} \text{ (positive definite)} \tag{6.1.29}$$

$$\operatorname{p\,lim} T^{-1}\boldsymbol{X}'\boldsymbol{\varepsilon}_m = \mathbf{0} \quad (m = 1, \ldots, M) \tag{6.1.30}$$

and

$$\lim_{T\to\infty} ET^{-1}\boldsymbol{X}'\boldsymbol{X} = \boldsymbol{\Sigma}_{\mathbf{xx}}. \tag{6.1.31}$$

If we define $\underset{M \times M}{\boldsymbol{W}} = (w_{mm'})$ and $\boldsymbol{Z} = \operatorname{diag}(\boldsymbol{X}, \ldots, \boldsymbol{X})$ the model has the form (see (17))

$$\left.\begin{aligned}
&\mathbf{y} = \boldsymbol{Z}\boldsymbol{\beta} + \boldsymbol{\varepsilon}, E\boldsymbol{\varepsilon} = \mathbf{0} \\
&\qquad E\boldsymbol{\varepsilon}\boldsymbol{\varepsilon}' = \boldsymbol{\phi} \\
&\qquad\qquad = \boldsymbol{W} \otimes \boldsymbol{I} \quad (\boldsymbol{W} \text{ positive definite}) \\
&\operatorname{p\,lim} T^{-1}\begin{pmatrix} \operatorname{tr}\boldsymbol{\varepsilon}_1\boldsymbol{\varepsilon}'_1 & \ldots & \operatorname{tr}\boldsymbol{\varepsilon}_1\boldsymbol{\varepsilon}'_M \\ \vdots & & \vdots \\ \operatorname{tr}\boldsymbol{\varepsilon}_M\boldsymbol{\varepsilon}'_1 & \ldots & \operatorname{tr}\boldsymbol{\varepsilon}_M\boldsymbol{\varepsilon}'_M \end{pmatrix} = \boldsymbol{W} \\
&\operatorname{p\,lim} T^{-1}\boldsymbol{Z}'\boldsymbol{\varepsilon} = \mathbf{0}, \quad \operatorname{p\,lim} T^{-1}\boldsymbol{Z}'\boldsymbol{Z} = \boldsymbol{\Sigma}_{\mathbf{xx}} \otimes \boldsymbol{I}
\end{aligned}\right\} \tag{6.1.32}$$

(here $\otimes$ denotes the Kronecker product, see Theorem A.31).

The assumption $\operatorname{p\,lim} T^{-1}\boldsymbol{Z}'\boldsymbol{\varepsilon} = \mathbf{0}$ is the formulation of the contemporaneously uncorrelatedness of the stochastic processes $\{\mathbf{x}(t)\}$ and $\{\boldsymbol{\varepsilon}(t)\}$. This ensures the possibility of deriving consistent estimators of the parameters. As $\boldsymbol{Z}$ fulfils the condition $\boldsymbol{X}_1 = \ldots = \boldsymbol{X}_M = \boldsymbol{X}$ we can use the result of Theorem 6.2 and make the statement that $\mathbf{b} = \hat{\boldsymbol{\beta}}$, i.e. the parameter vectors $\boldsymbol{\beta}_m$ of each equation of the system (32) may be estimated partially from the mth equation. This gives the OLS-estimators

$$\hat{\boldsymbol{\beta}}_m = (\boldsymbol{X}'\boldsymbol{X})^{-1}\boldsymbol{X}'\mathbf{y}_m \tag{6.1.33}$$

which have

$$\operatorname{p\,lim} \hat{\boldsymbol{\beta}}_m = \boldsymbol{\beta}_m. \tag{6.1.34}$$

In other words, $\hat{\boldsymbol{\beta}}_m$ $(m = 1, \ldots, M)$ is consistent (and asymptotically unbiased) and has the asymptotic correlation matrix (see (11))

$$\bar{V}(\hat{\boldsymbol{\beta}}_m, \hat{\boldsymbol{\beta}}_{m'}) = T^{-1} w_{mm'} \boldsymbol{\Sigma}_{\mathbf{xx}}^{-1} \tag{6.1.35}$$

(if $m = m'$ this is the asymptotic covariance matrix of $\hat{\boldsymbol{\beta}}_m$).

If we estimate $w_{mm'}$ consistently by

$$\hat{w}_{mm'} = (T-K)^{-1} (\mathbf{y}_m - \boldsymbol{X}\hat{\boldsymbol{\beta}}_m)' (\mathbf{y}_{m'} - \boldsymbol{X}\hat{\boldsymbol{\beta}}_{m'}) \tag{6.1.36}$$

we get the consistent estimator of $\bar{V}(\hat{\boldsymbol{\beta}}_m, \hat{\boldsymbol{\beta}}_{m'})$

$$\boldsymbol{S}(\hat{\boldsymbol{\beta}}_m, \hat{\boldsymbol{\beta}}_{m'}) = \hat{w}_{mm'} (\boldsymbol{X}'\boldsymbol{X})^{-1}. \tag{6.1.37}$$

Theorem 6.3

In the multivariate contemporaneously uncorrelated regression model (32) the classical OLS-estimates

$$\hat{\boldsymbol{\beta}} = \begin{pmatrix} \hat{\boldsymbol{\beta}}_1 \\ \vdots \\ \hat{\boldsymbol{\beta}}_M \end{pmatrix} = (\boldsymbol{Z}'\boldsymbol{Z})^{-1} \boldsymbol{Z}'\mathbf{y}, \tag{6.1.38}$$

$$\hat{\boldsymbol{W}} = (T-K)^{-1} (\mathbf{y} - \boldsymbol{Z}\hat{\boldsymbol{\beta}})' (\mathbf{y} - \boldsymbol{Z}\hat{\boldsymbol{\beta}})$$
$$\boldsymbol{S}(\hat{\boldsymbol{\beta}}, \hat{\boldsymbol{\beta}}) = \hat{\boldsymbol{W}} \otimes (\boldsymbol{X}'\boldsymbol{X})^{-1}$$

are consistent and asymptotically unbiased.

The consistency of an estimator is insufficient to justify its use, since many estimators may be consistent. If the sample size grows to infinity, an additional criterion is necessary in order to choose the 'best' estimator where 'best' may be related to the following definition.

Definition 6.2 (Asymptotic efficiency) An estimator $\hat{\boldsymbol{\beta}}$ of $\boldsymbol{\beta}$ is called asymptotically efficient if it is consistent and if its asymptotic covariance matrix satisfies the condition

$$\bar{V}(\tilde{\boldsymbol{\beta}}) - \bar{V}(\hat{\boldsymbol{\beta}}) \geq \boldsymbol{0} \tag{6.1.39}$$

where $\tilde{\boldsymbol{\beta}}$ is any arbitrary consistent estimator.

If there are two consistent estimators given, we may compare them according to Definition 6.3.

Definition 6.3 A consistent estimator $\hat{\boldsymbol{\beta}}_1$ is called asymptotically better than another consistent estimator $\hat{\boldsymbol{\beta}}_2$ if

$$\bar{V}(\hat{\boldsymbol{\beta}}_2) - \bar{V}(\hat{\boldsymbol{\beta}}_1) \geq \boldsymbol{0}. \tag{6.1.40}$$

(see the definition (11) of the asymptotic covariance matrix).

6.2 THE SIMULTANEOUS-EQUATION MODEL

6.2.1 Introduction—the Keynesian Model

As mentioned above, the typical situation in econometrics is that systems of interdependent relations have to be analysed and estimated, most of the difficulties being based on the correlation between exogenous variables and disturbances. The simplest model of any economy shall be used as an example for the nature of the problems. So we assume that we have the simplified Keynesian model in its statistical version

$$C_t = \alpha + \beta Y_t + \varepsilon_{1t}, \tag{6.2.1}$$

$$Y_t = C_t + I_t \quad (t = 1, \ldots, T) \tag{6.2.2}$$

where C_t, Y_t, I_t are, respectively, consumption, income, and investment at time t. The first equation describes the consumption behaviour, while the second states the national income identity. The variables C_t and Y_t are endogenous (common dependent variables) while I_t is taken as exogenous (predetermined variable). This leads to the specification

$$EI_t\varepsilon_{1t'} = 0 \quad (t, t' = 1, \ldots, T). \tag{6.2.3}$$

Arranging the system (1) and (2) to have the familiar explicit dependence of the endogenous variables on the exogenous variable yields the two-equation system at index t:

$$C_t = \frac{\alpha}{1-\beta} + \frac{\beta}{1-\beta} I_t + \frac{1}{1-\beta}\varepsilon_{1t}. \tag{6.2.4}$$

$$Y_t = \frac{\alpha}{1-\beta} + \frac{1}{1-\beta} I_t + \frac{1}{1-\beta}\varepsilon_{1t}. \tag{6.2.5}$$

From (5) we deduce

$$EY_t\varepsilon_{1t} = \frac{\alpha}{1-\beta}E\varepsilon_{1t} + \frac{\beta}{1-\beta}EI_t\varepsilon_{1t} + \frac{1}{1-\beta}E\varepsilon_{1t}^2$$

$$= \frac{\sigma^2}{1-\beta} \qquad \text{(see (1) and (3))}$$

and, moreover, we have under further weak assumptions the relation

$$\operatorname{plim} T^{-1}\sum_t \varepsilon_{1t}(Y_t - \bar{Y}) = E\varepsilon_{1t}Y_t = \frac{\sigma^2}{1-\beta}. \tag{6.2.6}$$

If we calculate the OLSE of β from the first equation (1) we get

$$\hat{\beta} = \frac{\sum_t C_t(Y_t - \bar{Y})}{\sum_t (Y_t - \bar{Y})^2}; \quad \bar{Y} = T^{-1}\sum_t Y_t$$

or, more explicitly,

$$\hat{\beta} = \frac{\alpha \sum_t (Y_t - \bar{Y}) + \beta \sum_t Y_t (Y_t - \bar{Y}) + \sum_t \varepsilon_{1t}(Y_t - \bar{Y})}{\sum_t (Y_t - \bar{Y})^2}$$

$$= \beta + \frac{T^{-1} \sum_t \varepsilon_{1t} (Y_t - \bar{Y})}{T^{-1} \sum_t (Y_t - \bar{Y})^2}.$$

If we use (6) and the relation

$$\operatorname{plim} T^{-1} \sum_{t=1}^{T} (Y_t - \bar{Y})^2 = E(Y_t - EY_t)^2 = \sigma_y^2$$

(which holds under weak assumptions) we get

$$\operatorname{plim} \hat{\beta} = \beta + \frac{\sigma^2}{(1-\beta)\sigma_y^2} > \beta \tag{6.2.7}$$

as $0 < \beta < 1$ has to be assumed as a natural restriction. Therefore the OLSE $\hat{\beta}$ of β is not consistent.

The main reason for this is the fact that Y_t is in (1) in the position of an explanatory variable, whereas in (2) it is a dependent variable. As may be seen from (4) and (5), C_t and Y_t are subject to the same disturbances. So, regressing C_t on Y_t does not give an estimator of β with desirable properties. Consequently, other procedures have to be developed which give consistent estimators.

6.2.2 Specification of the Model

Let the economic system be characterized at time t ($t = 1, \ldots, T$) by the M-equation model

$$y_{tm} = \sum_{j=1}^{M} y_{tj} \beta_{jm} + \sum_{i=1}^{K} x_{ti} \gamma_{im} + \varepsilon_{tm} \quad (m = 1, \ldots, M) \tag{6.2.8}$$

which describes the interrelation of M dependent variables $\boldsymbol{Y}_1, \ldots, \boldsymbol{Y}_M$, K predetermined variables $\boldsymbol{X}_1, \ldots, \boldsymbol{X}_K$, and M disturbances $\boldsymbol{\varepsilon}_1, \ldots, \boldsymbol{\varepsilon}_M$ at index t.

As pointed out in the context of multivariate regression, we also may assume in the simultaneous equation model (8) some of the coefficients β_{jm}, γ_{im} known to be zero. In particular, all β_{ii} are known to be zero.

As we are intending to estimate the parameters of any of the M equations of the system (8), without loss of generality let us concentrate on the first equation, that is

$$\begin{aligned} \mathbf{y}_1 &= \boldsymbol{Y\beta} + \boldsymbol{X\gamma} + \boldsymbol{\varepsilon}_1 \\ &= \boldsymbol{Y}_1 \boldsymbol{\beta}_1 + \boldsymbol{Y}_2 \boldsymbol{\beta}_2 + \boldsymbol{X}_1 \boldsymbol{\gamma}_1 + \boldsymbol{X}_2 \boldsymbol{\gamma}_2 + \boldsymbol{\varepsilon}_1. \end{aligned} \tag{6.2.9}$$

Here the T observations of all variables of this equation are contained. The variables and parameters in (9) are of the following dimensions:

$$\begin{aligned}
&\mathbf{y}_1 : T \times 1, \boldsymbol{Y}_1 : T \times m_1, \boldsymbol{Y}_2 : T \times m_2 \\
&\boldsymbol{\beta}_1 = m_1 \times 1, \boldsymbol{\beta}_2 : m_2 \times 1, \\
&\boldsymbol{X}_1 : T \times k_1, \boldsymbol{X}_2 : T \times k_2, \\
&\boldsymbol{\gamma}_1 : k_1 \times 1, \boldsymbol{\gamma}_2 : k_2 \times 1, \boldsymbol{\varepsilon}_1 : T \times 1 \\
&(m_1 + m_2 = M - 1, k_1 + k_2 = K).
\end{aligned}$$

Now, we assume that we know *a priori* that the variables contained in $\boldsymbol{Y}_2$ and $\boldsymbol{X}_2$, respectively, do not occur in the first equation. This information is used if the corresponding parameter vectors $\boldsymbol{\beta}_2$ and $\boldsymbol{\gamma}_2$ are equated to zero. The resulting equation is

$$\mathbf{y}_1 = \boldsymbol{Y}_1 \boldsymbol{\beta}_1 + \boldsymbol{X}_1 \boldsymbol{\gamma}_1 + \boldsymbol{\varepsilon}_1. \tag{6.2.10}$$

As far as the stochastic properties of the variables in (8) are concerned we make the following assumption (see, for example, Goldberger, 1964, p. 299; Dhrymes, 1974, p. 172):

$$\left.\begin{aligned}
&E\varepsilon_{tm} = 0, E(\varepsilon_{tm}\varepsilon_{t'm'}) = \delta_{tt'}\sigma_{mm'} \\
&\operatorname{plim} T^{-1}\boldsymbol{\varepsilon}'\boldsymbol{\varepsilon} = \boldsymbol{\Sigma} \\
&\operatorname{plim} T^{-1}\boldsymbol{X}'\boldsymbol{X} = \boldsymbol{\Sigma}_{\mathbf{xx}} \text{ (positive definite)} \\
&E(x_{tm}\varepsilon_{t'm'}) = 0 \text{ all } m, m', t, \text{ and } t' \\
&\operatorname{plim} T^{-1}\boldsymbol{X}'\boldsymbol{\varepsilon} = \mathbf{0}.
\end{aligned}\right\} \tag{6.2.11}$$

Due to the second assumption, the error variables ε_{tm} define a contemporaneously uncorrelated but equationwise correlated stochastic process.

Furthermore, we assume that the system (8) uniquely determines the y_{tm} in terms of the x_{tj} and the disturbances ε_{tm}. As a necessary conclusion of this assumption we may deduce that $\boldsymbol{I} - \boldsymbol{B}$ must be nonsingular. Then the variables y_m are referred to as the jointly dependent or endogenous variables. Note that, in general,

$$E(y_{tm}\varepsilon_{tm'}) \neq 0 \quad (m, m' = 1, \ldots, m). \tag{6.2.12}$$

This is the well-known reason for the fact that the OLSE yields no unbiased estimators of the parameters β_{jm}, γ_{im} (for a detailed discussion see Dhrymes, 1974, p. 174, as well as the arguments in Section 6.1.3).

Furthermore, the $\mathbf{x}_i$ are said to be the predetermined variables consisting of (nonstochastic) exogenous and (stochastic, but at time t realized) lagged endogenous variables. Their correlation with the disturbances shall be such that $E(X_{tm}\varepsilon_{t'm'}) = 0$ (see (11)). Additionally, we assume that

$$\operatorname{rank} \boldsymbol{X} = K. \tag{6.2.13}$$

Collecting the T observations of all the M model equations given in (8) yields the matrix form of the simultaneous equation model

$$Y = YB + XC + \varepsilon \tag{6.2.14}$$

where

$\underset{T,M}{Y} = (\mathbf{y}_1, Y_1, Y_2)$ is the matrix of all endogenous variables,

$\underset{T,K}{X} = (X_1, X_2)$ contains all predetermined variables of the system,

$\underset{M,M}{B} = (\beta_{jm})$, $\underset{K,M}{C} = (\gamma_{im})$, $\underset{T,M}{\varepsilon} = (\varepsilon_1, \ldots, \varepsilon_M)$.

According to (11) we have

$$E\varepsilon = \mathbf{0}, \; E\varepsilon_m \varepsilon_{m'} = \sigma_{mm'} I \tag{6.2.15}$$

$$E \begin{pmatrix} \varepsilon_1 \\ \vdots \\ \varepsilon_M \end{pmatrix} (\varepsilon'_1, \ldots, \varepsilon'_M) = \Sigma \otimes I.$$

Definition 6.4 Let the operation of an economic system be described by the equations in (8) (or summarized in (14)) and the assumptions (11) and (13). Such a system is said to be a simultaneous-equation model or a system of structural equations.

6.2.3 The Reduced Form

As pointed out above, we shall confine ourselves to systems which uniquely determine the jointly dependent variables in terms of the predetermined variables; that is, we assume

$$I - B \text{ is nonsingular.} \tag{6.2.16}$$

Based on this we may solve the system (14) for Y which gives the reduced form of the simultaneous-equation model

$$Y = XC(I - B)^{-1} + \varepsilon(I - B)^{-1}. \tag{6.2.17}$$

If we use the notation

$$(I - B)^{-1} = D, \quad CD = \Pi, \quad \varepsilon D = V$$

we get the reduced form more compactly as

$$\underset{T,M}{Y} = \underset{T,K}{X} \; \underset{K,M}{\Pi} + \underset{T,M}{V}. \tag{6.2.18}$$

The relevant assumptions (11) are transformed as follows

$$\operatorname{plim} T^{-1} \boldsymbol{V}'\boldsymbol{V} = \operatorname{plim} T^{-1} \boldsymbol{D}'\boldsymbol{\varepsilon}'\boldsymbol{\varepsilon}\boldsymbol{D} \tag{6.2.19}$$

$$= \boldsymbol{D}' \boldsymbol{\Sigma} \boldsymbol{D}$$

$$\operatorname{plim} T^{-1} \boldsymbol{X}'\boldsymbol{V} = (\operatorname{plim} T^{-1} \boldsymbol{X}'\boldsymbol{\varepsilon})\boldsymbol{D} = \boldsymbol{0}. \tag{6.2.20}$$

Now, the reduced form (18) looks like a multivariate regression model. For the mth dependent variable $\boldsymbol{Y}_m$ of $\boldsymbol{Y} = (Y_1, \ldots, Y_M)$ we have as an equivalent formulation of (18) the presentation as M univariate regression equations

$$\boldsymbol{Y}_m = \boldsymbol{X}\boldsymbol{\pi}_m + \boldsymbol{V}_m \qquad (m = 1, \ldots, M) \tag{6.2.21}$$

where $\boldsymbol{\pi}_m$ and $\boldsymbol{V}_m$ are the mth columns of $\boldsymbol{\pi}$ and $\boldsymbol{V}$, respectively. Now, let us estimate the $K \times 1$-dimensional parameter vectors $\boldsymbol{\pi}_m (m = 1, \ldots, M)$ of (21). The OLSE of $\boldsymbol{\pi}_m$ is defined as

$$\hat{\boldsymbol{\pi}}_m = (\boldsymbol{X}'\boldsymbol{X})^{-1} \boldsymbol{X}'\boldsymbol{Y}_m \tag{6.2.22}$$

$$= \boldsymbol{\pi}_m + (\boldsymbol{X}'\boldsymbol{X})^{-1} \boldsymbol{X}'\boldsymbol{V}_m.$$

Using (11) and (20) we prove immediately the consistency of the OLSE

$$\operatorname{plim} \hat{\boldsymbol{\pi}}_m = \boldsymbol{\pi}_m + (\operatorname{plim} T^{-1} \boldsymbol{X}'\boldsymbol{X})^{-1} \operatorname{plim} T^{-1} \boldsymbol{X}'\boldsymbol{V}_m \tag{6.2.23}$$

$$= \boldsymbol{\pi}_m + \boldsymbol{\Sigma}_{\mathbf{xx}}^{-1} \cdot 0 = \boldsymbol{\pi}_m.$$

Composing the M estimators $\hat{\boldsymbol{\pi}}_m$ into $\hat{\boldsymbol{\Pi}} = (\hat{\boldsymbol{\pi}}_1, \ldots, \hat{\boldsymbol{\pi}}_M)$ gives Theorem 6.4.

Theorem 6.4

In the reduced form model $\boldsymbol{Y} = \boldsymbol{X}\boldsymbol{\Pi} + \boldsymbol{V}$ the OLSE $\hat{\boldsymbol{\Pi}}$ of $\boldsymbol{\Pi}$ is consistent.

Now, is the econometrician grateful to this result? In general, not, for he is interested in the estimation of the structural parameters $\boldsymbol{B}$ and $\boldsymbol{C}$. To ensure that the estimator $\hat{\boldsymbol{\Pi}}$ allows the calculation of desirable estimators $\boldsymbol{B}$ and $\boldsymbol{C}$, the relation

$$\hat{\boldsymbol{\Pi}} = \hat{\boldsymbol{C}}\hat{\boldsymbol{D}} \tag{6.2.24}$$

must have a unique solution $(\hat{\boldsymbol{C}}, \hat{\boldsymbol{D}})$. Clearly, if $\hat{\boldsymbol{D}}$ is uniquely determined, $\hat{\boldsymbol{B}} = \boldsymbol{I} - \hat{\boldsymbol{D}}^{-1}$ may be calculated uniquely, too.

Theorem 6.5

Any nonsingular linear transformation of a simultaneous-equation model results in the same reduced form model.

Proof Let two simultaneous-equation models in the same variables $\boldsymbol{Y}$, $\boldsymbol{X}$, $\boldsymbol{\varepsilon}$, namely

$$\boldsymbol{Y} = \boldsymbol{Y}\boldsymbol{B} + \boldsymbol{X}\boldsymbol{C} + \boldsymbol{\varepsilon}$$

and

$$\tilde{Y} = Y\tilde{B} + X\tilde{C} + \tilde{\varepsilon},$$

be given, where $\tilde{B} = BA, \tilde{C} = CA,\ \tilde{\varepsilon} = \varepsilon A,\ \tilde{Y} = YA$ and A is an $M \times M$-nonsingular matrix. Then the reduced form of the second model is

$$Y = X\tilde{\Pi} + \tilde{V}$$

where

$$\begin{aligned}\tilde{\Pi} = \tilde{C}\tilde{D} &= \tilde{C}(A - \tilde{B})^{-1} \\ &= CAA^{-1}(I - B)^{-1} = \Pi.\end{aligned}$$

6.2.4 Identification

As a conclusion of Theorem 6.5 we have to ensure the unique solution of (24) for C and D. This problem is known as the question of identifiability of the structural parameters of C and D. For the purpose of this book it is sufficient to give the main rules for identification (for a more detailed discussion of the identification problem the reader is referred to standard literature, e.g. Fisher (1976), Duncan (1975), Leser (1974), and Koopmans (1953)).

To answer the question, let us concentrate on one structural equation, say the first (see (9)):

$$y_1 = Y_1\beta_1 + Y_2\beta_2 + X_1\gamma_1 + X_2\gamma_2 + \varepsilon_1$$

where β_2 and γ_2 are assumed known to be zero. Now partition the reduced form corresponding to the first structural equation by

$$\left(\underset{T\times 1}{y_1}\ \underset{T\times m_1}{Y_1}\ \underset{T\times m_2}{Y_2}\right) = \left(\underset{T\times k_1}{X_1}\ \underset{T\times k_2}{X_2}\right)\begin{pmatrix}\underset{k_1\times 1}{\pi_1} & \underset{k_1\times m_1}{\Pi_{11}} & \underset{k_1\times m_2}{\Pi_{12}} \\ \underset{k_2\times 1}{\pi_2} & \underset{k_2\times m_1}{\Pi_{21}} & \underset{k_2\times m_2}{\Pi_{22}}\end{pmatrix} + V. \tag{6.2.25}$$

The subset of relations in $\hat{\Pi} = \hat{C}(I - \hat{B})^{-1}$ that is relevant to our problem is

$$\hat{\Pi}\begin{pmatrix}1 \\ -\hat{\beta}_1 \\ 0\end{pmatrix} = \begin{pmatrix}\hat{\gamma}_1 \\ 0\end{pmatrix} \tag{6.2.26}$$

which may be written by the help of (25) as

$$\hat{\Pi}_{11}\hat{\beta}_1 + \hat{\gamma}_1 = \hat{\pi}_1 \tag{6.2.27}$$

and

$$\hat{\Pi}_{21}\hat{\beta}_1 = \hat{\pi}_2. \tag{6.2.28}$$

The first structural equation is identifiable if and only if there is a unique solution $\hat{\beta}_1$ in (28). This holds if and only if

$$\text{rank}\ \hat{\Pi}_{21} = m_1. \tag{6.2.29}$$

The proof follows from well-known results of linear algebra. The uniquely determined $\hat{\boldsymbol{\beta}}_1$ gives by (27) a unique $\hat{\gamma}_1$.

Definition 6.5 Let the rank condition (29) be fulfilled. Then the estimation procedure, which at first calculates the (consistent) OLSE $\hat{\boldsymbol{\Pi}}$ in the reduced form and, as the second step, calculates $\hat{\boldsymbol{\beta}}_1$ and $\hat{\gamma}_1$ from (28) and (27) is called *Indirect Least Squares Estimator* (ILSE) of the structural parameters.

From Slutsky's Theorem (Theorem A.57) it may be deduced that the ILSE is consistent.

Now, $\hat{\boldsymbol{\Pi}}_{21}$ is a $k_2 \times m_1$-matrix. The rank condition (29) implies $k_2 \geq m_1$.

Definition 6.6

(i) If $k_2 = m_1$ and rank $\hat{\boldsymbol{\Pi}}_{21} = m_1$, the first structural equation is just-identified;
(ii) if $k_2 > m_1$ and rank $\hat{\boldsymbol{\Pi}}_{21} = m_1$, the first structural equation is overidentified;
(iii) if $k_2 < m_1$ we name it the underidentified or nonidentified case.

Note Conditions of identifiability should refer to the true (but unknown) parameters of the structural model and not to their estimators.

Theorem 6.6

The first equation of a simultaneous-equation model is identifiable if and only if

$$\text{rank } \boldsymbol{\Pi}_{21} = m_1.$$

If, in addition, $k_2 = m_1$ holds, then the ILS estimator is unique.

6.3 SIMULTANEOUS-EQUATION TECHNIQUES USING PRIOR INFORMATION

6.3.1 The Two-Stage Least Squares (2 SLSE)

In estimating the parameters $\boldsymbol{\beta}_1$ and γ_1 in the first structural equation (6.2.10), which looks like a regression model, the first idea would be to use the familiar OLSE procedure. But, as mentioned above, the correlation between the right-hand dependent variables $\boldsymbol{Y}_1$ and the error term ε_1 (i.e. $\text{p lim } T^{-1}\boldsymbol{Y}'_1\varepsilon_1 \neq \boldsymbol{0}$ in general) is the reason why the OLSE is biased and inconsistent.

Let us assume the case of an overidentified first structural equation, that is $k_2 \geq m_1$. To overcome the difficulty described before, we would eliminate the right-hand stochastic variables $\boldsymbol{Y}_1$ with the help of the reduced form (6.2.18) which may be partitioned corresponding to the first-equation's notation by

$$(\boldsymbol{Y}_1, \boldsymbol{Y}_2) = \boldsymbol{X}(\boldsymbol{\Pi}_1, \boldsymbol{\Pi}_2) + (\boldsymbol{V}_1, \boldsymbol{V}_2). \tag{6.3.1}$$

The part of the reduced form model corresponding to the remaining dependent variables $\boldsymbol{Y}_1$ in (6.2.10) is

$$\boldsymbol{Y}_1 = \boldsymbol{X}\boldsymbol{\Pi}_1 + \boldsymbol{V}_1 \tag{6.3.2}$$

(here $\boldsymbol{\Pi}_1$ is a $K \times m_1$-matrix, $\boldsymbol{V}_1$ is of dimension $T \times m_1$). We note that $\boldsymbol{Y}_1$ is the sum of a nonstochastic component $\boldsymbol{X\Pi}_1$ and a random component $\boldsymbol{V}_1$. By using only the systematic part $\boldsymbol{X\Pi}_1$ of $\boldsymbol{Y}_1$ we could guarantee at least consistency. But $\boldsymbol{\Pi}_1$ is unknown and has to be estimated by its OLSE $\hat{\boldsymbol{\Pi}}_1$ which is consistent. If we replace $\boldsymbol{Y}_1$ in (6.2.10) by $\boldsymbol{X}\hat{\boldsymbol{\Pi}}_1$ we could hope that the resulting estimator of the structural parameters $(\boldsymbol{\beta}_1, \boldsymbol{\gamma}_1)$ would also be consistent.

Let us now formalize these heuristic arguments. We have to estimate the parameters $(\boldsymbol{\beta}_1, \boldsymbol{\gamma}_1)$ of the structural equation (6.2.10). The corresponding reduced form is given in relation (2). Thus the OLSE of $\boldsymbol{\Pi}_1$ is

$$\hat{\boldsymbol{\Pi}}_1 = (\boldsymbol{X}'\boldsymbol{X})^{-1}\boldsymbol{X}'\boldsymbol{Y}_1. \tag{6.3.3}$$

According to assumption (6.2.13) the inverse of $(\boldsymbol{X}'\boldsymbol{X})$ exists. If we define the matrix of OLS residuals as

$$\hat{\boldsymbol{V}}_1 = \boldsymbol{Y}_1 - \boldsymbol{X}\hat{\boldsymbol{\Pi}}_1 \tag{6.3.4}$$

we have

$$\boldsymbol{X}'\hat{\boldsymbol{V}}_1 = \boldsymbol{X}'\boldsymbol{Y}_1 - \boldsymbol{X}'\boldsymbol{X}(\boldsymbol{X}'\boldsymbol{X})^{-1}\boldsymbol{X}'\boldsymbol{Y}_1 = \boldsymbol{0}. \tag{6.3.5}$$

$$\hat{\boldsymbol{V}}_1'\hat{\boldsymbol{V}}_1 = (\boldsymbol{Y}_1' - \hat{\boldsymbol{\Pi}}_1\boldsymbol{X}')\hat{\boldsymbol{V}}_1 = \hat{\boldsymbol{Y}}_1'\hat{\boldsymbol{V}}_1 \tag{6.3.6}$$

and therefore

$$(\boldsymbol{Y}_1 - \hat{\boldsymbol{V}}_1)'(\boldsymbol{Y}_1 - \hat{\boldsymbol{V}}_1) = \boldsymbol{Y}_1'\boldsymbol{Y}_1 - \hat{\boldsymbol{V}}_1'\hat{\boldsymbol{V}}_1. \tag{6.3.7}$$

By a trivial re-formulation we get the equation (6.2.10) to be estimated as

$$\begin{aligned} \boldsymbol{y}_1 &= (\boldsymbol{Y}_1 - \hat{\boldsymbol{V}}_1)\boldsymbol{\beta}_1 + \boldsymbol{X}_1\boldsymbol{\gamma}_1 + (\boldsymbol{\varepsilon}_1 + \hat{\boldsymbol{V}}_1\boldsymbol{\beta}_1) \\ &= \boldsymbol{Z}_1\boldsymbol{\delta}_1 + \tilde{\boldsymbol{\varepsilon}}_1, \text{ say} \end{aligned} \tag{6.3.8}$$

where

$$\boldsymbol{Z}_1 = (\boldsymbol{Y}_1 - \hat{\boldsymbol{V}}_1, \boldsymbol{X}_1),$$
$$\boldsymbol{\delta}_1' = (\boldsymbol{\beta}_1', \boldsymbol{\gamma}_1')$$

and

$$\tilde{\boldsymbol{\varepsilon}}_1 = \boldsymbol{\varepsilon}_1 + \hat{\boldsymbol{V}}_1\boldsymbol{\beta}_1.$$

The covariance matrix of $\tilde{\boldsymbol{\varepsilon}}_1$ is not of the type $\sigma^2\boldsymbol{I}$. It is a function of the unknown $\boldsymbol{\beta}_1$, the matrix $\boldsymbol{B}$, and the matrix $\boldsymbol{\Sigma}$. Moreover, in general it is not of the type $\boldsymbol{\Sigma} \otimes \boldsymbol{W}$. Therefore, the theory of best linear unbiased (BLU) or asymptotic BLU estimation is not applicable.

So we are restricted to methods which give consistent estimators based on a more heuristic background. One of these methods is the 2 SLS estimation procedure.

The idea is as follows. Regardless of the exact covariance matrix of the disturbance term $\tilde{\boldsymbol{\varepsilon}}_1$ in (8) the parameter $\boldsymbol{\delta}_1$ is estimated by the OLSE applied to equation (8). The result is said to be the 2 SLSE of $\boldsymbol{\delta}_1$. Its first stage consists in estimating the 'systematic part' of $\boldsymbol{Y}_1$ by $\boldsymbol{X}\hat{\boldsymbol{\Pi}}_1$ and the second stage consists in estimating the 'regression' parameters $\boldsymbol{\beta}_1$ and γ_1 by OLSE applied to (8). Therefore the 2 SLSE of $\boldsymbol{\delta}_1$ is

$$\hat{\boldsymbol{\delta}}_1 = (\boldsymbol{Z}_1'\boldsymbol{Z}_1)^{-1}\boldsymbol{Z}_1'\mathbf{y}_1 \tag{6.3.9}$$

where the existence of the inverse is assured.

To write $\hat{\boldsymbol{\delta}}_1$ in a computationally efficient form we use equations (5)–(9) and get

$$\hat{\boldsymbol{\delta}}_1 = \begin{pmatrix} Y_1'Y_1 - \hat{V}_1'\hat{V}_1 & Y_1'X_1 \\ X_1'Y_1 & X_1'X_1 \end{pmatrix}^{-1} \begin{pmatrix} (Y_1 - \hat{V}_1)'\mathbf{y}_1 \\ X_1'\mathbf{y}_1 \end{pmatrix}. \tag{6.3.10}$$

If we use the partition $\boldsymbol{X} = (\boldsymbol{X}_1, \boldsymbol{X}_2)$ we have

$$\boldsymbol{X}_1 = \boldsymbol{X}\begin{pmatrix} \boldsymbol{I} \\ \boldsymbol{0} \end{pmatrix}$$

and therefore

$$X_1'Y_1 = X_1'X(X'X)^{-1}X'Y_1,$$
$$X_1'X_1 = X_1'X(X'X)^{-1}X'X_1.$$

Thus the 2 SLSE (10) may be equivalently written as

$$\hat{\boldsymbol{\delta}}_1 = \begin{pmatrix} Y_1'X(X'X)^{-1}X'Y_1 & Y_1'X(X'X)^{-1}X'X_1 \\ X_1'X(X'X)^{-1}X'Y_1 & X_1'X(X'X)^{-1}X'X_1 \end{pmatrix}^{-1} \begin{pmatrix} Y_1'X(X'X)^{-1}X'\mathbf{y}_1 \\ X_1'X(X'X)^{-1}X'\mathbf{y}_1 \end{pmatrix}. \tag{6.3.11}$$

6.3.2 Interpretation of 2 SLSE as Aitken Estimator

In the preceding section, the derivation of the 2 SLSE was based on well-known historical ideas which are arranged in a heuristic manner. In other words, this derivation leads to a practicable estimator for overidentified equations but, on the other hand, does not allow an interpretation in the sense of mathematical optimality (whatever we understand by this notion). Now we shall follow an idea of Dhrymes (1974, p. 183), which was to formalize the 2 SLSE as a GLS estimator. Besides this interesting interpretation of 2 SLSE the proposed method enables the transfer of usual regression techniques, including restricted least squares, to the simultaneous-equation model.

Thus let us regard again the first structural equation

$$\mathbf{y}_1 = Y_1\boldsymbol{\beta}_1 + X_1\gamma_1 + \boldsymbol{\varepsilon}_1. \tag{6.3.12}$$

Now, we transform this by premultiplying by the matrix $\boldsymbol{X}$ of all predetermined variables of the system, i.e.

$$\begin{aligned} X'\mathbf{y}_1 &= X'Y_1\boldsymbol{\beta}_1 + X'X_1\gamma_1 + X'\boldsymbol{\varepsilon}_1 \\ &= (X'Y_1, X'X_1)\boldsymbol{\delta}_1 + X'\boldsymbol{\varepsilon}_1. \end{aligned} \tag{6.3.13}$$

Notice that the right-hand (explanatory) variables are T times the sample cross moments between the current endogenous and the predetermined variables. Dividing (13) by T and increasing the sample size, we can expect the new explanatory variables $X'Y_1$ and $X'X_1$ to converge (in probability) to a non-stochastic limit and thus to be uncorrelated with the disturbances $X'\varepsilon_1$. Following the arguments on stochastic regression, we can expect to obtain desirable estimators if we apply least squares to (13).

In general, it is not true that

$$\operatorname{cov}(X'\varepsilon_1) = E(X'\varepsilon_1\varepsilon_1 X) = \sigma_{11}X'X \tag{6.3.14}$$

since the matrix X may contain stochastic elements (lagged endogenous variables). But under reasonable assumptions on the disturbances we have, at least approximately,

$$\operatorname{cov}(X'\varepsilon_1) = \sigma_{11}X'X \tag{6.3.15}$$

(for a full discussion of this problem see Dhrymes, 1974, p. 184).

Note By assumption (6.2.13) the inverse $(X'X)^{-1}$ exists.

Theorem 6.7

The 2 SLSE of the structural parameters $\delta_1' = (\beta_1', \gamma_1')$ of the first equation (12) of the structural model is equivalent to the GLSE of δ_1 in the model (13) provided that we take the covariance matrix of the error term in (13) to be $\sigma_{11}X'X$.

The idea of the proof is as follows. The Aitken-estimator (GLSE) of δ_1 in (13) is just

$$\delta_1^* = (\bar{Z}_1'(X'X)^{-1}\bar{Z}_1)^{-1}(\bar{Z}_1'(X'X)^{-1}X'y_1) \tag{6.3.16}$$

where $\bar{Z}_1 = (X'Y_1, X'X_1)$. To show the equivalence of δ_1^* and $\hat{\delta}_1$ (11) we only have to compare the corresponding matrices. (This is left to the reader.)

Thus we have the following advantages of the equivalence theorem:

(i) by transforming the structural equation to the form (13), the expense in calculating the 2 SLSE can be decreased in the sense that only one regression has to be calculated.

(ii) The asymptotic covariance matrix of the GLSE δ_1^* gives just the asymptotic covariance matrix of the 2 SLSE $\hat{\delta}_1$, i.e. the p lim of the sample 'covariance' matrix

$$\begin{aligned}\bar{V}(\hat{\delta}_1) &= \operatorname{p\,lim}\sigma_{11}(\bar{Z}_1'(X'X)^{-1}\bar{Z}_1)^{-1} \\ &= \sigma_{11}\operatorname{p\,lim}\begin{pmatrix} Y_1'X(X'X)^{-1}X'Y_1 & Y_1'X_1 \\ X_1'Y_1 & X_1'X_1 \end{pmatrix}^{-1} \\ &= T^{-1}\sigma_{11}\begin{pmatrix} \Pi_1'\Sigma_{xx}\Pi_1 & \Pi_1'\Sigma_{x_1x} \\ \Sigma_{x_1x}'\Pi_1 & \Sigma_{x_1x_1} \end{pmatrix}\end{aligned} \tag{6.3.17}$$

(Σ_{x_1x} and $\Sigma_{x_1x_1}$ are submatrices of Σ_{xx} corresponding to the partition of $X = (X_1, X_2)$).

(iii) The equivalence theorem opens up the possibility of applying some special topics of the theory of linear regression estimation and prediction to econometrics, e.g. the method of restricted estimation.

6.3.3 The Restricted 2 SLS Estimator

We assume that we have outside or prior information on the coefficients of the first structural equation (12), i.e. we assume

$$\mathbf{r} = \boldsymbol{R}_1\boldsymbol{\beta}_1 + \boldsymbol{R}_1\gamma_1 + \boldsymbol{\phi} = \boldsymbol{R}\boldsymbol{\delta}_1 + \boldsymbol{\phi} \tag{6.3.18}$$

to hold. Here $\mathbf{r}$ is a realized (i.e. known) $J \times 1$-vector, $\boldsymbol{R}$ a known $(J \times k_1 + m_1)$-matrix with $J < k_1 + m_1$, $\boldsymbol{R} = (\boldsymbol{R}_1, \boldsymbol{R}_2)$ with $\boldsymbol{R}_2$ a $J \times k_1$-matrix, and

$$\operatorname{rank} \boldsymbol{R}_2 = k_1. \tag{6.3.19}$$

This last assumption (19) is essential in order to ensure the tractability of the method (cf. matrix (29)). Furthermore, we assume

$$E\boldsymbol{\phi} = \mathbf{0}, \quad E\boldsymbol{\phi}\boldsymbol{\phi}' = \boldsymbol{V}, \quad E\boldsymbol{\phi}\boldsymbol{\varepsilon}_1' = \boldsymbol{0}. \tag{6.3.20}$$

Following the idea of Theil (1963) we shall use the prior information (18) by enlarging the structural equation (12), i.e.

$$\mathbf{y}_1 = \tilde{\boldsymbol{Z}}_1\boldsymbol{\delta}_1 + \boldsymbol{\varepsilon}_1 \quad \text{with} \quad \tilde{\boldsymbol{Z}}_1 = (\boldsymbol{Y}_1, \boldsymbol{X}_1).$$

This gives the 'restricted' structural equation

$$\begin{pmatrix} \mathbf{y}_1 \\ \mathbf{r} \end{pmatrix} = \begin{pmatrix} \tilde{\boldsymbol{Z}}_1 \\ \boldsymbol{R} \end{pmatrix}\boldsymbol{\delta}_1 + \begin{pmatrix} \boldsymbol{\varepsilon}_1 \\ \boldsymbol{\phi} \end{pmatrix} \tag{6.3.21}$$

or, written more concisely,

$$\tilde{\mathbf{y}}_1 = \tilde{\boldsymbol{Z}}\boldsymbol{\delta}_1 + \tilde{\boldsymbol{\varepsilon}}_1. \tag{6.3.22}$$

The corresponding enlargement of the whole system (6.2.14) is therefore

$$\begin{pmatrix} \boldsymbol{Y} & \boldsymbol{0} \\ \boldsymbol{0} & \mathbf{r} \end{pmatrix} = \begin{pmatrix} \boldsymbol{Y} & \boldsymbol{0} \\ \boldsymbol{0} & \boldsymbol{R}_1 \end{pmatrix}\begin{pmatrix} \boldsymbol{B} & \boldsymbol{0} \\ \boldsymbol{0} & \boldsymbol{\beta}_1 \end{pmatrix} + \begin{pmatrix} \boldsymbol{X} & \boldsymbol{0} \\ \boldsymbol{0} & \boldsymbol{R}_2 \end{pmatrix}\begin{pmatrix} \boldsymbol{C} & \boldsymbol{0} \\ \boldsymbol{0} & \gamma_1 \end{pmatrix} + \begin{pmatrix} \boldsymbol{\varepsilon} & \boldsymbol{0} \\ \boldsymbol{0} & \boldsymbol{\phi} \end{pmatrix}. \tag{6.3.23}$$

If we remember the idea of the equivalence theorem, especially equation (13), we may conclude that transforming equation (21) by

$$\tilde{\boldsymbol{X}} = \begin{pmatrix} \boldsymbol{X} & 0 \\ 0 & \boldsymbol{R}_2 \end{pmatrix}$$

is equivalent to transforming equation (12) by $\boldsymbol{X}$ to give (13). In other words, the matrix of all predetermined variables now also contains the matrix $\boldsymbol{R}_2$ of restrictions on the parameters of the first equation's predetermined variables. Thus the transformation depends on the restrictions.

The transformed equation is

$$\tilde{\boldsymbol{X}}'\tilde{\mathbf{y}}_1 = \tilde{\boldsymbol{X}}'\tilde{\boldsymbol{Z}}\boldsymbol{\delta}_1 + \tilde{\boldsymbol{X}}'\tilde{\boldsymbol{\varepsilon}}_1. \tag{6.3.24}$$

Let us now define the restricted 2 SLS estimation procedure. The covariance matrix of $\tilde{\varepsilon}_1$ in (22) may be given by

$$\psi = \begin{pmatrix} \sigma_{11} I & 0 \\ 0 & V \end{pmatrix}.$$

Therefore the covariance matrix of $\tilde{X}'\tilde{\varepsilon}_1$ in (24) may be approximated by $(\tilde{X}'\psi\tilde{X})$. As rank $\tilde{X} = K + k_1$, the inverse $(\tilde{X}'\psi\tilde{X})^{-1}$ exists. Now, the formal Aitken estimator of δ_1 in the generalized regression model (24) under the assumption $E(\tilde{X}'\tilde{\varepsilon}_1\tilde{\varepsilon}_1\tilde{X}) = \tilde{X}'\psi\tilde{X}$ may be derived as

$$\mathbf{d}_1 = [\tilde{Z}'\tilde{X}(\tilde{X}'\psi\tilde{X})^{-1}\tilde{X}'\tilde{Z}]^{-1}\,\tilde{Z}'\tilde{X}(\tilde{X}'\psi\tilde{X})^{-1}\tilde{X}'\tilde{\mathbf{y}}_1 \tag{6.3.25}$$

where

$$\begin{aligned} &(\tilde{Z}'\tilde{X}(\tilde{X}'\psi\tilde{X})^{-1}\tilde{X}'\tilde{Z}) \\ &= \begin{pmatrix} \tilde{Z}_1 \\ R \end{pmatrix}\tilde{X}\begin{pmatrix} \sigma_{11}X'X & 0 \\ 0 & R_2'VR_2 \end{pmatrix}^{-1}\tilde{X}'\begin{pmatrix} \tilde{Z}_1 \\ R \end{pmatrix} \\ &= \sigma_{11}^{-1}\tilde{Z}_1'X(X'X)^{-1}X'\tilde{Z}_1 + R'R_2(R_2'VR_2)^{-1}R_2'R \\ &= D_0, \text{ say} \end{aligned} \tag{6.3.26}$$

$$\tilde{Z}'\tilde{X}(\tilde{X}'\psi\tilde{X})^{-1}\tilde{X}'\tilde{\mathbf{y}}_1 = \sigma_{11}^{-1}\tilde{Z}_1'X(X'X)^{-1}X'\mathbf{y}_1 + R'R_2(R_2'VR_2)^{-1}R_2'\mathbf{r}. \tag{6.3.27}$$

Thus the restricted 2 SLSE $\mathbf{d}_1$ is of the form

$$\mathbf{d}_1 = D_0^{-1}[\sigma_{11}^{-1}\tilde{Z}_1'X(X'X)^{-1}X'\mathbf{y}_1 + R'R_2(R_2'VR_2)^{-1}R_2'\mathbf{r}]. \tag{6.3.28}$$

The consistency of $\mathbf{d}_1$ is proved in a similar way to the proof for the usual 2 SLSE, namely by interpretation of $\mathbf{d}_1$ as an IVE (see (6.1.9)). We now investigate whether the restricted 2 SLSE $\mathbf{d}_1$ decreases the asymptotic risk of estimation compared with the unrestricted 2 SLSE $\hat{\delta}_1$.

Theorem 6.8 (Toutenburg and Wargowske, 1978)

Without loss of generality let the first structural equation of an econometric system and linear stochastic restrictions on the parameter of this equation be given:

$$\begin{aligned} \mathbf{y}_1 &= Y_1\beta_1 + X_1\gamma_1 + \varepsilon_1 \\ \mathbf{r} &= R_1\beta_1 + R_2\gamma_1 + \phi \end{aligned}$$

with

$$\text{rank } R_2 = k_1.$$

Then the restricted 2-stage-least-squares-estimator is $\mathbf{d}_1$ (28) and for the asymptotic covariance matrices the following relation holds:

$$\bar{V}^{-1}(\mathbf{d}_1) - \bar{V}^{-1}(\hat{\delta}_1) = R'R_2(R_2'VR_2)^{-1}R_2'R \geq 0. \tag{6.3.29}$$

Therefore, $\mathbf{d}_1$ is asymptotically better than $\hat{\delta}_1$.

In other words, the use of restrictions as in (18) decreases the asymptotic covariance matrix of 2 SLSE.

Proof of (29) As $\mathbf{d}_1$ is an Aitken-type estimator its sample matrix of second moments is just D_0^{-1} (see (26)); the asymptotic covariance matrix of $\mathbf{d}_1$ may therefore be calculated as

$$\begin{aligned}\bar{V}(\mathbf{d}_1) &= T^{-1}\operatorname{p\,lim} T D_0^{-1} \\ &= T^{-1}\{[\bar{V}(\hat{\delta}_1)]^{-1} + R' R_2 (R_2' V R_2)^{-1} R' R_2^{-1}\}\end{aligned} \tag{6.3.30}$$

where $\bar{V}(\hat{\delta}_1)$ is the asymptotic covariance matrix (17) of the 2 SLSE $\hat{\delta}_1$. Due to Theorem A.12 it follows from (30) that the difference $\bar{V}^{-1}(\mathbf{d}_1) - \bar{V}^{-1}(\hat{\delta}_1)$ is just as given in (29) and this matrix $R' R_2 (R_2' V R_2)^{-1} R_2' R$ is nonnegative definite.

Remark It is evident from the sample moment matrices $\sigma_{11}(\bar{Z}_1'(X'X)^{-1}\bar{Z}_1)^{-1}$ of $\hat{\delta}_1$ and D_0 of $\mathbf{d}_1$, respectively, that for finite samples

$$D_0^{-1} - \sigma_{11}(\bar{Z}_1'(X'X)^{-1}\bar{Z}_1)^{-1}$$

is nonnegative definite. Therefore a gain in efficiency is also ensured for finite samples.

Note For alternative set-ups in using linear and other restrictions the reader is referred to Rothenberg (1973) and Pollock (1979). Fukuhara (1976) gave a generalization of the restricted least squares to structural parameter estimation where difficulties arise caused by the identification problem. Kelly (1975) investigates cross-equation constraints and develops a strategy for using these constraints for single-equation identification. For another restricted 2 SLSE and its practical employment see Kakwani and Court (1970).

6.3.4 A Numerical Example

The following example demonstrates the gain in efficiency by using the proposed restricted 2 SLS technique. We assume the following 2-equations submodel of the econometric model of the German Democratic Republic (Wölfling, Biebler, and Schiele, 1975):

$$\text{FIPC} = \beta_{11}\,\text{BIPC} + \gamma_{10}\mathbf{i} + \gamma_{11}\,\text{UIPC}_{-1} + \varepsilon_1 \tag{6.3.31}$$

$$\text{BIPC} = \gamma_{20}\mathbf{i} + \gamma_{21}\,\text{BIPE} + \varepsilon_2 \tag{6.3.32}$$

where

BIPC	gross investment
FIPC	realized investment
UIPC_{-1}	partially realized investment (with lag 1)
BIPE	gross investment, measured in effective prices.

Let us concentrate on estimating the first relation's parameters with the two alternative methods.

We have

$$\begin{aligned}
&\mathbf{y}_1 = \text{FIPC}, \quad Y_1 = \text{BIPC}, \quad X_1 = (\mathbf{i}, \text{UIPC}_{-1}) \\
&\boldsymbol{\beta}_1 = (\beta_{11}), \quad \gamma_1' = (\gamma_{10}, \gamma_{11}), \\
&K = 3, M = 2, m_1 = 1, k_1 = 2, k_2 = 1.
\end{aligned}$$

Assuming rank $\Pi_{21} = m_1 = 1$ (which is true for all $\Pi_{21} \neq 0$) we have by $k_2 = m_1 = 1$ the just-identified case (see Definition 6.6), for which, as is well known, the 2 SLSE and the ILSE coincide. We shall use the formula of 2 SLSE to compare it with the restricted 2 SLSE. Given samples of size $T = 9$ we calculate

$$\hat{\sigma}_{11} = (\mathbf{y}_1 - X\bar{\boldsymbol{\delta}}_1)'(\mathbf{y} - X\bar{\boldsymbol{\delta}}_1) = 0.522$$

where $\bar{\boldsymbol{\delta}}_1$ is the OLSE of $\boldsymbol{\delta}_1' = (\beta_{11}, \gamma_{10}, \gamma_{11})$. The estimation of the sample moment matrix of the 2 SLSE $\hat{\boldsymbol{\delta}}_1$ gives

$$\operatorname{cov} \hat{\boldsymbol{\delta}}_1 = \hat{\sigma}_{11} \begin{pmatrix} Y_1' X(X'X)^{-1} X' Y_1 & Y_1' X_1 \\ X_1' Y_1 & X_1' X_1 \end{pmatrix} \tag{6.3.33}$$

$$= \begin{pmatrix} 0.00136 & -0.00956 & -0.00127 \\ -0.00956 & 1.05152 & 0.04646 \\ -0.00127 & 0.04646 & 0.00483 \end{pmatrix}.$$

The following restrictions on $\boldsymbol{\delta}$ are given:

$$\gamma_{10} \in (1.2; 1.8)$$

and

$$\beta_{11} - 3\gamma_{11} \in (-0.2; 0.2).$$

Both relations hold 'almost certainly' in the sense of a 3σ-rule, i.e. ± 0.3 and ± 0.2 are the 3σ-limits of the components of $\boldsymbol{\phi}$. Then the restrictions may be written as (see Section 4.5.2).

$$\mathbf{r} = \boldsymbol{R}\boldsymbol{\delta} + \boldsymbol{\phi} = \boldsymbol{R}_1 \boldsymbol{\beta}_1 + \boldsymbol{R}_2 \gamma_1 + \boldsymbol{\phi}, \quad E\boldsymbol{\phi}\boldsymbol{\phi}' = \boldsymbol{V}$$

where

$$\mathbf{r} = \begin{pmatrix} 0 \\ 1.5 \end{pmatrix}, \quad \boldsymbol{R} = \begin{pmatrix} 1 & 0 & -3 \\ 0 & 1 & 0 \end{pmatrix},$$

$$\boldsymbol{R}_1 = \begin{pmatrix} 1 \\ 0 \end{pmatrix}, \quad \boldsymbol{R}_2 = \begin{pmatrix} 0 & -3 \\ 1 & 0 \end{pmatrix},$$

$$\boldsymbol{V} = \begin{pmatrix} 0.0044 & 0 \\ 0 & 0.0100 \end{pmatrix}.$$

Thus we may calculate the sample moment matrix of the restricted 2 SLSE $\mathbf{d}_1$ as

$$\operatorname{cov} \mathbf{d}_1 = ([\operatorname{cov} \hat{\boldsymbol{\delta}}_1]^{-1} + \boldsymbol{R}'\boldsymbol{R}_2(\boldsymbol{R}_2' \boldsymbol{V} \boldsymbol{R}_2)^{-1} \boldsymbol{R}_2' \boldsymbol{R})^{-1} \tag{6.3.34}$$

$$= \begin{pmatrix} 0.00019434 & -0.00028425 & -0.00016807 \\ -0.00024425 & 0.00987058 & -0.00013766 \\ -0.00016807 & -0.00013766 & 0.00030380 \end{pmatrix}.$$

The elements of the diagonal in (33) and (34), respectively, are the estimated variances of the components of $\boldsymbol{\delta}$. We may conclude that in this model the use of the restricted 2 SLSE decreases the estimated variances of the estimated parameters $(\beta_{11}, \gamma_{10}, \gamma_{11})$ compared with the usual 2 SLSE:

for β_{11} by 99.06%
for γ_{10} by 85.71%
for γ_{11} by 93.71%.

If we calculate an ex-post-predictor $p = \mathbf{a}'\hat{\boldsymbol{\delta}}_1$ or $p = \mathbf{a}'\mathbf{d}_1$ where $\mathbf{a}$ is the vector of last year's realization of (BIPC, **i**, UIPC_{-1}), its variance is estimated by $\mathbf{a}' \operatorname{cov} \hat{\boldsymbol{\delta}}_1 \mathbf{a} = 4.80461$ or $\mathbf{a}' \operatorname{cov} \mathbf{d}_1 \mathbf{a} = 0.095338$. The use of restrictions therefore decreases the ex-post-predictor risk by 98.02%.

These gains in efficiency clearly depend on the actual data of the finite sample, but they do show the great possibilities of the proposed method.

Appendix A Matrix Results

The intention of this appendix, which is a revised version of the appendix in Bibby and Toutenburg, 1978, is to give a choice of important theorems and definitions from matrix theory which are useful for linear statistical methods. Most of the following results are referred to within this book. Where the theorems are standard, proofs are omitted. For a fuller discussion of the relevant theory the reader is referred to Graybill (1961), Rao (1973), to the early chapters of Johnston (1972), and to Mardia, Kent, and Bibby (1979).

A.1 QUADRATIC FORMS, DEFINITENESS, ETC.

A matrix A is a rectangular array of numbers. If A has m rows and n columns we say that A is of order $m \times n$. For abbreviation we use the notation $\underset{m \times n}{A}$ or '$m \times n$-matrix A' or $A = (a_{ij})$ where a_{ij} $(i = 1, \ldots, m; j = 1, \ldots, n)$ are the elements of the matrix. All over this book we assume a_{ij} to be real-valued.

Definition A.1 The homogeneous quadratic function in the variables $x_1, \ldots, x_n$

$$Q = \sum_{i=1}^{n} \sum_{j=1}^{n} a_{ij} x_i x_j = \mathbf{x}' A \mathbf{x}$$

with $\mathbf{x}' = (x_1, \ldots, x_n)$ and the $n \times n$-matrix A symmetric is said to be a quadratic form in $\mathbf{x}$. A is called the matrix of the quadratic form.

Theorem A.2 (Spectral decomposition theorem)

(i) Let $\underset{n,n}{A}$ be a symmetric matrix. Then there exists an orthogonal matrix P such that

$$P'AP = \Lambda \quad \text{or } A = P\Lambda P'$$

with $P'P = PP' = I$, P the matrix whose columns $P_1, \ldots, P_n$ are the standardized eigenvalues of A, and $\Lambda = \text{diag}(\lambda_1, \ldots, \lambda_n)$ the matrix of the eigenvalues of A.

Proof See, for example, Mardia, Kent, and Bibby, 1979, p. 469

(ii) If $\boldsymbol{A} > \boldsymbol{0}$, or at least $\boldsymbol{A} \geq \boldsymbol{0}$, then $\boldsymbol{A}$ can be written as

$$\boldsymbol{A} = \boldsymbol{A}^{1/2} \boldsymbol{A}^{1/2}$$

where $\boldsymbol{A}^{1/2} = \boldsymbol{P} \boldsymbol{\Lambda}^{1/2} \boldsymbol{P}'$ is symmetric.

Proof Follows from (i), immediately

(iii) If $\boldsymbol{A}$ is symmetric, then

$$\operatorname{tr} \boldsymbol{A} = \sum \lambda_i.$$

Proof $\operatorname{tr} \boldsymbol{A} = \operatorname{tr} \boldsymbol{P}' \boldsymbol{\Lambda} \boldsymbol{P} = \operatorname{tr} \boldsymbol{\Lambda} \boldsymbol{P} \boldsymbol{P}' = \operatorname{tr} \boldsymbol{\Lambda} = \Sigma \lambda_i$ (see Definition A.18).

The properties of a quadratic form depend on the matrix $\boldsymbol{A}$ which classifies the quadratic form. This classification is known as the definiteness of matrices and quadratic forms, respectively.

*Definition A.*3 A matrix $\underset{n \times n}{\boldsymbol{A}}$ is positive definite when it is symmetric and satisfies $Q = \mathbf{x}' \boldsymbol{A} \mathbf{x} > 0$ for all $\mathbf{x} \neq \mathbf{0}$. As an abbreviation we write $\boldsymbol{A} > \boldsymbol{0}$.

Theorem A.4

If $\underset{n \times n}{\boldsymbol{A}} > \boldsymbol{0}$ then

(i) $\boldsymbol{A}$ has full rank n
(ii) all eigenvalues of $\boldsymbol{A}$ are positive
(iii) $\operatorname{tr} \boldsymbol{A} = \sum_{i=1}^{n} a_{ii} > 0.$
(*iv*) $\det \boldsymbol{A} = |\boldsymbol{A}| > 0.$

Notation A rectangular matrix $\underset{m,n}{\boldsymbol{P}}$ with $m \leq n$ is said to be of full rank if rank $\boldsymbol{P} = m$.

Theorem A.5

If $\underset{n,n}{\boldsymbol{A}} > \boldsymbol{0}$ and $\underset{n,m}{\boldsymbol{P}}$ has full rank $m \leq n$, then $\boldsymbol{P}' \boldsymbol{A} \boldsymbol{P} > \boldsymbol{0}$. In particular (taking $\boldsymbol{A} = \boldsymbol{I}$), the matrix $\boldsymbol{P}' \boldsymbol{P} > \boldsymbol{0}$ for all matrices $\boldsymbol{P}$ of full rank.

Theorem A.6

If $\boldsymbol{A}$ is positive definite then the inverse $\boldsymbol{A}^{-1}$ is positive definite, too. We write this as $\boldsymbol{A} > \boldsymbol{0} \leftrightarrow \boldsymbol{A}^{-1} > \boldsymbol{0}$.

Proof Put $\boldsymbol{P} = \boldsymbol{A}^{-1}$ in Theorem A.5.

Definition A.7 A matrix $\boldsymbol{A}$ is nonnegative definite when it is symmetric and satisfies

$$Q = \mathbf{x}'\boldsymbol{A}\mathbf{x} \geq 0 \text{ for all } \mathbf{x}.$$

As an abbreviation we write $\boldsymbol{A} \geq \boldsymbol{0}$.

Theorem A.8

If $\boldsymbol{A} \geq \boldsymbol{0}$ then $\lambda_i \geq 0$ $(i = 1, \ldots, n)$.

Proof If $0 \leq \mathbf{x}'\boldsymbol{A}\mathbf{x} = \mathbf{x}'\boldsymbol{P}\boldsymbol{\Lambda}\boldsymbol{P}'\mathbf{x} = \lambda_1 y_1{}^2 + \ldots + \lambda_n y_n^2$ (see Theorem A.2) for all $\mathbf{y} = \boldsymbol{P}'\mathbf{x}$, then this is true especially for $\mathbf{y}' = (1, 0, \ldots, 0)$, $\mathbf{y}' = (0, 1, 0, \ldots)$, and so on. Thus $\lambda_i \geq 0$ $(i = 1, \ldots, n)$.

Theorem A.9

If $\boldsymbol{A}$ is positive definite then $\boldsymbol{P}'\boldsymbol{A}\boldsymbol{P}$ is nonnegative definite for all $\boldsymbol{P}$. If $\boldsymbol{P}$ is square and regular, then $\boldsymbol{P}'\boldsymbol{A}\boldsymbol{P}$ is positive definite.

Proof Since $\mathbf{x}'\boldsymbol{P}'\boldsymbol{A}\boldsymbol{P}\mathbf{x} = \mathbf{y}'\boldsymbol{A}\mathbf{y}$ where $\mathbf{y} = \boldsymbol{P}\mathbf{x}$, then $\mathbf{x}'\boldsymbol{P}'\boldsymbol{A}\boldsymbol{P}\mathbf{x} \geq 0$, for all $\mathbf{x}$. If $\boldsymbol{P}$ is square and of full rank then $\mathbf{y} = \mathbf{0}$ only when $\mathbf{x} = \mathbf{0}$, so $\mathbf{x}'\boldsymbol{P}'\boldsymbol{A}\boldsymbol{P}\mathbf{x} > 0$ for all nonzero $\mathbf{x}$.

Theorem A.10

For any matrix $\boldsymbol{P}$, $\boldsymbol{P}'\boldsymbol{P} \geq \boldsymbol{0}$.

Proof Put $\boldsymbol{A} = \boldsymbol{I}$ in Theorem A.9.

Theorem A.11

If $\boldsymbol{A} > \boldsymbol{0}$ and $\boldsymbol{B} \geq \boldsymbol{0}$, then $(\boldsymbol{A} + \boldsymbol{B}) > \boldsymbol{0}$.

Proof

$$\mathbf{x}'(\boldsymbol{A} + \boldsymbol{B})\mathbf{x} = \mathbf{x}'\boldsymbol{A}\mathbf{x} + \mathbf{x}'\boldsymbol{B}\mathbf{x} > 0 \text{ for } \mathbf{x} \neq \mathbf{0}.$$

Theorem A.12

Using the notation in Theorem A.11,

(i) $|\boldsymbol{A}| \leq |\boldsymbol{A} + \boldsymbol{B}|$

(ii) $\boldsymbol{A}^{-1} - (\boldsymbol{A} + \boldsymbol{B})^{-1}$ is nonnegative definite.

Proof See Goldberger (1964, p. 38).

Theorem A.13

Let $\underset{n,n}{A}$ be a symmetric matrix with eigenvalues $\lambda_1 \geq \ldots \geq \lambda_n$ and the corresponding eigenvectors $P_1, \ldots, P_n$. Then

(i) $\sup\limits_{\beta} \dfrac{\beta' A \beta}{\beta' \beta} = \lambda_1$

(ii) $\inf\limits_{\beta} \dfrac{\beta' A \beta}{\beta' \beta} = \lambda_n$

Proof see Rao (1973).

Theorem A.14

For any matrix $A \geq 0$

$$0 \leq \lambda_i \leq 1 \quad (i = 1, \ldots, n) \quad \text{iff } (I - A) \geq 0.$$

Proof Using Theorem A.2 we have

$$(I - A) = P(I - \Lambda)P' \geq 0 \text{ iff}$$

$$P'P(I - \Lambda)P'P = (I - \Lambda) \geq 0.$$

(i) Let $(I - \Lambda) \geq 0$, then the eigenvalues of this diagonal matrix $(1 - \lambda_1), \ldots, (1 - \lambda_n)$ are nonnegative (Theorem A.8).
(ii) Let $0 \leq \lambda_i \leq 1$, then

$$\mathbf{x}'(I - \Lambda)\mathbf{x} = \sum_{i=1}^{n} x_i^2 (1 - \lambda_i) \geq 0.$$

Theorem A.15 (Theobald, 1974)

A symmetric $n \times n$-matrix D is nonnegative definite if and only if tr $CD \geq 0$ for all nonnegative definite C, i.e.

$$D \geq 0 \leftrightarrow \text{tr } CD \geq 0 \text{ for all } C \geq 0$$

Proof Following Theorem A.2 we have the representation

$$D = P \Lambda P'$$

$$= \sum_{i=1}^{n} \lambda_i P_i P_i'$$

where $\lambda_1, \ldots, \lambda_n$ are the eigenvalues of D ($\Lambda = \text{diag}(\lambda_1, \ldots, \lambda_n)$) and $P = (P_1, \ldots, P_n)$ is the matrix of the corresponding orthogonal eigenvectors.

From the above it follows that

$$\operatorname{tr}(\boldsymbol{CD}) = \operatorname{tr}\left(\sum_{i=1}^{n} \lambda_i \boldsymbol{C P}_i \boldsymbol{P}_i'\right)$$
$$= \sum_{i=1}^{n} \lambda_i \boldsymbol{P}_i' \boldsymbol{C P}_i .$$

The terms $\boldsymbol{P}_i' \boldsymbol{CP}_i$ $(i = 1, \ldots, n)$ are nonnegative for all $\boldsymbol{C} \geq \boldsymbol{0}$. If $\boldsymbol{D} \geq \boldsymbol{0}$, all λ_i are nonnegative and thus we have tr $\boldsymbol{CD} \geq 0$.

Conversely, if tr $\boldsymbol{CD} \geq 0$ for all $\boldsymbol{C} \geq \boldsymbol{0}$ this is true especially for $\boldsymbol{C} = \boldsymbol{P}_i \boldsymbol{P}_i' \geq \boldsymbol{0}$ $(i = 1, \ldots, n)$. Then we have

$$\operatorname{tr} \boldsymbol{CD} = \operatorname{tr}\left[\boldsymbol{P}_i \boldsymbol{P}_i' \left(\sum_{j=1}^{n} \lambda_j \boldsymbol{P}_j \boldsymbol{P}_j'\right)\right]$$
$$= \lambda_i \geq 0 \qquad (i = 1, \ldots, n).$$

This completes the proof.

Theorem A.16 (Teräsvirta, 1980)

Let $\mathbf{h}$ and $\mathbf{g}$ be two linearly independent $m \times 1$ vectors. Then $\mathbf{hh}' - \mathbf{gg}'$ is indefinite and its nonzero eigenvalues are

$$\lambda = -(\mathbf{g}'\mathbf{g} - \mathbf{h}'\mathbf{h}) \pm [(\mathbf{g}'\mathbf{g} - \mathbf{h}'\mathbf{h})^2 + \mathbf{g}'\mathbf{g}\mathbf{h}'\mathbf{h} - (\mathbf{g}'\mathbf{h})^2]^{1/2}.$$

Proof We have

$$(\mathbf{hh}' - \mathbf{gg}')\mathbf{z} = \lambda \mathbf{z} \tag{A.1.1}$$

where λ is an eigenvalue of $\mathbf{hh}' - \mathbf{gg}'$ and $\mathbf{z}$ is the corresponding eigenvector. Left-multiplying (1) by $\mathbf{h}'$ and $\mathbf{g}'$ yields

$$(\mathbf{h}'\mathbf{h} - \lambda)\mathbf{h}'\mathbf{z} = \mathbf{h}'\mathbf{g}\mathbf{g}'\mathbf{z} \tag{A.1.2}$$

and

$$(\mathbf{g}'\mathbf{g} + \lambda)\mathbf{g}'\mathbf{z} = \mathbf{g}'\mathbf{h}\mathbf{h}'\mathbf{z}. \tag{A.1.3}$$

Solving (3) for $\mathbf{h}'\mathbf{z}$ and inserting the result into (2) gives for $\mathbf{g}'\mathbf{z} \neq 0$

$$(\mathbf{h}'\mathbf{h} - \lambda)(\mathbf{g}'\mathbf{g} + \lambda) = (\mathbf{h}'\mathbf{g})^2. \tag{A.1.4}$$

From (4) we have

$$\lambda = -(\mathbf{g}'\mathbf{g} - \mathbf{h}'\mathbf{h}) \pm [(\mathbf{g}'\mathbf{g} - \mathbf{h}'\mathbf{h})^2 + \mathbf{g}'\mathbf{g}\mathbf{h}'\mathbf{h} - (\mathbf{g}'\mathbf{h})^2]^{1/2}.$$

By the Cauchy–Schwarz inequality

$$\mathbf{g}'\mathbf{g}\mathbf{h}'\mathbf{h} \geq (\mathbf{g}'\mathbf{h})^2$$

so that the roots of (4) are of different sign.

Theorem A.17

Let $\boldsymbol{I}$ be the $n \times n$ identity matrix and $\mathbf{g}$ a $n \times 1$-vector. Then

$$\boldsymbol{I} - \mathbf{g}\mathbf{g}' \geq \boldsymbol{0} \quad \text{iff} \quad \mathbf{g}'\mathbf{g} \leq 1.$$

Proof Let $\boldsymbol{C}$ be an orthogonal matrix such that

$$\boldsymbol{C}\mathbf{g} = \begin{pmatrix} \alpha \\ 0 \\ \vdots \\ 0 \end{pmatrix}$$

Now, $\boldsymbol{I} - \mathbf{g}\mathbf{g}' \geq \boldsymbol{0}$ iff $\boldsymbol{C}(\boldsymbol{I} - \mathbf{g}\mathbf{g}')\boldsymbol{C}' \geq \boldsymbol{0}$, i.e.

$$\boldsymbol{I} - \begin{pmatrix} \alpha^2 & & 0 \\ & 0 & \\ & & \ddots \\ 0 & & 0 \end{pmatrix} \geq \boldsymbol{0} \quad \text{iff} \quad \alpha^2 = \mathbf{g}'\mathbf{g} \leq 1.$$

(see also Yancey *et al.*, 1974).

A.2 TRACE OF A MATRIX

Definition A.18 The trace of a matrix $\boldsymbol{A}$, tr $\boldsymbol{A}$, is the sum of the diagonal elements i.e.

$$\text{tr } \boldsymbol{A} = \sum_i a_{ii}.$$

Theorem A.19

If $\boldsymbol{A}$ and $\boldsymbol{B}$ are $(n \times n)$ matrices and c is a scalar, then

(a) $\text{tr } (\boldsymbol{A} + \boldsymbol{B}) = \text{tr } \boldsymbol{A} + \text{tr } \boldsymbol{B}$
(b) $\text{tr } \boldsymbol{A}' = \text{tr } \boldsymbol{A}$
(c) $\text{tr } c\boldsymbol{A} = c \text{ tr } \boldsymbol{A}$
(d) $\text{tr } \boldsymbol{A}\boldsymbol{B} = \text{tr } \boldsymbol{B}\boldsymbol{A}$

Proof Follows directly from the definition of trace. (Part (d) is also true if $\boldsymbol{A}$ is $n \times m$ while $\boldsymbol{B}$ is $m \times n$.)

Theorem A.20

For any two $(n \times 1)$ vectors $\mathbf{a}$ and $\mathbf{b}$

$$\mathbf{a}'\mathbf{b} = \sum a_i b_i = \text{tr } \mathbf{a}\mathbf{b}'.$$

A.3 IDEMPOTENT MATRICES

Definition A.21 A matrix $\boldsymbol{A}$ is idempotent when it is symmetric and satisfies

$$\boldsymbol{A}^2 = \boldsymbol{A}\boldsymbol{A} = \boldsymbol{A}.$$

Theorem A.22

The eigenvalues of an idempotent matrix are all 1 or 0.

Theorem A.23

If $\boldsymbol{A}$ is idempotent and of full rank then $\boldsymbol{A} = \boldsymbol{I}$.

Theorem A.24

If $\boldsymbol{A}$ is idempotent and of rank r, then there exists an orthogonal matrix $\boldsymbol{P}$ such that $\boldsymbol{P}'\boldsymbol{A}\boldsymbol{P} = \boldsymbol{E}_r$, where $\boldsymbol{E}_r$ is a diagonal matrix with r ones on the main diagonal and zeros elsewhere.

Theorem A.25

If $\boldsymbol{A}$ is idempotent and of rank r then $\operatorname{tr} \boldsymbol{A} = r$.

Theorem A.26

If $\boldsymbol{A}$ and $\boldsymbol{B}$ are idempotent and $\boldsymbol{A}\boldsymbol{B} = \boldsymbol{B}\boldsymbol{A}$, then $\boldsymbol{A}\boldsymbol{B}$ is also idempotent.

Theorem A.27

If $\boldsymbol{A}$ is idempotent and $\boldsymbol{P}$ is orthogonal, then $\boldsymbol{P}'\boldsymbol{A}\boldsymbol{P}$ is idempotent.

Theorem A.28

If $\boldsymbol{A}$ is idempotent and $\boldsymbol{B} = \boldsymbol{I} - \boldsymbol{A}$, then $\boldsymbol{B}$ is idempotent and $\boldsymbol{A}\boldsymbol{B} = \boldsymbol{B}\boldsymbol{A} = \boldsymbol{0}$.

Proof of Theorems A.21–A.28 See Graybill (1961).

A.4 ORTHOGONAL MATRICES

Definition A.29 A $n \times n$-matrix $\boldsymbol{A}$ is orthogonal if $\boldsymbol{A}\boldsymbol{A}' = \boldsymbol{I}$.

Theorem A.30

For an orthogonal matrix $\boldsymbol{A}$

(i) $A^{-1} = A'$
(ii) $A'A = I$
(iii) $|A| = \pm 1$
(iv) AB is orthogonal if A and B are orthogonal $n \times n$-matrices,

Proof Follows easily from Definition A.29.

A.5 KRONECKER PRODUCT

Theorem A.31

If $A\,(m \times n)$ and $B(p \times q)$ are matrices, then their Kronecker product $A \otimes B$ is the $(mp \times nq)$ matrix defined by

$$A \otimes B = \begin{pmatrix} a_{11}B & \dots & a_{1n}B \\ \vdots & & \vdots \\ a_{m1}B & \dots & a_{mn}B \end{pmatrix}$$

The Kronecker product has the following properties:

(i) $c(A \otimes B) = (cA) \otimes B = A \otimes (cB)$;
(ii) $A \otimes (B \otimes C) = (A \otimes B) \otimes C$;
(iii) $A \otimes (B + C) = (A \otimes B) + (A \otimes C)$;
(iv) $(A \otimes B)' = A' \otimes B'$.

A.6 INVERSE OF A MATRIX

Definition A.32 The inverse of a square and nonsingular matrix A is the unique matrix A^{-1} satisfying

$$AA^{-1} = A^{-1}A = I.$$

Theorem A.33

For partitioned A

$$A = \begin{pmatrix} E & F \\ G & H \end{pmatrix}$$

where E and $D = H - GE^{-1}F$ are invertible we have the partitioned inverse

$$A^{-1} = \begin{pmatrix} E^{-1} + E^{-1}FD^{-1}GE^{-1} & -E^{-1}FD^{-1} \\ -D^{-1}GE^{-1} & D^{-1} \end{pmatrix}.$$

Proof Check that the product of the matrix and its inverse reduces to the identity matrix.

Theorem A.34 (see, for example, Mardia, Kent, and Bibby, 1979, p. 459)

If all the necessary inverses exist, then for $A(n \times n)$, $B(n \times m)$, $C(m \times m)$, and $D(m \times n)$ the inverse of the following sum is

$$(A + BCD)^{-1} = A^{-1} - A^{-1}B(C^{-1} + DA^{-1}B)^{-1}DA^{-1}.$$

Theorem A.35

If $\underset{n \times n}{A}$ is nonsingular and $\mathbf{a}$ and $\mathbf{b}$ are $n \times 1$-vectors, then

$$(A + \mathbf{a}\mathbf{b}')^{-1} = A^{-1} - \frac{(A^{-1}\mathbf{a})(\mathbf{b}'A^{-1})}{1 + \mathbf{b}'A^{-1}\mathbf{a}}.$$

Proof Use Theorem A.34.

A.7 GENERALIZED INVERSE

Definition A.36 Let A be a $m \times n$-matrix of any rank. Then the matrix A^- is called a g-inverse of A if $\mathbf{x} = A^-\mathbf{y}$ is a solution of $A\mathbf{x} = \mathbf{y}$ for any fixed vector $\mathbf{y}$. (*Note* A generalized inverse always exists, but in general it is not unique.)

Theorem A.37

A^{-1} exists iff $AA^-A = A$.

Theorem A.38

If $\underset{n \times n}{A}$ is nonsingular, then $A^- = A^{-1}$.

Theorem A.39

If $\underset{n \times n}{A}$ is of rank $m \leq n$, rearrange A as

$$A = \begin{pmatrix} A_{11} & A_{12} \\ A_{21} & A_{22} \end{pmatrix}$$

where A_{11} is a $(m \times m)$ nonsingular matrix. Then

$$A^- = \begin{pmatrix} A_{11}^{-1} & 0 \\ 0 & 0 \end{pmatrix}$$

is a g-inverse.

A.8 FUNCTIONS OF NORMALLY DISTRIBUTED VARIABLES

Definition A.40 Let $\mathbf{x}' = (x_1, \ldots, x_p)$ be a p-dimensional random vector. $\mathbf{x}$ is said to be (p-dimensional) normally distributed with mean vector μ and covariance matrix Σ (abbreviated as $\mathbf{x} \sim N_p(\mu, \Sigma)$) if it has the common density function

$$\mathbf{f}(\mathbf{x};\mu,\Sigma) = \{(2\pi)^p |\Sigma|\}^{-1/2} \exp\{-\tfrac{1}{2}(\mathbf{x}-\mu)'\Sigma^{-1}(\mathbf{x}-\mu)\}.$$

Theorem A.41

If $\mathbf{x} \sim N_p(\mu, \Sigma)$, A is a $(p \times p)$ matrix, and $\mathbf{b}$ is a $(p \times 1)$ vector (both A and $\mathbf{b}$ nonstochastic), then

$$\mathbf{y} = A\mathbf{x} + \mathbf{b} \sim N_p(A\mu + \mathbf{b},\ A\Sigma A').$$

Theorem A.42

If $\mathbf{x} \sim N_p(0, I)$, then $w = \mathbf{x}'\mathbf{x}$ has a (central) chi squared distribution with p degrees of freedom. We write

$$w \sim \chi^2_p.$$

Theorem A.43

If $\mathbf{x} \sim N_p(\mu, I)$ then $w = \mathbf{x}'\mathbf{x}$ has a noncentral χ^2_p distribution with noncentrality parameter $\lambda = \mu'\mu$. We write

$$w \sim \chi^2_p(\lambda).$$

Theorem A.44

If $\mathbf{x} \sim N_p(\mu, \Sigma)$ and Σ is positive definite then

(i) $\mathbf{x}'\Sigma^{-1}\mathbf{x} \sim \chi^2_p(\mu'\Sigma^{-1}\mu)$,

(ii) $(\mathbf{x}-\mu)'\Sigma^{-1}(\mathbf{x}-\mu) \sim \chi^2_p$.

Proof As $\Sigma > 0$ is symmetric we may write $\Sigma = P\Lambda^{1/2}\Lambda^{1/2}P' = RR'$ (see Theorem A.2) where $R = P\Lambda^{1/2}$ is regular. Then we have

(i)
$$\begin{aligned}\mathbf{x}'\Sigma^{-1}\mathbf{x} &= [R^{-1}\mathbf{x}]'[R^{-1}\mathbf{x}]\\ &= \mathbf{y}'\mathbf{y} \sim \chi^2_p(\mu'\Sigma^{-1}\mu)\end{aligned}$$

as $\mathbf{y} = R^{-1}\mathbf{x} \sim N(R^{-1}\mu, I)$ (see Theorem A.41).

(ii) Similarly, we calculate

$$\begin{aligned}(\mathbf{x}-\mu)'\Sigma^{-1}(\mathbf{x}-\mu) &= [R^{-1}(\mathbf{x}-\mu)]'\,[R^{-1}(\mathbf{x}-\mu)]\\ &= \tilde{\mathbf{y}}'\tilde{\mathbf{y}} \sim \chi^2_p\end{aligned}$$

where $\tilde{\mathbf{y}} = R^{-1}(\mathbf{x}-\mu) \sim N_p(0, I)$ (see Theorem A.41).

Theorem A.45

If $Q_1 \sim \chi^2_m(\lambda)$ and $Q_2 \sim \chi^2_n$, and Q_1 and Q_2 are independent, then

(i) the quotient

$$F = \frac{Q_1/m}{Q_2/n}$$

has a noncentral F distribution with m and n degrees of freedom and noncentrality parameter λ. We write $F \sim F_{m,n}(\lambda)$.

(ii) If $\lambda = 0$ then $F \sim F_{m,n}$ (the central F distribution).

(iii) if $m = 1$ then $\sqrt{F}$ has a noncentral $t_n(\sqrt{\lambda})$ distribution, or a central t_n distribution if $\lambda = 0$.

Theorem A.46

If $\mathbf{x} \sim N_p(\mu, I)$ and A is an idempotent $(p \times p)$ matrix with rank r, then

$$\mathbf{x}' A \mathbf{x} \sim \chi^2_r(\mu' A \mu).$$

Proof From Theorem A.24 there exists an orthogonal matrix P such that (without loss of generality)

$$P' A P = E_r = \begin{pmatrix} I_r & 0 \\ 0 & 0 \end{pmatrix}.$$

We write $P = \left(\underset{p \times r}{P_1}, \underset{p \times (p-r)}{P_2} \right)$ and get

$$P' \mathbf{x} = \mathbf{y} = \begin{pmatrix} \mathbf{y}_1 \\ \mathbf{y}_2 \end{pmatrix} = \begin{pmatrix} P_1' \mathbf{x} \\ P_2' \mathbf{x} \end{pmatrix}.$$

As $\mathbf{y} \sim N_p(P'\mu, I_p)$ (Theorem A.41) it follows that $\mathbf{y}_1 \sim N_r(P_1'\mu, I_r)$ and (by Theorem A.43) $\mathbf{y}_1' \mathbf{y}_1 \sim \chi^2_r(\mu' P_1 P_1' \mu)$.

As P is orthogonal we have $P'P = PP' = I$ and therefore

$$A = (PP')A(PP') = P(P'AP)P'$$
$$= (P_1, P_2)\begin{pmatrix} I_r & 0 \\ 0 & 0 \end{pmatrix}\begin{pmatrix} P_1' \\ P_2' \end{pmatrix} = P_1 P_1',$$

so that

$$\mathbf{x}' A \mathbf{x} = \mathbf{x}' P_1 P_1' \mathbf{x} = \mathbf{y}_1' \mathbf{y}_1 \sim \chi^2_r(\mu' A \mu).$$

Theorem A.47

If $\mathbf{x} \sim N_p(\mu, I)$, A is an idempotent $(p \times p)$ matrix with rank r, B is a $(p \times n)$ matrix, and $AB = 0$, then the linear form $B\mathbf{x}$ is independent of the quadratic form $\mathbf{x}' A \mathbf{x}$.

Proof Consider the matrix $\boldsymbol{P}$ as in Theorem A.46. Then $\boldsymbol{BPP'AP} = \boldsymbol{BAP} = \boldsymbol{0}$, as $\boldsymbol{BA} = \boldsymbol{0}$ was assumed. On the other hand, if $\boldsymbol{BP} = \boldsymbol{D} = (\boldsymbol{D}_1, \boldsymbol{D}_2)$, then we have

$$\boldsymbol{BPP'AP} = (\boldsymbol{D}_1, \boldsymbol{D}_2)\begin{pmatrix} \boldsymbol{I}_r & \boldsymbol{0} \\ \boldsymbol{0} & \boldsymbol{0} \end{pmatrix} = (\boldsymbol{D}_1, \boldsymbol{0}) = (\boldsymbol{0}, \boldsymbol{0})$$

and, therefore, $\boldsymbol{D}_1 = \boldsymbol{0}$. This gives

$$\boldsymbol{B}\mathbf{x} = \boldsymbol{BPP'}\mathbf{x} = \boldsymbol{D}\mathbf{y} = (\boldsymbol{0}, \boldsymbol{D}_2)\begin{pmatrix} \mathbf{y}_1 \\ \mathbf{y}_2 \end{pmatrix} = \boldsymbol{D}_2 \mathbf{y}_2$$

where $\mathbf{y} = \boldsymbol{P}'\mathbf{x}$ and $\mathbf{y}_2$ is as in Theorem A.46.

Since $\mathbf{y} \sim N(\boldsymbol{P}'\mu, \boldsymbol{I})$ all components of $\mathbf{y}$ are independent. Now, we may conclude that

$$\boldsymbol{B}\mathbf{x} = \boldsymbol{D}_2 \mathbf{y}_2 \quad \text{and} \quad \mathbf{x}'\boldsymbol{A}\mathbf{x} = \mathbf{y}_1'\mathbf{y}_1$$

are distributed independently.

Theorem A.48

If $\mathbf{x} \sim N_p(\boldsymbol{0}, \boldsymbol{I})$ and $\boldsymbol{A}$ and $\boldsymbol{B}$ are idempotent $(p \times p)$ matrices of rank r and s respectively, and if $\boldsymbol{BA} = \boldsymbol{0}$, then the quadratic forms $\mathbf{x}'\boldsymbol{A}\mathbf{x}$ and $\mathbf{x}'\boldsymbol{B}\mathbf{x}$ are independently distributed.

Proof Consider the matrix $\boldsymbol{P}$ as in Theorem A.46. If we set $\boldsymbol{C} = \boldsymbol{P}'\boldsymbol{BP}$($\boldsymbol{C}$ symmetric) we get with the assumption $\boldsymbol{BA} = \boldsymbol{0}$

$$\begin{aligned} \boldsymbol{CP'AP} &= \boldsymbol{P'BPP'AP} \\ &= \boldsymbol{P'BAP} = \boldsymbol{0}. \end{aligned}$$

This relation may be written as

$$\boldsymbol{CP'AP} = \begin{pmatrix} \boldsymbol{C}_1 & \boldsymbol{C}_2 \\ \boldsymbol{C}_2' & \boldsymbol{C}_3 \end{pmatrix}\begin{pmatrix} \boldsymbol{I} & \boldsymbol{0} \\ \boldsymbol{0} & \boldsymbol{0} \end{pmatrix} = \begin{pmatrix} \boldsymbol{C}_1 & \boldsymbol{0} \\ \boldsymbol{C}_2' & \boldsymbol{0} \end{pmatrix} = \begin{pmatrix} \boldsymbol{0} & \boldsymbol{0} \\ \boldsymbol{0} & \boldsymbol{0} \end{pmatrix}$$

where

$$\boldsymbol{C} = \begin{pmatrix} \boldsymbol{C}_1 & \boldsymbol{C}_2 \\ \boldsymbol{C}_2' & \boldsymbol{C}_3 \end{pmatrix}.$$

So $\boldsymbol{C}_1 = \boldsymbol{0}$ and $\boldsymbol{C}_2 = \boldsymbol{0}$, and

$$\begin{aligned} \mathbf{x}'\boldsymbol{B}\mathbf{x} &= \mathbf{x}'(\boldsymbol{PP}')\boldsymbol{B}(\boldsymbol{PP}')\mathbf{x} \\ &= \mathbf{x}'\boldsymbol{P}(\boldsymbol{P}'\boldsymbol{BP})\boldsymbol{P}'\mathbf{x} = \mathbf{x}'\boldsymbol{PCP}'\mathbf{x} \\ &= (\mathbf{y}_1', \mathbf{y}_2')\begin{pmatrix} \boldsymbol{0} & \boldsymbol{0} \\ \boldsymbol{0} & \boldsymbol{C}_3 \end{pmatrix}\begin{pmatrix} \mathbf{y}_1 \\ \mathbf{y}_2 \end{pmatrix} = \mathbf{y}_2'\boldsymbol{C}_3\mathbf{y}_2. \end{aligned}$$

As shown in Theorem A.46, $\mathbf{x}'\boldsymbol{A}\mathbf{x} = \mathbf{y}_1'\mathbf{y}_1$, so the quadratic forms $\mathbf{x}'\boldsymbol{A}\mathbf{x}$ and $\mathbf{x}'\boldsymbol{B}\mathbf{x}$ are independent.

A.9 DIFFERENTIATION OF SCALAR FUNCTIONS OF MATRICES (See also Appendix B)

Definition A.49. If $f(\boldsymbol{X})$ is a real scalar function of an $(m \times n)$ matrix $\boldsymbol{X} = (x_{ij})$, then the partial differential of f with respect to $\boldsymbol{X}$ is defined as the $(m \times n)$ matrix of partial differentials $\partial f / \partial x_{ij}$. That is,

$$\frac{\partial f}{\partial \boldsymbol{X}} = \begin{pmatrix} \dfrac{\partial f}{\partial x_{11}} & \cdots & \dfrac{\partial f}{\partial x_{1n}} \\ \vdots & & \vdots \\ \dfrac{\partial f}{\partial x_{m1}} & \cdots & \dfrac{\partial f}{\partial x_{mn}} \end{pmatrix}.$$

Theorem A.50

(i) $\dfrac{\partial}{\partial \mathbf{x}} \mathbf{x}' \boldsymbol{A} \mathbf{x} = (\boldsymbol{A} + \boldsymbol{A}')\mathbf{x}$.

(ii) If $\boldsymbol{A}$ is symmetric then

$$\frac{\partial}{\partial \mathbf{x}} \mathbf{x}' \boldsymbol{A} \mathbf{x} = 2\boldsymbol{A}\mathbf{x}.$$

Theorem A.51

$$\frac{\partial}{\partial \boldsymbol{C}} \mathbf{x}' \boldsymbol{C} \mathbf{y} = \mathbf{x}\mathbf{y}'.$$

Theorem A.52

(i) $\dfrac{\partial}{\partial \boldsymbol{C}} \mathbf{x}' \boldsymbol{C}' \boldsymbol{A} \boldsymbol{C} \mathbf{y} = \boldsymbol{A}\boldsymbol{C}\mathbf{y}\mathbf{x}' + \boldsymbol{A}'\boldsymbol{C}\mathbf{x}\mathbf{y}'$.

(ii) If $\boldsymbol{A}$ is symmetric and $\mathbf{x} = \mathbf{y}$ then

$$\frac{\partial}{\partial \boldsymbol{C}} \mathbf{x}' \boldsymbol{C}' \boldsymbol{A} \boldsymbol{C} \mathbf{x} = 2\boldsymbol{A}\boldsymbol{C}\mathbf{x}\mathbf{x}'.$$

Theorem A.53

If $\boldsymbol{A}$ and $\boldsymbol{B}$ are matrices which depend on a scalar x, then

$$\frac{\partial}{\partial x} \operatorname{tr} \boldsymbol{A}\boldsymbol{B} = \operatorname{tr}\left(\frac{\partial \boldsymbol{A}}{\partial x} \boldsymbol{B}\right) + \operatorname{tr}\left(\boldsymbol{A} \frac{\partial \boldsymbol{B}}{\partial x}\right).$$

Theorem A.54

If $y = \operatorname{tr} AC'WC$ then

(i) $\frac{\partial y}{\partial C} = WCA + W'CA'$.

(ii) If W and A are symmetric then

$$\frac{\partial y}{\partial C} = 2WCA.$$

A.10 MISCELLANEOUS RESULTS, STOCHASTIC CONVERGENCE

Theorem A.55 (Tschebyschev Inequality)

For any n-dimensional random vector X and a given scalar $\varepsilon > 0$ we have

$$P\{|X| \geq \varepsilon\} \leq \frac{E|X|^2}{\varepsilon^2}.$$

Proof Let $F(\mathbf{x})$ be the common distribution function of $X = (X_1, \ldots, X_n)$. Then

$$\begin{aligned} E|X|^2 &= \int |\mathbf{x}|^2 \, \mathrm{d}F(\mathbf{x}) \\ &= \int_{\{\mathbf{x}:|\mathbf{x}| \geq \varepsilon\}} |\mathbf{x}|^2 \, \mathrm{d}F(\mathbf{x}) + \int_{\{\mathbf{x}:|\mathbf{x}| < \varepsilon\}} |\mathbf{x}|^2 \, \mathrm{d}F(\mathbf{x}) \\ &\geq \varepsilon^2 \int_{\{\mathbf{x}:|\mathbf{x}| \geq \varepsilon\}} \mathrm{d}F(\mathbf{x}) = \varepsilon^2 P(|X| \geq \varepsilon). \end{aligned}$$

Definition A.56 If $\{\mathbf{x}(t)\}, t = 1, 2, \ldots$ is a multivariate stochastic process which satisfies

$$\lim_{t \to \infty} P\{|\mathbf{x}(t) - \tilde{\mathbf{x}}| \geq \delta\} = 0$$

where $\delta > 0$ is any given scalar and $\tilde{\mathbf{x}}$ is a finite vector, then $\tilde{\mathbf{x}}$ is called the probability limit of $\{\mathbf{x}(t)\}$ and we write

$$\operatorname{plim} \mathbf{x} = \tilde{\mathbf{x}}.$$

Theorem A.57 (Slutsky's Theorem)

Using Definition A.56 we have

(i) if $\operatorname{plim} \mathbf{x} = \tilde{\mathbf{x}}$ then $\lim_{t \to \infty} E\{\mathbf{x}(t)\} = \bar{E}(\mathbf{x}) = \tilde{\mathbf{x}}$;

(ii) if $\mathbf{c}$ is a vector of constants then $\operatorname{plim} \mathbf{c} = \mathbf{c}$;
(iii) (Slutsky's Theorem) if $\operatorname{plim} \mathbf{x} = \tilde{\mathbf{x}}$ and $\mathbf{y} = \mathbf{f}(\mathbf{x})$ is any continuous vector function of $\mathbf{x}$, then $\operatorname{plim} \mathbf{y} = \mathbf{f}(\tilde{\mathbf{x}})$;
(iv) if A and B are random matrices, then when the following limits exist:

$$\operatorname{plim}(AB) = (\operatorname{plim} A)(\operatorname{plim} B)$$

and

$$\operatorname{plim}(A^{-1}) = (\operatorname{plim} A)^{-1};$$

(v) if $\operatorname{plim}[\sqrt{T}(\mathbf{x}(t) - E\mathbf{x}(t))]'[\sqrt{T}(\mathbf{x}(t) - E\mathbf{x}(t))] = V$ then the asymptotic covariance matrix is

$$\bar{V}(\mathbf{x}, \mathbf{x}) = \bar{E}[\mathbf{x} - \bar{E}(\mathbf{x})]'[\mathbf{x} - \bar{E}(\mathbf{x})] = T^{-1} V.$$

Definition A.58 If $\{\mathbf{x}(t)\}, t = 1, 2, \ldots$ is a multivariate stochastic process which satisfies

$$\lim_{t \to \infty} E|\mathbf{x}(t) - \tilde{\mathbf{x}}|^2 = 0$$

then $\{\mathbf{x}(t)\}$ is called convergent in the quadratic mean, and we write

$$\text{l.i.m.}\, \mathbf{x} = \tilde{\mathbf{x}}.$$

Theorem A.59

If l.i.m. $\mathbf{x} = \tilde{\mathbf{x}}$, then $\operatorname{plim} \mathbf{x} = \tilde{\mathbf{x}}$.

Proof Using Theorem A.55 we get

$$0 \leq \lim_{t \to \infty} P(|\mathbf{x}(t) - \tilde{\mathbf{x}}| \geq \varepsilon) \leq \lim_{t \to \infty} \frac{E|\mathbf{x}(t) - \tilde{\mathbf{x}}|^2}{\varepsilon^2} = 0.$$

Theorem A.60

If l.i.m. $(\mathbf{x}(t) - E\mathbf{x}(t)) = \mathbf{0}$ and $\lim_{t \to \infty} E\mathbf{x}(t) = \mathbf{c}$, then $\operatorname{plim} \mathbf{x}(t) = \mathbf{c}$.

Proof $\lim_{t \to \infty} P(|\mathbf{x}(t) - \mathbf{c}| \geq \varepsilon) \leq \varepsilon^{-2} \lim_{t \to \infty} E|\mathbf{x}(t) - \mathbf{c}|^2$

$$= \varepsilon^{-2} \lim_{t \to \infty} E|\mathbf{x}(t) - E\mathbf{x}(t) + E\mathbf{x}(t) - \mathbf{c}|^2$$

$$= \varepsilon^{-2} \lim_{t \to \infty} E|\mathbf{x}(t) - E\mathbf{x}(t)|^2 + \varepsilon^{-2} \lim_{t} |E\mathbf{x}(t) - \mathbf{c}|^2$$

$$+ 2\varepsilon^{-2} \lim_{t \to \infty} E\{(E\mathbf{x}(t) - \mathbf{c})'(\mathbf{x}(t) - E\mathbf{x}(t))\}$$

$$= 0.$$

Theorem A.61

l.i.m. $\mathbf{x} = \mathbf{c}$ iff

l.i.m. $(\mathbf{x}(t) - E\mathbf{x}(t)) = \mathbf{0}$ and $\lim_{t \to \infty} E\mathbf{x}(t) = \mathbf{c}$.

Proof As in Theorem A.60, we may write

$$\begin{aligned}\lim_{t \to \infty} E|\mathbf{x}(t) - \mathbf{c}|^2 &= \lim_{t \to \infty} E|\mathbf{x}(t) - E\mathbf{x}(t)|^2 \\ &\quad + \lim_{t \to \infty} |E\mathbf{x}(t) - \mathbf{c}|^2 \\ &\quad + 2 \lim_{t \to \infty} E(E\mathbf{x}(t) - \mathbf{c})'(\mathbf{x}(t) - E\mathbf{x}(t)) \\ &= 0.\end{aligned}$$

Theorem A.62

Using Theorem A.61, where $\mathbf{x}(t)$ shall be an estimator of a parameter vector $\boldsymbol{\theta}$, we have the result

$$\lim_{t \to \infty} E\mathbf{x}(t) = \boldsymbol{\theta} \quad \text{if} \quad \text{l.i.m.}\ (\mathbf{x}(t) - \boldsymbol{\theta}) = \mathbf{0}.$$

That is, $\mathbf{x}(t)$ is an asymptotically unbiased estimator for $\boldsymbol{\theta}$ if $\mathbf{x}(t)$ converges to $\boldsymbol{\theta}$ in the quadratic mean.

Appendix B* Matrix Differentiation

B.1 INTRODUCTION

In this appendix we develop a general theory, from which the results cited in Theorems A.50–A.54 follow as special cases. These theorems concern scalar functions such as

$$y = \operatorname{tr} \boldsymbol{AX} + \operatorname{tr} \boldsymbol{XBX}' \tag{B.1.1}$$

which have to be maximized or minimized with respect to the matrix $\boldsymbol{X}$. Under certain conditions such turning points may be found by differentiating the function y with respect to each of the elements of $\boldsymbol{X}$ in turn. If $\boldsymbol{X}$ has n rows and p columns then as in Definition A.49 we define the differential $\partial y/\partial \boldsymbol{X}$ to be the $(n \times p)$ matrix whose (i, j)th element is $\partial y/\partial x_{ij}$ $(i = 1, \ldots, n; j = 1, \ldots, p)$.

In many cases y may be expressed not as a function of $\boldsymbol{X}$ directly, but as a function of a matrix $\boldsymbol{Z}$ which itself depends upon $\boldsymbol{X}$. For instance, when $\boldsymbol{X}$ is square we may write (B.1.1) as

$$y = \operatorname{tr} \boldsymbol{Z}, \text{ where } \boldsymbol{Z} = \boldsymbol{AX} + \boldsymbol{XBX}'.$$

Now in general if s is any scalar and

$$y = f(\boldsymbol{Z}) \text{ then } \frac{\partial y}{\partial s} = \sum_{\alpha, \beta} \frac{\partial y}{\partial Z_{\alpha\beta}} \frac{\partial Z_{\alpha\beta}}{\partial s}.$$

This formula can also be written

$$\frac{\partial y}{\partial s} = \operatorname{tr} \frac{\partial y}{\partial \boldsymbol{Z}} \frac{\partial \boldsymbol{Z}}{\partial s}, \tag{B.1.2}$$

where $\partial y/\partial \boldsymbol{Z}$ is defined according to Definition A.49 and $\partial \boldsymbol{Z}/\partial s$ is the matrix whose (α, β)th element is $\partial Z_{\alpha\beta}/\partial s$. Note that $\partial \boldsymbol{Z}/\partial s$ is the differential of a matrix with respect to a scalar, in contrast to $\partial y/\partial \boldsymbol{Z}$ which is the differential of a scalar with respect to a matrix. Both concepts will be required in what follows, and they are brought together in the important result cited above as (B.1.2).

Equation (B.1.2) takes particularly simple forms if y is, say, the trace or

* This appendix is taken from Bibby and Toutenburg, 1978.

determinant of $\boldsymbol{Z}$. For it is easily shown (Fisk, 1967, pp. 147–154; Anderson, 1958, pp. 346–349) that

$$\frac{\partial}{\partial \boldsymbol{Z}} \operatorname{tr} \boldsymbol{Z} = \boldsymbol{I}, \tag{B.1.3}$$

$$\frac{\partial}{\partial \boldsymbol{Z}} |\boldsymbol{Z}| = |\boldsymbol{Z}|(\boldsymbol{Z}')^{-1}, \tag{B.1.4}$$

and

$$\frac{\partial}{\partial \boldsymbol{Z}} \log |\boldsymbol{Z}| = (\boldsymbol{Z}')^{-1}. \tag{B.1.5}$$

Hence as special cases of (B.1.2) we get

$$\frac{\partial}{\partial s} \operatorname{tr} \boldsymbol{Z} = \operatorname{tr} \frac{\partial \boldsymbol{Z}}{\partial s}, \tag{B.1.6}$$

$$\frac{\partial}{\partial s} |\boldsymbol{Z}| = |\boldsymbol{Z}| \operatorname{tr} \boldsymbol{Z}^{-1} \frac{\partial \boldsymbol{Z}}{\partial s}, \tag{B.1.7}$$

and

$$\frac{\partial}{\partial s} \log |\boldsymbol{Z}| = \operatorname{tr} \boldsymbol{Z}^{-1} \frac{\partial \boldsymbol{Z}}{\partial s}. \tag{B.1.8}$$

In addition, matrix differentials have the usual linearity properties of differential operators, as well as the following extensions of the well known scalar formulae for the differential of a product:

$$\frac{\partial}{\partial \boldsymbol{X}} ab = \frac{\partial a}{\partial \boldsymbol{X}} b + a \frac{\partial b}{\partial \boldsymbol{X}}, \tag{B.1.9}$$

and

$$\frac{\partial}{\partial s} \boldsymbol{AB} = \frac{\partial \boldsymbol{A}}{\partial s} \boldsymbol{B} + \boldsymbol{A} \frac{\partial \boldsymbol{B}}{\partial s}. \tag{B.1.10}$$

B.2 THE TRACE FUNCTION

We now return for the moment to (B.1.1), which may be written as $y = y_1 + y_2$, where

$$y_1 = \operatorname{tr} \boldsymbol{AX} \text{ and } y_2 = \operatorname{tr} \boldsymbol{XBX}'. \tag{B.2.1}$$

Let us first find the differentials $\partial y_1 / \partial \boldsymbol{X}$ and $\partial y_2 / \partial \boldsymbol{X}$. Clearly

$$y_1 = \sum_{\alpha, \beta} a_{\alpha\beta} x_{\alpha\beta}. \tag{B.2.2}$$

Therefore

$$\frac{\partial y_1}{\partial x_{ji}} = a_{ji}, \tag{B.2.3}$$

since all the terms in (B.2.2) have a zero differential, except the one where $\beta = i$ and $\alpha = j$. Now a_{ji} is the (i, j)th element of $\boldsymbol{A}'$. Hence, collecting together the results represented by (B.2.3), we deduce that

$$\frac{\partial y_1}{\partial \boldsymbol{X}} = \boldsymbol{A}'. \tag{B.2.4}$$

Having obtained this differential, we may now seek the differential of $y_2 = \operatorname{tr} \boldsymbol{XBX}'$. Since this question is somewhat trickier, it will pay to be more pedantic in seeking an answer. Note that

$$y_2 = \operatorname{tr} \boldsymbol{XBX}' = \sum_{\alpha,\beta,\gamma} x_{\alpha\beta} b_{\beta\gamma} x_{\alpha\gamma}.$$

In evaluating $\partial y_2/\partial x_{ij}$, all the terms in this summation may be ignored except when either $\alpha = i$ and $\beta = j$ or when $\alpha = i$ and $\gamma = j$. These requirements may be abbreviated in the following table:

α	β	γ
i	j	—
i	—	j

Hence the terms in y_2 which have nonzero differential with respect to x_{ij} may be written

$$\sum_{\gamma} x_{ij} b_{j\gamma} x_{i\gamma} + \sum_{\beta} x_{i\beta} b_{\beta j} x_{ij}.$$

The required differential is therefore

$$\frac{\partial y_2}{\partial x_{ij}} = \sum_{\gamma} b_{j\gamma} x_{i\gamma} + \sum_{\beta} x_{i\beta} b_{\beta j} = (\boldsymbol{XB}')_{ij} + (\boldsymbol{XB})_{ij}.$$

Hence we deduce that

$$\frac{\partial y_2}{\partial \boldsymbol{X}} = \boldsymbol{XB}' + \boldsymbol{XB}.$$

Taking this equation along with (B.2.4) it is clear that the differential of y given by (B.1.1) is

$$\frac{\partial y}{\partial \boldsymbol{X}} = \frac{\partial y_1}{\partial \boldsymbol{X}} + \frac{\partial y_2}{\partial \boldsymbol{X}} = \boldsymbol{A}' + \boldsymbol{XB}' + \boldsymbol{XB}. \tag{B.2.5}$$

Turning points in the value of y are given by equating this differential to zero. The general result is given by

$$X = -A'(B+B')^{-1}.$$

When B is symmetric this equals $-\frac{1}{2}A'B^{-1}$, which may be compared with the value $x = -\frac{1}{2}a/b$, which minimizes the scalar analogue of (B.1.1), namely $ax+bx^2$.

An alternative way of deriving (B.2.5) is to use (B.1.2) where

$$y = \mathrm{tr}\, Z \quad \text{and} \quad Z = AX + XBX'.$$

Now

$$\frac{\partial X}{\partial x_{ij}} = E_{ij} = \mathbf{e}_i\mathbf{e}_j' \quad \text{and} \quad \frac{\partial X'}{\partial x_{ij}} = E_{ji} = \mathbf{e}_j\mathbf{e}_i', \tag{B.2.6}$$

where E_{ij} is the matrix with a one in the (i, j)th position and zeros elsewhere, and the vectors $\mathbf{e}_i$ and $\mathbf{e}_j$ are defined similarly. Using (B.2.6) and (B.1.10) we note that

$$\frac{\partial Z}{\partial x_{ij}} = AE_{ij} + E_{ij}BX' + XBE_{ji}.$$

Now, using (B.1.6),

$$\frac{\partial y}{\partial x_{ij}} = \mathrm{tr}\frac{\partial Z}{\partial x_{ij}} = \mathrm{tr}(A\mathbf{e}_i\mathbf{e}_j' + \mathbf{e}_i\mathbf{e}_j'BX' + XB\mathbf{e}_j\mathbf{e}_i').$$

Using the commutativity property of the trace operator this equals

$$\mathrm{tr}(\mathbf{e}_j'A\mathbf{e}_i + \mathbf{e}_j'BX'\mathbf{e}_i + \mathbf{e}_i'XB\mathbf{e}_j) = (A)_{ji} + (BX')_{ji} + (XB)_{ij}.$$

This equals the (i, j)th element of $(A' + XB' + XB)$, thus confirming (B.2.5).

This latter method of deriving matrix differentials in fact has more general application. For instance, consider

$$y = \mathrm{tr}\, Z, \quad \text{where } Z = AXBXCX'. \tag{B.2.7}$$

Now, using (B.2.6),

$$\frac{\partial Z}{\partial x_{ij}} = AE_{ij}BXCX' + AXBE_{ij}CX' + AXBXCE_{ji}. \tag{B.2.8}$$

Therefore from (B.1.6)

$$\frac{\partial y}{\partial x_{ij}} = \mathrm{tr}\frac{\partial Z}{\partial x_{ij}}.$$

But taking the final term from (B.2.8) and using the commutativity of the trace operator we know for instance that,

$$\mathrm{tr}\, AXBXCE_{ji} = \mathrm{tr}\, AXBXC\mathbf{e}_j\mathbf{e}_i' = \mathbf{e}_i'AXBXC\mathbf{e}_j = (AXBXC)_{ij}.$$

Developing the other terms in (B.2.8) similarly we see that

$$\frac{\partial y}{\partial x_{ij}} = \operatorname{tr}\frac{\partial \boldsymbol{Z}}{\partial x_{ij}} = (\boldsymbol{BXCX'A})_{ji} + (\boldsymbol{CX'AXB})_{ji} + (\boldsymbol{AXBXC})_{ij}.$$

Therefore

$$\frac{\partial y}{\partial \boldsymbol{X}} = \boldsymbol{A'XC'X'B'} + \boldsymbol{B'X'A'XC'} + \boldsymbol{AXBXC}.$$

This result may be confirmed by noting the scalar-valued special case, that when $y = abcx^3$ then $\partial y/\partial x = 3abcx^2$.

Further results which may be obtained in a manner similar to the above are given in Figure B.2.1. Note that Theorem A.50 is a special case of result (b), obtained when $\boldsymbol{A}$ is symmetric and $\boldsymbol{X}$ is a vector. Theorem A.51 corresponds to (a), and Theorems A.52 and A.54 to (f).

	y	$\partial y/\partial \boldsymbol{X}$
(a)	$\operatorname{tr}\boldsymbol{AX}$	$\boldsymbol{A'}$
(b)	$\operatorname{tr}\boldsymbol{X'AX}$	$(\boldsymbol{A}+\boldsymbol{A'})\boldsymbol{X}$
(c)	$\operatorname{tr}\boldsymbol{XAX}$	$\boldsymbol{X'A'}+\boldsymbol{A'X'}$
(d)	$\operatorname{tr}\boldsymbol{XAX'}$	$\boldsymbol{X}(\boldsymbol{A}+\boldsymbol{A'})$
(e)	$\operatorname{tr}\boldsymbol{X'AX'}$	$\boldsymbol{AX'}+\boldsymbol{X'A}$
(f)	$\operatorname{tr}\boldsymbol{X'AXB}$	$\boldsymbol{AXB}+\boldsymbol{A'XB'}$

Figure B.2.1 Matrix differentials obtained as described in Section B.2

B.3 DIFFERENTIATING INVERSE MATRICES

The methods outlined in Section B.2 are insufficient to allow us to differentiate, for instance,

$$y = \operatorname{tr}\boldsymbol{AX}^{-1}. \tag{B.3.1}$$

To do this we need the formula (B.1.10) for differentiating a product. Clearly

$$\frac{\partial}{\partial s}\boldsymbol{T}^{-1}\boldsymbol{T} = \frac{\partial \boldsymbol{T}^{-1}}{\partial s}\boldsymbol{T} + \boldsymbol{T}^{-1}\frac{\partial \boldsymbol{T}}{\partial s},$$

where s is any scalar. In other words,

$$\frac{\partial \boldsymbol{T}^{-1}}{\partial s} = -\boldsymbol{T}^{-1}\frac{\partial \boldsymbol{T}}{\partial s}\boldsymbol{T}^{-1}, \tag{B.3.2}$$

because $\boldsymbol{T}^{-1}\boldsymbol{T}$ is constant, and therefore has zero differential.

This result may be extended by noting that

$$\frac{\partial}{\partial s}\boldsymbol{RT}^{-1} = \frac{\partial \boldsymbol{R}}{\partial s}\boldsymbol{T}^{-1} - \boldsymbol{RT}^{-1}\frac{\partial \boldsymbol{T}}{\partial s}\boldsymbol{T}^{-1}, \tag{B.3.3}$$

a formula which generalizes the well-known scalar result

$$\frac{\partial}{\partial x}\left(\frac{u}{v}\right) = \frac{v\partial u/\partial x - u\partial v/\partial x}{v^2}.$$

As a particular case of (B.3.2) note that

$$\frac{\partial \boldsymbol{X}^{-1}}{\partial x_{ij}} = -\boldsymbol{X}^{-1}\frac{\partial \boldsymbol{X}}{\partial x_{ij}}\boldsymbol{X}^{-1} = -\boldsymbol{X}^{-1}\boldsymbol{E}_{ij}\boldsymbol{X}^{-1}. \tag{B.3.4}$$

Therefore, differentiating (B.3.1),

$$\frac{\partial}{\partial x_{ij}}\operatorname{tr}\boldsymbol{A}\boldsymbol{X}^{-1} = \operatorname{tr}\boldsymbol{A}\frac{\partial \boldsymbol{X}^{-1}}{\partial x_{ij}} = -\operatorname{tr}\boldsymbol{A}\boldsymbol{X}^{-1}\boldsymbol{E}_{ij}\boldsymbol{X}^{-1} = -(\boldsymbol{X}^{-1}\boldsymbol{A}\boldsymbol{X}^{-1})_{ji}.$$

Hence we have the result that

$$\frac{\partial}{\partial \boldsymbol{X}}\operatorname{tr}\boldsymbol{A}\boldsymbol{X}^{-1} = -(\boldsymbol{X}^{-1}\boldsymbol{A}\boldsymbol{X}^{-1})'.$$

Other results which may be derived in a similar manner are given in Figure B.3.1.

y	$\partial y/\partial \boldsymbol{X}$
$\operatorname{tr}\boldsymbol{A}\boldsymbol{X}^{-1}$	$-(\boldsymbol{X}^{-1}\boldsymbol{A}\boldsymbol{X}^{-1})'$
$\operatorname{tr}\boldsymbol{X}^{-1}\boldsymbol{A}\boldsymbol{X}^{-1}\boldsymbol{B}$	$-(\boldsymbol{X}^{-1}\boldsymbol{A}\boldsymbol{X}^{-1}\boldsymbol{B}\boldsymbol{X}^{-1}+\boldsymbol{X}^{-1}\boldsymbol{B}\boldsymbol{X}^{-1}\boldsymbol{A}\boldsymbol{X}^{-1})'$

Figure B.3.1 Matrix differentials involving the inverse matrix

B.4 THE DETERMINANT

Suppose that we seek

$$\frac{\partial}{\partial \boldsymbol{X}}\log|\boldsymbol{X}'\boldsymbol{A}\boldsymbol{X}|.$$

Writing $\boldsymbol{Z} = \boldsymbol{X}'\boldsymbol{A}\boldsymbol{X}$ we know that

$$\frac{\partial \boldsymbol{Z}}{\partial x_{ij}} = \boldsymbol{E}_{ji}\boldsymbol{A}\boldsymbol{X} + \boldsymbol{X}'\boldsymbol{A}\boldsymbol{E}_{ij}.$$

Therefore, substituting in (B.1.8),

$$\begin{aligned}\frac{\partial}{\partial x_{ij}}\log|\boldsymbol{X}'\boldsymbol{A}\boldsymbol{X}| &= \operatorname{tr}(\boldsymbol{Z}')^{-1}[\boldsymbol{E}_{ji}\boldsymbol{A}\boldsymbol{X} + \boldsymbol{X}'\boldsymbol{A}\boldsymbol{E}_{ij}] \\ &= [\boldsymbol{A}\boldsymbol{X}(\boldsymbol{Z}')^{-1} - \boldsymbol{A}'\boldsymbol{X}\boldsymbol{Z}^{-1}]_{ij},\end{aligned}$$

where $\boldsymbol{Z} = \boldsymbol{X}'\boldsymbol{A}\boldsymbol{X}$. In particular, when $\boldsymbol{A}$ is symmetric, this simplifies to

$$2\boldsymbol{A}\boldsymbol{X}(\boldsymbol{X}'\boldsymbol{A}\boldsymbol{X})^{-1}.$$

Further results involving the determinant function are given in Figure B.4.1.

y	$\partial y/\partial \mathbf{X}$
$\lvert\mathbf{X}\rvert$	$\lvert\mathbf{X}\rvert(\mathbf{X}')^{-1}$
$\log\lvert\mathbf{X}\rvert$	$(\mathbf{X}')^{-1}$
$\log\lvert\mathbf{X}'\mathbf{A}\mathbf{X}\rvert$	$\mathbf{A}\mathbf{X}(\mathbf{X}'\mathbf{A}'\mathbf{X})^{-1}+\mathbf{A}'\mathbf{X}(\mathbf{X}'\mathbf{A}\mathbf{X})^{-1}$
$\log\lvert\mathbf{X}'\mathbf{X}\rvert$	$2\mathbf{X}(\mathbf{X}'\mathbf{X})^{-1}$
$\log\lvert\mathbf{Z}\rvert$	$\mathbf{A}\mathbf{X}\mathbf{Z}^{-1}+\mathbf{A}'\mathbf{X}(\mathbf{Z}^{-1})'+(\mathbf{Z}^{-1})\mathbf{B}'+\mathbf{Z}^{-1}\mathbf{C}$
where $\mathbf{Z}=\mathbf{X}'\mathbf{A}\mathbf{X}+\mathbf{X}\mathbf{B}+\mathbf{C}\mathbf{X}'+\mathbf{D}$	

Figure B.4.1 Matrix differentials involving the determinant function

Glossary

This glossary presents a description of the main notational conventions and abbreviations adopted throughout the book.

NOTATIONAL CONVENTIONS

$\mathbf{X} \sim (\boldsymbol{\mu}, \boldsymbol{\Sigma})$	The random vector $\mathbf{X}$ has mean vector $\boldsymbol{\mu}$ and dispersion (variance–covariance) matrix $\boldsymbol{\Sigma}$
$\mathbf{X} \sim N_p(\boldsymbol{\mu}, \boldsymbol{\Sigma})$	The random vector $\mathbf{X}$ has the p-variable normal distribution with mean $\boldsymbol{\mu}$ and dispersion matrix $\boldsymbol{\Sigma}$
χ_m^2	The central chi squared distribution with m degrees of freedom
$\chi_m^2(\lambda)$	The noncentral χ_m^2-distribution with noncentrality parameter λ
$F_{i,j}$	The central F distribution with i and j degrees of freedom
$F_{i,j}(\lambda)$	The noncentral F_{ij}-distribution with noncentrality parameter λ
$\boldsymbol{A} \geq \boldsymbol{0}$	The matrix $\boldsymbol{A}$ is nonnegative definite
$\boldsymbol{A} > \boldsymbol{0}$	The matrix $\boldsymbol{A}$ is positive definite

ABBREVIATIONS

LSE	Least squares estimator
OLSE	Ordinary least squares estimator
GLSE	Generalized least squares estimator
MILE	Minimax–linear estimator
RLSE	Restricted least squares estimator
MLE	Maximum likelihood estimator
MSE	Mean square error
SMSE	Scalar mean square error
MSEP	Mean square error of prediction
BLUE	Best linear unbiased estimator
IVE	Instrumental variable estimator
ILSE	Indirect least squares estimator
2SLSE	Two-stage least squares estimator

MATRICES AND VECTORS

$\boldsymbol{S}$	$\boldsymbol{S} = \boldsymbol{X}'\boldsymbol{W}^{-1}\boldsymbol{X}$
U	$U = (X'X)^{-1}$
$\mathbf{i}$	$\mathbf{i}' = (1, \ldots, 1)$ is the vector which introduces a constant into regression
E^K	Euclidean space of dimension K
$\Delta(\hat{\boldsymbol{\beta}}_1, \hat{\boldsymbol{\beta}}_2)$	Difference of the mean square errors of two estimators $\hat{\boldsymbol{\beta}}_1, \hat{\boldsymbol{\beta}}_2$
$\Delta^k(\hat{\boldsymbol{\beta}}_1, \hat{\boldsymbol{\beta}}_2)$	Difference of the minimax risks of estimators $\hat{\boldsymbol{\beta}}_1$ and $\hat{\boldsymbol{\beta}}_2$, defined in (4.2.25) as $$\Delta^k(\hat{\boldsymbol{\beta}}_1, \hat{\boldsymbol{\beta}}_2) = \sup_{\boldsymbol{\beta}'T\boldsymbol{\beta} \leq k} R(\hat{\boldsymbol{\beta}}_1, \mathbf{aa}') - \sup_{\boldsymbol{\beta}'T\boldsymbol{\beta} \leq k} R(\hat{\boldsymbol{\beta}}_2, \mathbf{aa}')$$
$\boldsymbol{E}_{ij}$	The matrix with a one in the (i, j)th position and zeros elsewhere
$\mathbf{e}_i$	The vector with a one as ith component and zeros elsewhere
$R(\hat{\boldsymbol{\beta}}, \boldsymbol{A})$ or $R(\hat{\boldsymbol{\beta}})$	Quadratic risk of an estimator $\hat{\boldsymbol{\beta}}$
$\boldsymbol{R}, \mathbf{r}$	Linear restriction $\mathbf{r} = \boldsymbol{R\beta}$
$\lambda_{\max}(\boldsymbol{A})$	Maximal eigenvalue of the matrix $\boldsymbol{A}$
s^2	Estimator of σ^2, defined in (2.1.51) and (2.2.9), respectively
$\boldsymbol{D}$	$\boldsymbol{D} = k^{-1}\sigma^2\boldsymbol{T} + \boldsymbol{S}$ (defined in (4.2.6))
$\boldsymbol{D}$	$\boldsymbol{D} = (\boldsymbol{I} - \boldsymbol{B})^{-1}$, the matrix which transforms the simultaneous model into its reduced form
$\boldsymbol{Z}_k$	Defined in (3.3.10)
$\boldsymbol{\Pi}$	Parameter matrix of the reduced form (see (6.2.18))
$V(\boldsymbol{\beta})$	Dispersion matrix of an estimator $\hat{\boldsymbol{\beta}}$

ESTIMATORS OF $\boldsymbol{\beta}$

$\mathbf{b}_0$	Ordinary least squares estimator, defined in (2.1.15) as $\mathbf{b}_0 = (\boldsymbol{X}'\boldsymbol{X})^{-1}\boldsymbol{X}'\mathbf{y}$
$\mathbf{b}$	Generalized least squares estimator, defined in (2.2.4) as $\mathbf{b} = (\boldsymbol{X}'\boldsymbol{W}^{-1}\boldsymbol{X})^{-1}\boldsymbol{X}'\boldsymbol{W}^{-1}\mathbf{y} = \boldsymbol{S}^{-1}\boldsymbol{X}'\boldsymbol{W}^{-1}\mathbf{y}$
$\hat{\boldsymbol{\beta}}_1$	Heterogeneous R-optimal estimator, defined in (2.4.5) as $\hat{\boldsymbol{\beta}}_1 = \boldsymbol{\beta}$ (also called trivial estimator)
$\hat{\boldsymbol{\beta}}_2$	Homogeneous biased R-optimal estimator, defined in (2.4.6) as $\hat{\boldsymbol{\beta}}_2 = \boldsymbol{\beta\beta}'\boldsymbol{X}'(\boldsymbol{X\beta\beta}'\boldsymbol{X}' + \sigma^2\boldsymbol{W})^{-1}\mathbf{y}$
$\mathbf{b}(k)$	Ridge estimator, defined in (2.3.18) as $\mathbf{b}(k) = (\boldsymbol{X}'\boldsymbol{X} + k\boldsymbol{I})^{-1}\boldsymbol{X}'\mathbf{y}$
$\hat{\boldsymbol{\beta}}(\rho)$	Shrunken estimator, defined in (2.3.40) as $\hat{\boldsymbol{\beta}}(\rho) = (\rho + 1)^{-1}\mathbf{b}$, $\rho \geq 0$
$\tilde{\boldsymbol{\beta}}_*$	General shrunken estimator defined in (2.3.46)
$\mathbf{b}_{\boldsymbol{R}}$	Restricted least squares estimator under constraint $\mathbf{0} = \boldsymbol{R\beta}$, defined in (2.3.15) as $\mathbf{b}_{\boldsymbol{R}} = (\boldsymbol{S} + \boldsymbol{R}'\boldsymbol{R})^{-1}\boldsymbol{X}'\boldsymbol{W}^{-1}\mathbf{y}$

$\tilde{\mathbf{b}}_R$	Conditional restricted least squares estimator (Conditional RLSE), defined in (2.3.71) as $\tilde{\mathbf{b}}_R = \mathbf{b} + S^{-1}R'[RS^{-1}R']^{-1}(\mathbf{r} - R\mathbf{b})$
$\mathbf{b}_R(V)$	Generalized restricted least squares estimator, defined in (3.5.6) as $\mathbf{b}_R(V) = \mathbf{b} + S^{-1}R'[\sigma^{-2}V + RS^{-1}R']^{-1}(\mathbf{r} - R\mathbf{b})$
$\mathbf{b}_R(\mathbf{r})$	General conditional restricted LSE under constraint $\mathbf{r} = R\boldsymbol{\beta}$, defined in (2.3.5) as $\mathbf{b}_R(\mathbf{r}) = (S + R'R)^{-1}(X'W^{-1}\mathbf{y} + R'\mathbf{r})$
$\mathbf{b}^* = \begin{pmatrix} \mathbf{b}_1^* \\ \mathbf{b}_2^* \end{pmatrix}$	Piecewise estimator, defined in (3.5.22)
$\mathbf{b}^*$	Minimax linear estimator under constraint $\boldsymbol{\beta}'T\boldsymbol{\beta} \le k$, defined in (4.2.7) as $\mathbf{b}^* = (k^{-1}\sigma^2 T + S)^{-1}X'W^{-1}\mathbf{y} = D^{-1}X'W^{-1}\mathbf{y}$
$\mathbf{b}^*(\boldsymbol{\beta}_0)$	Minimax linear estimator under constraint $(\boldsymbol{\beta} - \boldsymbol{\beta}_0)'T(\boldsymbol{\beta} - \boldsymbol{\beta}_0) \le k$, defined in (4.2.11) as $\mathbf{b}^*(\boldsymbol{\beta}_0) = \boldsymbol{\beta}_0 + D^{-1}X'W^{-1}(\mathbf{y} - X\boldsymbol{\beta}_0)$
$\mathbf{b}_c^*$	Two-stage minimax linear estimator (σ^2 in $\mathbf{b}^*$ replaced by the constant c), defined in (4.3.29) as $\mathbf{b}_c^* = (k^{-1}cT + S)^{-1}X'W^{-1}\mathbf{y}$
$\mathbf{b}_R^*$	So-to-speak restricted minimax linear estimator, defined in (4.5.48)
$\hat{\boldsymbol{\beta}}_r, \hat{\boldsymbol{\beta}}_{r_1}$	Estimator of $\boldsymbol{\beta}$ in multivariate regression with r and r_1 replications of the input matrix, respectively (see definitions (5.3.2) and (5.3.4))
$\hat{\boldsymbol{\beta}}(B^*)$	Two-stage estimator of $\boldsymbol{\beta}$, defined in (5.3.14)
$\hat{\boldsymbol{\delta}}_1$	Two-stage LSE of parameters in an overidentified structural equation (see (6.3.9))
$\mathbf{d}_1$	Restricted 2 SLSE, defined in (6.3.25)

Bibliography

Aitchison, J., and Dunsmore, I. R. (1968). Linear-loss interval estimation of location and scale parameters. *Biometrika*, **55**, 141–148.

Aitken, A. C. (1935). On least squares and linear combinations of observations. *Proc. R. Soc. Edinburgh*, **55**, 42–48.

Almon, S. (1965). The distributed lag between capital approximations and expenditures. *Econometrica*, **33**, 178–196.

Anderson, T. W. (1958). *An Introduction to Multivariate Statistical Analysis*. Wiley, London.

Baldwin, F., and Hoerl, A. E. (1978). Bounds on minimum mean squared error in ridge regression. *Commun. Statist.—Theor. Meth.*, A7(13), 1209–1218.

Barten, A. P. (1962). Note on unbiased estimation of the squared multiple correlation coefficient. *Statist. Neerlandica*, **16**, 151–163.

Basmann, R. L. (1957). A generalized classical method of linear estimation of coefficients in a structural equation. *Econometrica*, **25**, 77–83.

Bibby, J. (1972). Minimum mean square error estimation, ridge regression, and some unanswered questions. *Proc. 9th European Meeting of Statisticians, Budapest*. Reprinted in J. Gani, K. Sarkadi, and I. Vincze (Eds) (1974). *Progress in Statistics, Colloquia Mathematica*, Janos Bolyai, p. 107–121.

Bibby, J., and Toutenburg, H. (1978). *Prediction and Improved Estimation in Linear Models*, Wiley, Chichester.

Brook, R. J. (1976). On the use of a regret function to set significance points in prior tests of estimation. *J. Amer. Statist. Assoc.*, **71**, 126–131.

Bunke, O. (1975). Minimax linear, ridge and shrunken estimators for linear parameters. *Math. Operationsforschung Statistik*, **6**, 697–701.

Bunke, O. (1977). On optimal prediction. *Math. Operationsforschung Statistik*, *Series Statistics*, **8**, 453–455.

Chipman, J. S., and Rao, M. M. (1964). The treatment of linear restrictions in regression analysis. *Econometrica*, **32**, 198–209.

Cohen, A. (1966). All admissible estimates of the mean vector. *Ann. Math. Statist.*, **2**, 458–463.

Dhrymes, P. J. (1974). *Econometrics*, Springer-Verlag, New York.

Duncan, O. D. (1975). *Introduction to structural models*, Academic Press, New York.

Durbin, J., and Watson, G. S. (1950). Testing for serial correlation in least squares regression I. *Biometrika*, **37**, 409–428.

Durbin, J., and Watson, G. S. (1951). Testing for serial correlation in least squares regression II. *Biometrika*, **38**, 159–178.

Durbin, J. (1953). A note on regression when there is extraneous information about one of the coefficients. *J. Amer. Statist. Assoc.*, **48**, 799–808.

Farebrother, R. W. (1975). The minimum mean square error linear estimator and ridge regression. *Technometrics*, **17**, 127–128.

Farebrother, R. W. (1976). Further results on the mean square error of ridge regression. *J. R. Statist. Soc.*, B, **38**, 248–250.

Farebrother, R. W. (1978a). Estimating regression coefficients under conditional

specification: comment. *Commun. Statist.*, A7(2), 193–196.
Farebrother, R. W. (1978b). A class of shrinkage estimators. *J. R. Statist. Soc.*, B, **40**, 47–49.
Farebrother, R. W. (1978c). Partitioned ridge regression. *Technometrics*, **20**, 121–122.
Ferguson, T. S. (1967). *Mathematical statistics*, Academic Press, New York.
Fisher, F. M. (1976). *The identification problem in econometrics*, Krieger, New York.
Fisk, P. R. (1967). *Stochastically Dependent Equations: An Introductory Text for Econometricians.* Griffin Monographs, London.
Fukuhara, F. (1976). A method of restricted two-stage least squares and the treatment of qualitative factors. *Unpublished paper*, Aoyama Gakuin University, Tokyo.
Gilchrist, W. (1976). *Statistical forecasting*, Wiley, London.
Goldberger, A. S. (1962). Best linear unbiased prediction in the generalized linear regression model. *J. Amer. Statist. Assoc.*, **57**, 369–375.
Goldberger, A. S. (1964). *Econometric Theory*, Wiley, London.
Goldstein, M., and Smith, F. N. (1974). Ridge-type estimators for regression analysis. *J. R. Statist. Soc.*, **36**, 284–291.
Goodnight, J. H., and Wallace, T. D. (1972). Operational techniques and tables for making weak MSE tests for restrictions in regressions. *Econometrika*, **40**, 699–709.
Graybill, F. A. (1961). *An Introduction to Linear Statistical Models*, McGraw-Hill, New York.
Gunst, R. F., and Mason, R. L. (1977). Biased estimation in regression: An evaluation using mean squared error. *J. Amer. Statist. Assoc.*, **72**, 616–628.
Guttman, I. (1970). *Statistical tolerance regions*, Griffin Monographs, London.
Hartung, J. (1978). Zur Verwendung von Vorinformation in der Regressionsanalyse. (Using prior information in regression.) *Research Report*, Department of Applied Statistics, University of Bonn.
Hemmerle, W. J., and Brantle, T. F. (1978). Explicit and constrained generalized ridge estimation. *Technometrics*, **20**, 109–120.
Hocking, R. R. (1976). The analysis and selection of variables in linear regression. *Biometrika*, **32**, 1–49.
Hocking, R. R., Speed, F. M., and Lynn, M. J. (1976). A class of biased estimators in linear regression. *Technometrics*, **18**, 425–437.
Hoerl, A. E., and Kennard, R. W. (1970a). Ridge regression: Biased estimation for nonorthogonal problems. *Technometrics*, **12**, 55–67.
Hoerl, A. S., and Kennard, R. W. (1970b). Ridge Regression: Applications to nonorthogonal problems. *Technometrics*, **12**, 69–82.
Hoffmann, K. (1977). Admissibility of linear estimators with respect to restricted parameter sets. *Math. Operationsforschung Statistik*, **8**, 425–438.
Huang, D. S. (1970). *Regression and econometric methods*, Wiley, London.
Humak, K. M. S. (1977). *Statistische Modellbildung*, Bd. I, Akademie-Verlag, Berlin.
Johnston, J. (1972) *Econometric Methods*, 2nd Ed. McGraw-Hill, New York.
Judge, G. G., and Bock, M. E. (1978). *The statistical implications of pre-test and Stein-rule estimators in econometrics*, North-Holland, Amsterdam.
Kakwani, N. C., and Court, R. (1970). Simultaneous-equation estimation under linear restrictions with application to New Zealand meat demand. *Rev. Int. Statist. Inst.*, **38**, 244–257.
Kelly, J. S. (1975). Linear cross-equation constraints and the identification problem. *Econometrica*, **43**, 125–140.
Kendall, M. G., and Babington-Smith, B. (1951). *Tables of random sampling numbers. Tracts for computers*, XXIV, Cambridge University Press.
Koopmans, C. (1953). Identification problems in economic model constructions. *Cowles Commission Monograph*, **14**, 27–48.
Kruskal, W. (1968). When are Gauss–Markov and least squares estimators identical? A coordinate-free approach. *Ann. Math. Statist.*, **39**, 70–75.

Kuks, J. (1972). Minimaksnaja ozenka koeffizientow regressii. (A minimax estimator of regression coefficients.) *Iswestija Akademija Nauk Estonskoj SSR*, **21**, 73–78.

Kuks, J., and Olman, W. (1971). Minimaksnaja linejnaja ozenka koefficientow regressii. (Minimax linear estimation of regression coefficients.) *Iswestija Akademija Nauk Estonskoj SSR*, **20**, 480–482.

Kuks, J., and Olman, W. (1972). Minimaksnaja linejnaja ozenka koeffizientow regressii, II. (Minimax linear estimation of regression coefficients, II.) *Iswestija Akademija Nauk Estonskoj SSR*, **21**, 66–72.

Läuter, H. (1975). A minimax linear estimator for linear parameters under restrictions in form of inequalities. *Math. Operationsforschung Statistik*, **6**, 689–696.

Leamer, E. E. (1979). *Specification searches*, Wiley, New York.

Leser, C. E. V. (1974). *Econometric techniques and problems*, 2nd Ed. Griffin, London.

Lieberman, G. L., and Miller, R. G. (1963). Simultaneous tolerance intervals in regression. *Biometrika*, **50**, 155–168.

Liski, E. (1979). On reduced risk estimation in linear models. *Tampere, Acta Universitatis Tamperensis*, A, **105**.

Mantel, N. (1969). Restricted least squares regression and convex quadratic programming. *Technometrics*, **11**, 763–773.

Mardia, K. V., Kent, J. T., and Bibby, J. M. (1979). *Multivariate Analysis*. Academic Press, London.

Marquardt, W., and Snee, D. (1975). Ridge regression in practice. *Amer. Statist.*, **29**, 3–20.

Mayer, L. S., Singh, J., and Willke, T. A. (1974). Utilizing initial estimates in estimating the coefficients in a linear model. *J. Amer. Statist. Assoc.*, **69**, 219–222.

Mayer, L. S., and Willke, T. A. (1973). On biased estimation in linear models. *Technometrics*, **15**, 497–508.

McElroy, F. W. (1967). A necessary and sufficient condition that ordinary least-squares estimators be best linear unbiased. *J. Amer. Statist. Assoc.*, **62**, 1302–1304.

Menges, G. (1959). Zur statistischen Grundlegung der Ökonometrie. (On the statistical foundation of econometrics.) *Zeitschrift für die gesamte Staatswissenschaft*, **115**, 611–625.

Menges, G. (1971). Some decision- and information-theoretical considerations about the econometric problem of specification and identification. *Statistische Hefte*, **12**, 23.

Möller, H. D. (1976). *Probleme der statistischen Spezifikation der Struktur von Einzelgleichungsmodellen* (*Specifying the structure of single—equation models*), Vandenhoeck and Ruprecht, Göttingen.

Nagar, A. L., and Kakwani, N. C. (1964). The bias and moment matrix of a mixed regression estimator. *Econometrica*, **32**, 174–182.

Pollock, D. S. G. (1979). *The algebra of econometrics*, Wiley, Chichester.

Ramsey, J. B. (1969). Tests for specification errors in classical linear least-squares regression analysis. *J. R. Statist. Soc.*, B, **31**, 350–371.

Rao, C. R. (1971). Unified theory of linear estimation. *Sankhya*, A **33**, 370–396; and *Sankhya*, A **34**, 477.

Rao, C. R. (1973). *Linear Statistical Inference and Its Applications*, 2nd Ed. Wiley, London.

Rao, C. R. (1976). The 1975 Wald memorial lectures. Estimation of parameters in a linear model. *Ann. Statist.*, **4**, 1023–1037.

Roeder, B. (1978). Minimax-Schätzungen im linearen Modell. (Minimax estimation in the linear model.) *Unpublished research paper*, Department of Mathematics, Humboldt University, Berlin.

Rothenberg, T. J. (1973). Efficient estimation with a priori information. *Cowles foundation monograph*, Vol. 23, New Haven and London.

Schiele, K. (1980). Ein Verfahren zur Konstruktion modelladäquater Schätzmethoden für ökonometrische Modelle der sozialistischen Volkswirtschaft. (On model-adequate estimators of econometric macro-models of socialist countries.) *Ph.D Thesis*, Academy of Sciences of the GDR, Berlin.

Schneeweiss, H. (1971). *Ökonometrie* (*Econometrics*), Physica-Verlag, Würzburg-Wien.

Schönfeld, P. (1969). *Methoden der Ökonometrie (Econometric Methods)*, Verlag Franz Vahlen, Berlin.

Schönfeld, P. (1973) Descriptive regression. *Mimeopaper*, Institute of Econometrics and Operations Research, University of Bonn.

Seber, G. A. F. (1966). *The linear hypothesis*, Griffin, London.

Shiller, R. J. (1973). A distributed lag estimator derived from smoothness priors. *Econometrica*, **41**, 775–788.

Shinozaki, N. (1975). A study of generalized inverse of matrix and estimation with quadratic loss. *Ph.D Thesis*, Keio University, Japan.

Sprent, P. (1969). *Models in regression and related topics*, Methuen, London.

Swamy, P. A. V. B. (1971). Statistical inference in random coefficient models, Springer-Verlag, Berlin.

Swamy, P. A. V. B., and Mehta, J. S. (1977). A note on minimum average risk estimators for coefficients in linear models. *Commun. Statist. Theor. Math.*, A **6**(12), 1181–1186.

Swamy, P. A. V. B., Mehta, J. S., and Rappoport, P. N. (1978). Two methods of evaluating Hoerl and Kennard's ridge regression. *Commun. Statist.*, **12**, 1133–1155.

Swindel, B. F. (1968). On the bias of least-squares estimators of variance in a general linear model. *Biometrika*, **55**, 313–316.

Tan, W. Y. (1971). Note on an extension of the GM-theorems to multivariate linear regression models. *SIAM J. Appl. Math.*, **1**, 24–28.

Taylor, W. E. (1974). Smoothness priors and stochastic prior restrictions in distributed lag estimation. *Internat. Econom. Rev.*, **15**, 803–804.

Teräsvirta, T. (1979a). Some results on improving the least squares estimation of linear models by mixed estimation. *Discussion Paper 7914*, Louvain, CORE.

Teräsvirta, T. (1979b). The polynomial distributed lag revisited. *Discussion Paper 7919*, Louvain, CORE.

Teräsvirta, T. (1980). A comparison of mixed and minimax estimators of linear models. *Research Report No. 13*, Department of Statistics, University of Helsinki.

Teräsvirta, T., and Toutenburg, H. (1980). A note on the limits of a modified Theil estimator. *Biometrical J.*, **22**, 561–562.

Theil, H. (1963). On the use of incomplete prior information in regression analysis. *J. Amer. Statist. Assoc.*, **58**, 401–414.

Theil, H. (1971). *Principles of econometrics*, Wiley, New York.

Theil, H., and Goldberger, A. S. (1961). On pure and mixed estimation in economics. *Int. Economic Rev.*, **2**, 65–78.

Theobald, C. M. (1974). Generalizations of mean square error applied to ridge regression. *J. R. Statist. Soc.*, B, **36**, 103–106.

Till, R. (1973). The use of linear regression in geomorphology. *Area*, **5**(4), 303–308.

Toro-Vizcarrondo, C., and Wallace, T. D. (1968). A test of the mean square error criterion for restrictions in linear regression. *J. Amer. Statist. Assoc.*, **63**, 558–572.

Toutenburg, H. (1968). Vorhersage im allgemeinen linearen Regressionsmodell mit Zusatzinformation über die Koeffizienten. (Prediction in the generalized linear regression model with prior information on the coefficients.) *Operationsforschung Mathematische Statistik, I*, Akademie-Verlag, Berlin. pp. 107–120.

Toutenburg, H. (1970a). Über die Wahl zwischen erwartungstreuen oder nichterwartungstreuen Vorhersagen. (On the choice between biased and unbiased predictors.) *Operationsforschung Mathematische Statistik, II*, Akademie-Verlag Berlin, pp. 107–118.

Toutenburg, H. (1970b). Optimale Vorhersage von endogenen Variablen in einem linearen System von strukturellen Gleichungen. (Optimal prediction of endogenous

variables in a linear system of structural equations.) *Mathematische Operationsforschung Statistik*, **2**, 69–75.
Toutenburg, H. (1970c). Vorhersage im allgemeinen linearen Regressionsmodell mit stochastischen Regressoren. (Prediction in generalized linear regression with stochastic regressors.) *Mathematische Operationsforschung Statistik*, **2**, 105–116.
Toutenburg, H. (1970d). Vorhersagebereiche im allgemeinen linearen Regressionsmodell. (Prediction regions in generalized linear regression.) *Biometrische Zeitschrift*, **12**, 1–13.
Toutenburg, H. (1970e). Probleme linearer Vorhersagen im allgemeinen linearen Regressionsmodell. (Problems of linear prediction in generalized linear regression.) *Biometrische Zeitschrift*, **12**, 242–252.
Toutenburg, H. (1971). Probleme der Intervallvorhersage von normalverteilten Variablen. (Problems of interval prediction of normally distributed variables.) *Biometrische Zeitschrift*, **13**, 261–273.
Toutenburg, H. (1973). Lineare Restriktionen und Modellwahl im allgemeinen linearen Regressionsmodell. (Linear restrictions and model choice in generalized regression.) *Biometrische Zeitschrift*, **15**, 325–342.
Toutenburg, H. (1975a). *Vorhersage in linearen Modellen* (*Prediction in linear models*), Akademie-Verlag, Berlin.
Toutenburg, H. (1975b). The use of mixed prior information in regression analysis. *Biometrische Zeitschrift*, **17**, 365–372.
Toutenburg, H. (1975c). Minimax-linear estimation (MMLE) and 2-phase MMLE in a restricted linear regression model. *Math. Operationsforschung Statistik*, **6**, 703–706.
Toutenburg, H. (1976). Minimax-linear and MSE-estimators in generalized regression. *Biometrische Zeitschrift*, **18**, 91–100.
Toutenburg, H. (1977a). The combination of restrictions on the variance and the mean in regression. *Proc. 5th Conf. on Probability Theory, Brasov*. pp. 347–354.
Toutenburg, H. (1977b). Two-step procedure for estimation and model choice in regression. *Biometrische Zeitschrift*, **19**, 237–244.
Toutenburg, H. (1977c) Vorhersage, Schätzung und Modellwahl unter Zusatzinformation in linearen Modellen mit Anwendungen auf ein ökonometrisches Modell der Volkswirtschaft der DDR. (Prediction, estimation and model choice under prior information in linear models with application to an econometric model of the GDR.) *D.Sc. (Econ.) Thesis*, Academy of Sciences of the GDR, Berlin.
Toutenburg, H. (1980a). Nonlinear constraints for coefficients of linear models. *Proc. 6th Conf. Probability Theory and Statistics, Brasov.*
Toutenburg, H. (1980b). On the combination of equality and inequality restrictions on regression coefficients. *Biometrical J.*, **22**, 271–274.
Toutenburg, H., and Rödel, E. (1978). *Mathematisch-Statistische Methoden in der Ökonomie* (*Mathematical and statistical methods in economy*), Akademie-Verlag, Berlin.
Toutenburg, H., and Roeder, B. (1978). Minimax-linear and Theil estimator for restrained regression coefficients. *Mathematische Operationsforschung und Statistik, Series Statistics*, **9**, 499–505.
Toutenburg, H., and Wargowske, B. (1978). On restricted 2-stage-least-squares (2 SLSE) in a system of structural equations. *Math. Operationsforschung Statistik, Series Statistics*, **9**, 167–177.
Toyoda, T., and Wallace, F. D. (1976). Optimal critical values for pre-testing in regression. *Econometrica*, **44**, 365–375.
Vinod, H. S. (1978). A survey of ridge regression and related techniques for improvements over ordinary least squares. *Review of economics and statistics*, **60**, 121–131.
Wallace, T. D. (1972). Weaker criteria and tests for linear restrictions in regression. *Econometrica*, **40**, 689–698.
Wallace, T. D., and Toro-Vizcarrondo, C. (1969). Tables for the mean square error test for exact linear restrictions in regression. *J. Amer. Statist. Assoc.*, **64**,1649–1663.

Watson, G. S. (1955). Serial correlation in regression analysis, I. *Biometrika*, **42**, 327–341.

Wölfling, M., Biebler, E., and Schiele, K. (1975). Ein ökonometrisches Modell der Volkswirtschaft der DDR. (An econometric model of the GDR.) *Unpublished research report.*

Yancey, T. A., Judge, G. G., and Bock, M. E. (1973). Wallace's weak mean square error criterion for testing linear restrictions in regression: a tighter bound. *Econometrica*, **41**, 1203–1206.

Yancey, T. A., Judge, G. G., and Bock, M. E. (1974). A mean square error test when stochastic restrictions are used in regression. *Commun. Statist.*, **3**, 755–768.

Index

Applied Probability and Statistics (Continued)

FLEISS • Statistical Methods for Rates and Proportions, *Second Edition*
FRANKEN • Queues and Point Processes
GALAMBOS • The Asymptotic Theory of Extreme Order Statistics
GIBBONS, OLKIN, and SOBEL • Selecting and Ordering Populations: A New Statistical Methodology
GNANADESIKAN • Methods for Statistical Data Analysis of Multivariate Observations
GOLDBERGER • Econometric Theory
GOLDSTEIN and DILLON • Discrete Discriminant Analysis
GROSS and CLARK • Survival Distributions: Reliability Applications in the Biomedical Sciences
GROSS and HARRIS • Fundamentals of Queueing Theory
GUPTA and PANCHAPAKESAN • Multiple Decision Procedures: Theory and Methodology of Selecting and Ranking Populations
GUTTMAN, WILKS, and HUNTER • Introductory Engineering Statistics, *Second Edition*
HAHN and SHAPIRO • Statistical Models in Engineering
HALD • Statistical Tables and Formulas
HALD • Statistical Theory with Engineering Applications
HAND • Discrimination and Classification
HARTIGAN • Clustering Algorithms
HILDEBRAND, LAING, and ROSENTHAL • Prediction Analysis of Cross Classifications
HOEL • Elementary Statistics, *Fourth Edition*
HOLLANDER and WOLFE • Nonparametric Statistical Methods
JAGERS • Branching Processes with Biological Applications
JESSEN • Statistical Survey Techniques
JOHNSON and KOTZ • Distributions in Statistics
Discrete Distributions
Continuous Univariate Distributions—1
Continuous Univariate Distributions—2
Continuous Multivariate Distributions
JOHNSON and KOTZ • Urn Models and Their Application: An Approach to Modern Discrete Probability Theory
JOHNSON and LEONE • Statistics and Experimental Design in Engineering and the Physical Sciences, Volumes I and II, *Second Edition*
JUDGE, GRIFFITHS, HILL and LEE • The Theory and Practice of Econometrics
KALBFLEISCH and PRENTICE • The Statistical Analysis of Failure Time Data
KEENEY and RAIFFA • Decisions with Multiple Objectives
LAWLESS • Statistical Models and Methods for Lifetime Data
LEAMER • Specification Searches: Ad Hoc Inference with Nonexperimental Data
McNEIL • Interactive Data Analysis
MANN, SCHAFER and SINGPURWALLA • Methods for Statistical Analysis of Reliability and Life Data
MEYER • Data Analysis for Scientists and Engineers
MILLER • Survival Analysis
MILLER, EFRON, BROWN, and MOSES • Biostatistics Casebook
MONTGOMERY and PECK • Introduction to Linear Regression
NELSON • Applied Life Data Analysis
OTNES and ENOCHSON • Applied Time Series Analysis: Volume I, Basic Techniques
OTNES and ENOCHSON • Digital Time Series Analysis
POLLOCK • The Algebra of Econometrics
PRENTER • Splines and Variational Methods
RAO and MITRA • Generalized Inverse of Matrices and Its Applications
RIPLEY • Spatial Statistics
SCHUSS • Theory and Applications of Stochastic Differential Equations
SEAL • Survival Probabilities: The Goal of Risk Theory
SEARLE • Linear Models
SPRINGER • The Algebra of Random Variables
UPTON • The Analysis of Cross-Tabulated Data
WEISBERG • Applied Linear Regression
WHITTLE • Optimization Under Constraints